MATHEMATICS IN CHEMISTRY

MATHEMATICS IN CHEMISTRY

AN INTRODUCTION TO MODERN METHODS

Harry G. Hecht
South Dakota State University

Prentice Hall, Englewood Cliffs, New Jersey 07632

Library of Congress Cataloging-in-Publication Data

HECHT, HARRY G.
Mathematics in chemistry: an introduction to modern methods/
Harry G. Hecht
p. cm.
Includes bibliographical references.
ISBN 0-13-561069-9
1. Chemistry—Mathematics. I. Title.
QD39.3.M3H43 1990 89-16369
540′.15′1—dc20 CIP

Editorial/production supervision
and interior design: *Kathleen M. Lafferty*
Cover design: *Wanda Lubelska*
Manufacturing buyer: *Paula Massenaro*

Printed in the United States of America

10 9 8 7 6 5 4 3 2 1

ISBN 0-13-561069-9

Prentice-Hall International (UK) Limited, *London*
Prentice-Hall of Australia Pty. Limited, *Sydney*
Prentice-Hall Canada Inc., *Toronto*
Prentice-Hall Hispanoamericana, S.A., *Mexico*
Prentice-Hall of India Private Limited, *New Delhi*
Prentice-Hall of Japan, Inc., *Tokyo*
Simon & Schuster Asia Pte. Ltd., *Singapore*
Editora Prentice-Hall do Brasil, Ltda., *Rio de Janeiro*

CONTENTS

chapter five

SOME SPECIAL TOPICS IN CHEMOMETRICS 256

APPENDICES 336

PREFACE

The Spring 1986 newsletter of the American Chemical Society Committee on Professional Training contained an article indicating that several universities had either adopted or were considering the adoption of a degree program in computational chemistry. This is but one of the many attestations that the practice of modern chemistry relies on a firm grasp of theoretical principles. Its mastery demands a rather high degree of mathematical sophistication.

It seems ironic, but I think it is nevertheless true, that at the same time we have seen our discipline become more theoretical in nature, we have in many cases decreased our demands for formal mathematical training. Part of the reason for this dilemma lies in the fact that a large number of students are also considering the option of alternative professional careers, and this has led to greater diversity in the background of the students taking traditional chemistry courses.

Mathematics in Chemistry: An Introduction to Modern Methods is an outgrowth of an attempt to address this problem. It is intended for a one-semester, upper-division course which reviews basic mathematical concepts and introduces new ones with which the student may not be familiar. It has been written with the hope that it may also be useful as a supplemental text, or for individual study. The intent is not to maintain mathematical rigor throughout, but rather to rely on intuitive development of those concepts that are of most use to the chemist. Examples have been selected which illustrate the methods in a chemical context wherever possible; this will, I hope, assist the reader in bridging that formidable gap which separates the abstract world of x, y, z from the real world of p, V, T. The discussion is meant to be introductory in nature, providing the background necessary for a general understanding of the various topics considered. They are treated with more depth and rigor in the suggested readings found at the end of each chapter. To keep the book within reasonable limits, the material included in the appendices has been kept to a level appropriate to the material presented. Good reference collections of mathematical and statistical tables are available and should be used in conjunction with this book for extended applications.

Many of the methods treated lend themselves to computer solution, and this has been made an integral part of the presentation. The question of which language to use is always subject to debate. I have chosen BASIC for several reasons. It is universally available on PCs; they are in general use and adequate for the applications given. Essentials of computer programming are covered, but they are not the major emphasis. It has been my experience that teaching the important concepts is possible in a few lectures using BASIC, but attempts to cover them in FORTRAN get bogged down (for example, in complications such as FORMAT statements), leaving beginning students with a feeling of frustration. At the present time many chemistry students take an entire course in FORTRAN programming before they complete their degree, but they may not have yet done so when they encounter a need to utilize the skills developed in this text. Converting some of the programs to FORTRAN or to one of the other more advanced languages, such as Pascal, can be a very useful exercise for those who already have programming experience.

Some of the material was written while I was an exchange professor at the Yunnan Normal University in the People's Republic of China. I am grateful to my colleagues there who generously made their resources available to me. I also acknowledge the comments of the following reviewers: Lawrence Bottomley, Georgia Institute of Technology; Peter Lykos, Illinois Institute of Technology; Michele Mandrioli, Southeastern Massachusetts University; and J. D. Wimeforder, University of Florida.

Harry G. Hecht

chapter one

SUMMARY OF MATHEMATICAL FUNDAMENTALS

Practically the whole of chemistry, or for that matter of any natural science, is based upon an understanding of the relationship between variables. If we increase the temperature, what happens to the solubility of salt in water? If we decrease the pressure, what happens to the boiling point of alcohol? If we add acid to a given solution, what happens to the mobility of the ions in an electric field?

The relationships are often defined in terms of algebraic expressions involving real variables. Many times a sum of terms is required, each of which is of the form $a_n x^n$. The expression is then termed a *polynomial*.

Sometimes the polynomial involves more than one independent variable. If a replacement of each of the independent variables by a constant times the variable results in the same polynomial multiplied by the constant to the nth power,

$$f(cx, cy, \ldots) = c^n f(x, y, \ldots) \tag{1.1}$$

the polynomial is said to be *homogeneous and of degree n*. For example, $ax^2y + bxy^2 + cxyz$ is a homogeneous polynomial of degree 3, but $dx^2y + exy$ is not homogeneous.

Many of the fundamental quantities with which the chemist deals may be classified as homogeneous functions, and this provides the basis for mathematical manipulation of them to provide some very useful relationships, particularly in the study of thermodynamics. For example, we find it useful to distinguish between intensive properties and extensive properties of matter. Intensive properties do not depend on the amount of material present but are "inherent" quantities that define its nature. Temperature and density are examples of intensive properties. Extensive properties, on the other hand, do depend on the amount of material present, and we determine the total quantity for any system by summing that of all the parts. Examples are internal energy and enthalpy, or heat content. We typically use the mole as our fundamental measure of the quantity of matter, so in a mathematical sense we classify intensive properties as those that are zero-degree homogeneous functions of the mole numbers, and extensive properties are first-degree homogeneous functions of the mole numbers.

1.1 SOME SPECIAL FUNCTIONS

There are relationships encountered by the chemist that cannot be expressed in polynomial form. Some examples are trigonometric functions, logarithms, and exponentials. These are called *transcendental functions*. They are usually expressed in terms of an infinite series, and one must sum varying numbers of terms in such a series to calculate the desired quantity to a specified level of accuracy. When we use our calculator or computer to generate one of these functions, it performs the actual calculation using such a truncated infinite series. In Section 1.8 we will learn how any arbitrary function can be expressed in terms of a series; here we introduce some of the more common transcendental functions and define their characteristics.

1.1.1 Exponential Functions

Any number can be expressed as some base b raised to an appropriate power or exponent m:

$$n = b^m \tag{1.2}$$

Here m represents the number of times b must be multiplied by itself to give the number n. For example, if $n = 27$ and $b = 3$, then $m = 3$ since $3 \times 3 \times 3 = 27$. If the exponent is negative, the quantity is to be reciprocated:

$$b^{-m} = \frac{1}{b^m} \tag{1.3}$$

As an example we might write in this case,

$$10^{-3} = \frac{1}{10^3} = \frac{1}{10 \times 10 \times 10} = \frac{1}{1000} = 0.001$$

Several important *laws of exponents* may readily be verified if one is dealing with integral exponents m and n:

$$b^n b^m = b^{n+m} \tag{1.4}$$

$$\frac{b^m}{b^n} = b^{m-n} \tag{1.5}$$

$$(b^m)^n = b^{mn} \tag{1.6}$$

$$(ab)^m = a^m b^m \tag{1.7}$$

$$\left(\frac{a}{b}\right)^m = \frac{a^m}{b^m} = a^m b^{-m} \tag{1.8}$$

From Eq. 1.5 it follows that for $m = n$,

$$\frac{b^m}{b^m} = b^{m-m} = b^0 = 1 \tag{1.9}$$

and we have the important result that *any number (except zero) raised to the zero power is equal to 1.*

It is not required that m and n be integers. For a fractional exponent,

$$b^{m/n} = (b^{1/n})^m$$

where $b^{1/n}$ is the positive nth root of b if b is positive, or the negative nth root of b if b is negative and n is odd. It follows that $b^{m/n} = (b^m)^{1/n}$, and the five laws of exponents (Eqs. 1.4 to 1.8) are valid as long as m and n are integers or fractions, that is, *rational numbers*. In those cases where the exponent is irrational, for example π or $\sqrt{3}$, the quantity can still be evaluated using a close rational approximation to the true exponent, and this process may be carried to any desired degree of accuracy.

1.1.2 Logarithms

The relationship given by Eq. 1.2 may be placed in an alternative form. If $b^m = n$, we may state that m is the *logarithm* of n to the base b:

$$m = \log_b n \tag{1.10}$$

The base b of *common logarithms* is 10. The number 1000, for example, can be written as 10^3, so the common logarithm of 1000 is 3. Similarly, the common logarithm of 0.001 is -3. We often write log without specification of the base, in which case the common logarithm, or base 10, is assumed.

Logarithms have some very useful properties for mathematical manipulation. Suppose that we have the product of two numbers x and y. We write $x = b^m$ and $y = b^n$. Then, according to Eq. 1.4,

$$x \cdot y = b^m \cdot b^n = b^{m+n}$$

and

$$\log_b(x \cdot y) = m + n$$

In other words,

$$\log_b(x \cdot y) = \log_b x + \log_b y \tag{1.11}$$

and we *multiply* two numbers by *adding* their logarithms. Similarly, by Eq. 1.5

$$\frac{x}{y} = \frac{b^m}{b^n} = b^{m-n}$$

so that

$$\log_b \frac{x}{y} = m - n$$

In order to *divide* two numbers we therefore *subtract* their logarithms:

$$\log_b \frac{x}{y} = \log_b x - \log_b y \tag{1.12}$$

It is also very useful to note that we can perform exponentiations (i.e., raising a number to a certain power) by simple multiplication using logarithms. We can see this by assuming some number expressed as $x = b^m$ for which we wish to compute x^p. This, according to Eq. 1.6, is equivalent to

$$x^p = (b^m)^p = b^{mp}$$

Taking "logs" gives

$$\log_b x^p = m \cdot p$$

and since $m = \log_b x$, we have

$$\log_b x^p = p \cdot \log_b x \tag{1.13}$$

Thus we get the logarithm of the number x raised to the power p by *multiplying* the logarithm of x by the factor p.

Sometimes a change of base is required. The relationship between logarithms of a number to different bases is derived as follows: Assume that we have $\log_b x = m$. Then $x = b^m$ and taking the logarithm to the base a gives

$$\log_a x = \log_a b^m = m \cdot \log_a b$$

Since $m = \log_b x$, we have

$$\log_a x = (\log_a b)(\log_b x) \tag{1.14}$$

The real power of logarithms is that they apply to fractional as well as whole-number exponents. It may not be obvious at this point how such logarithms are calculated, but as mentioned above, they can be evaluated by summing an appropriate series as will be described later. Tables of logarithms are available in most reference works.

Another base that is often chosen is the irrational number e, which correct to five decimal places is 2.71828. Such logarithms are called *natural* or *Napierian* logs [after J. Napier (1550–1610), the inventor of the logarithm], and they are represented by $\log_e$ or more simply by ln. Although it may appear at first that the use of an *irrational* number as a logarithmic base is an *irrational* choice, there is a very good reason for it, which will become clear in Section 1.5.3. Some sources give tables for common logarithms only, but the conversion to natural logarithms is readily made using Eq. 1.14, which for this particular case becomes

$$\ln x = \ln 10 \log x$$
$$= 2.303 \log x \tag{1.15}$$

Most modern calculators give values for both log and ln, as well as the antilogs (10^x and e^x). When such calculators are available the use of tables may be totally unnecessary, but we should be familiar with their construction and use. The number e is of particular importance, as it appears often in scientific calculations. Based on the definitions above it should be recognized that raising e to a certain power x, which is often represented by exp x, is just the inverse of taking the logarithm $\ln x$. Thus $\exp(\ln x) = \ln(\exp x) = x$. Figure 1.1 shows plots of both $\exp x$ and $\ln x$, from which the inverse nature of the two functions can be seen. A similar inverse relationship exists between 10^x and $\log x$.

Example 1.1

A buffer solution contains 0.0100 mol of ammonium chloride and 0.0200 mol of ammonia in 100 mL of solution. What is the pOH?

Solution. The equilibrium with which we are concerned in this case is

$$NH_3 + H_2O \rightleftharpoons NH_4^+ + OH^-$$

for which the equilibrium constant may be written,

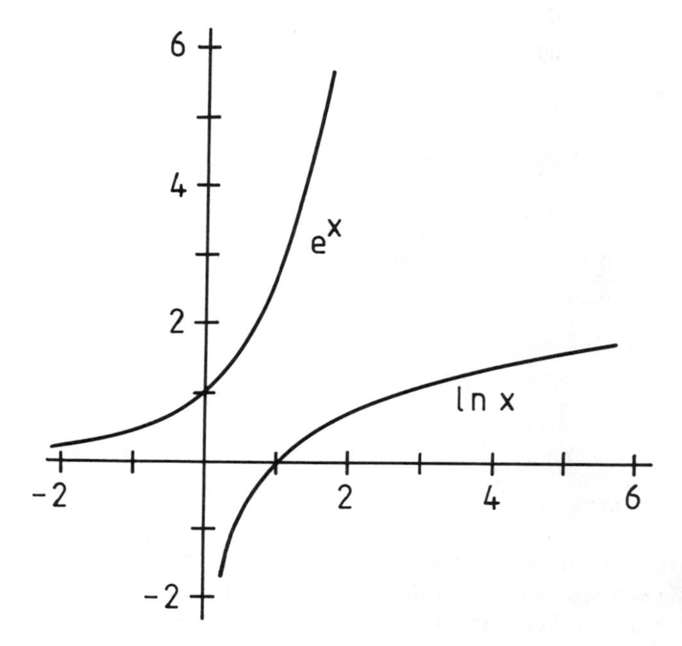

Figure 1.1 Plot of the logarithmic and exponential functions.

$$K_b = \frac{[NH_4^+][OH^-]}{[NH_3]}$$

Taking logs of both sides of the equation gives the following:

$$\log K_b = \log \frac{[NH_4^+]}{[NH_3]} + \log [OH^-]$$

$$pOH = pK_b + \log \frac{[NH_4^+]}{[NH_3]}$$

The acid dissociation constant K_a of the ammonium ion is 5.5×10^{-10}, from which we calculate the pK_a to be 9.26. This may be related to the pK_b of ammonia as follows:

$$K_b = \frac{[NH_4^+][OH^-]}{[NH_3]} \times \frac{[H^+]}{[H^+]} = \frac{[NH_4^+]}{[NH_3][H^+]} \times [H^+][OH^-] = \frac{1}{K_a} \times K_w$$

$$\log K_b = \log K_w - \log K_a$$

$$pK_b = pK_w - pK_a = 14.00 - 9.26 = 4.74$$

If the base hydrolysis reaction proceeds to a certain small extent x, then the equilibrium concentration of NH_3 is

$$\frac{0.0200 \text{ mol}}{0.100 \text{ L}} - x = 0.200 - x$$

and that of NH_4^+ is

$$\frac{0.0100 \text{ mol}}{0.100 \text{ L}} + x = 0.100 + x$$

Thus the pOH is given by

$$pOH = 4.74 + \log \frac{0.100 + x}{0.200 - x}$$

and if we assume that $x \ll 0.1$, then

$$\frac{0.100 + x}{0.200 - x} \approx \frac{0.100}{0.200} = \frac{1}{2.00}$$

and

$$\begin{aligned} pOH &= 4.74 - \log 2.00 = 4.74 - 0.30 \\ &= 4.44 \end{aligned}$$

1.1.3 Trigonometric Functions

Trigonometry deals with the measurement of angles. There are several *trigonometric functions*, and they can be defined in terms of the coordinates of a circle. If we have some arbitrary point P at a distance r from the origin, with x-coordinate a and

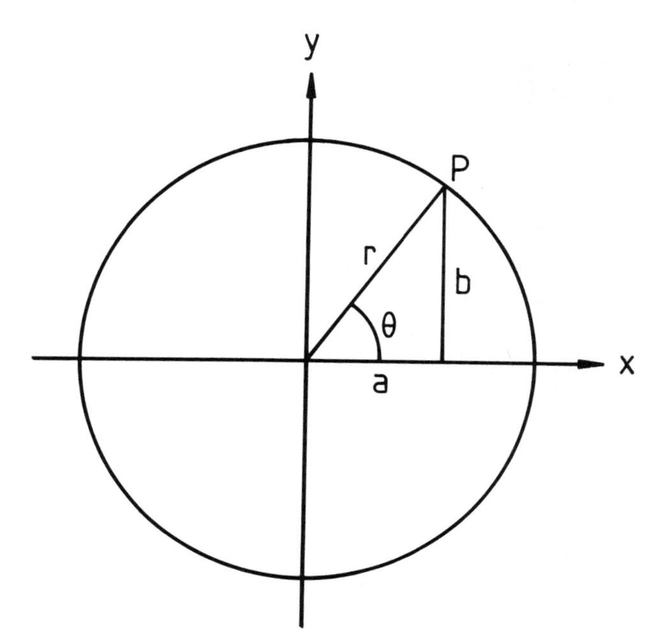

Figure 1.2 Relation of the parameters used to define the trigonometric functions in Eqs. 1.16 and 1.17 to the coordinates of a point P that lies on a circle.

y-coordinate b (see Figure 1.2), then the *sine, cosine*, and *tangent* of the angle θ which the line OP makes with respect to the x-axis are defined by

$$\sin\theta = \frac{b}{r} \qquad \cos\theta = \frac{a}{r} \qquad \tan\theta = \frac{\sin\theta}{\cos\theta} = \frac{b}{a} \tag{1.16}$$

The reciprocals of these functions are also in common use. They are called the *cosecant, secant*, and *cotangent*, respectively, and they are written

$$\csc\theta = \frac{1}{\sin\theta} = \frac{r}{b} \qquad \sec\theta = \frac{1}{\cos\theta} = \frac{r}{a} \qquad \cot\theta = \frac{1}{\tan\theta} = \frac{a}{b} \tag{1.17}$$

Each of these trigonometric functions traces out a certain characteristic curve as the angle goes from 0 to 360 degrees. If the value of r remains constant, the point P defines a circle, and the sine function for example starts at 0 for $\theta = 0$, rises to a maximum value of 1 at $\theta = 90$, goes back to 0 at $\theta = 180$, falls to a minimum value of -1 at $\theta = 270$, and then again returns to 0 for $\theta = 360$. The behavior of each of the other trigonometric functions may be similarly derived from the defining relations Eqs. 1.16 and 1.17, and plots of them may be found in many reference works. For many purposes it is more convenient to use the *radian* rather than the degree as the angular measure. A full circle of 360 degrees is defined to be 2π radians, so the conversion factor is

$$\frac{360 \text{ degrees}}{2\pi \text{ radians}} = 57.29578 \text{ deg rad}^{-1}$$

There are important *trigonometric relations* among the trigonometric functions that we have defined. Among the more important of these are the following:

$$\sin^2\theta + \cos^2\theta = 1 \tag{1.18}$$

$$\tan^2\theta + 1 = \sec^2\theta \tag{1.19}$$

$$\cot^2\theta + 1 = \csc^2\theta \tag{1.20}$$

which may be readily verified using the *Pythagorean theorem* ($r^2 = a^2 + b^2$).

We make frequent use of formulas for composite angles that are sums or differences of two angles. We will sketch how some of these are derived by reference to Figure 1.3. Consider two lines OR and OP which are at angles α and $\alpha + \beta$ with respect to the x-axis, respectively. Note that the lengths are chosen such that the line PR is perpendicular to OR, so that the angle SPR is also equal to α. By definition,

$$\cos(\alpha + \beta) = \frac{OQ}{OP} = \frac{OT - QT}{OP} = \frac{OT}{OP} - \frac{QT}{OP}$$

Multiplying the first term by OR/OR and the second term by PR/PR gives

$$\cos(\alpha + \beta) = \frac{OT}{OP}\frac{OR}{OR} - \frac{QT}{OP}\frac{PR}{PR}$$

$$= \frac{OT}{OR}\frac{OR}{OP} - \frac{QT}{PR}\frac{PR}{OP}$$

From the figure it can be seen that

$$\frac{OT}{OR} = \cos\alpha \qquad \frac{PR}{OP} = \sin\beta$$

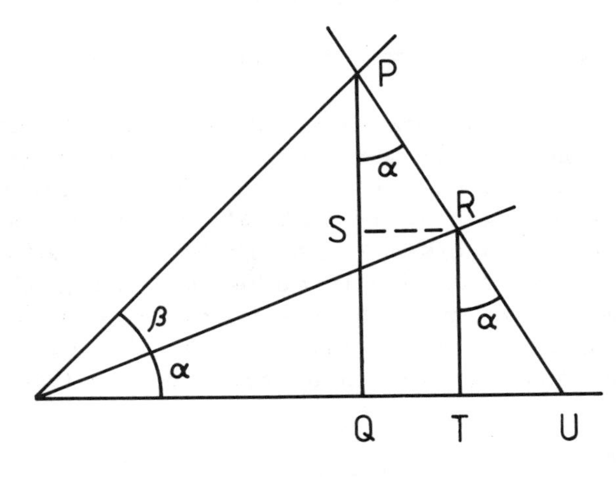

Figure 1.3 Figure used to derive the multiple angle formulas for trigonometric quantities as described in the text. The line PR is perpendicular to the line OR, so that the angles TRU and SPR are both equal to α.

$$\frac{OR}{OP} = \cos\beta \qquad \frac{QT}{PR} = \sin\alpha$$

Inserting these values in the equation above gives the important result

$$\cos(\alpha + \beta) = \cos\alpha\cos\beta - \sin\alpha\sin\beta$$

If we now replace β with $-\beta$, we have

$$\cos(\alpha - \beta) = \cos\alpha\cos(-\beta) - \sin\alpha\sin(-\beta)$$

and since

$$\cos(-\beta) = \cos\beta$$
$$\sin(-\beta) = -\sin\beta$$

we have the following formula for the cosine of the sum or difference of two angles:

$$\cos(\alpha \pm \beta) = \cos\alpha\cos\beta \mp \sin\alpha\sin\beta \tag{1.21}$$

Figure 1.3 can be used to derive a similar formula for the sine of the sum or difference of two angles. According to the definitions given in Eq. 1.16, the sine of $\alpha + \beta$ is

$$\sin(\alpha + \beta) = \frac{PQ}{OP} = \frac{QS + SP}{OP} = \frac{QS}{OP} + \frac{SP}{OP} = \frac{TR}{OP} + \frac{SP}{OP}$$

We multiply the first term by OR/OR and the second by PR/PR to obtain

$$\sin(\alpha + \beta) = \frac{TR}{OP}\frac{OR}{OR} + \frac{SP}{OP}\frac{PR}{PR}$$
$$= \frac{TR}{OR}\frac{OR}{OP} + \frac{SP}{PR}\frac{PR}{OP}$$

Since

$$\frac{TR}{OR} = \sin\alpha \qquad \frac{SP}{PR} = \cos\alpha$$
$$\frac{OR}{OP} = \cos\beta \qquad \frac{PR}{OP} = \sin\beta$$

we have

$$\sin(\alpha + \beta) = \sin\alpha\cos\beta + \cos\alpha\sin\beta$$

Once again we replace β by $-\beta$ to get

$$\sin(\alpha - \beta) = \sin\alpha\cos(-\beta) + \cos\alpha\sin(-\beta)$$

and with

$$\cos(-\beta) = \cos\beta$$
$$\sin(-\beta) = -\sin\beta$$

we obtain the following result for the sine of the sum or difference of two angles:

$$\sin(\alpha \pm \beta) = \sin\alpha\cos\beta \pm \cos\alpha\sin\beta \tag{1.22}$$

From Eqs. 1.21 and 1.22 many other important trigonometric relations may be derived. We will give but one example, which will be needed for our discussion of Section 1.5.5. Using Eq. 1.22 twice, once for the sum and once for the difference, we get

$$\sin(\alpha + \beta) - \sin(\alpha - \beta) = 2\cos\alpha\sin\beta$$

We now define $\gamma = \alpha + \beta$ and $\delta = \alpha - \beta$, and the result becomes

$$\sin\gamma - \sin\delta = 2\cos\frac{\gamma + \delta}{2}\sin\frac{\gamma - \delta}{2} \tag{1.23}$$

These and numerous other trigonometric identities will be found in standard reference works, so we do not give an exhaustive list here.

Example 1.2

Prove that

$$\csc\theta = \sqrt{1 + \cot^2\theta}$$

Solution. Using the definitions of the trigonometric functions given in Eq. 1.17, we have

$$\sqrt{1 + \cot^2\theta} = \sqrt{1 + \left(\frac{a}{b}\right)^2} = \sqrt{\frac{b^2 + a^2}{b^2}}$$

Since $a^2 + b^2 = r^2$, this is equivalent to

$$\sqrt{\frac{r^2}{b^2}} = \frac{r}{b} = \csc\theta$$

Example 1.3

Calculate the angle between the bonds for a tetrahedral molecule such as CH_4.

Solution. The geometry of the molecule is most readily visualized by placing the central atom at the center of a cube, with the four atoms to which it is bonded at alternate corners:

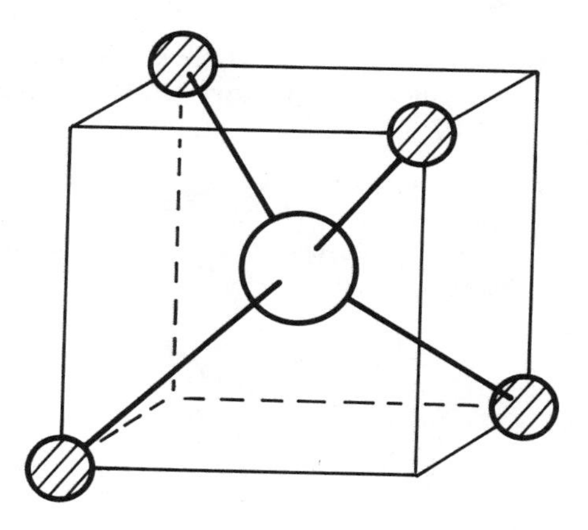

If the length of the cube is a, the distance between any pair of the four corner atoms, represented by a diagonal drawn across a face of the cube, is $\sqrt{a^2 + a^2} = \sqrt{2}\,a$. We slice the cube in two, by cutting along the diagonal. What we see is

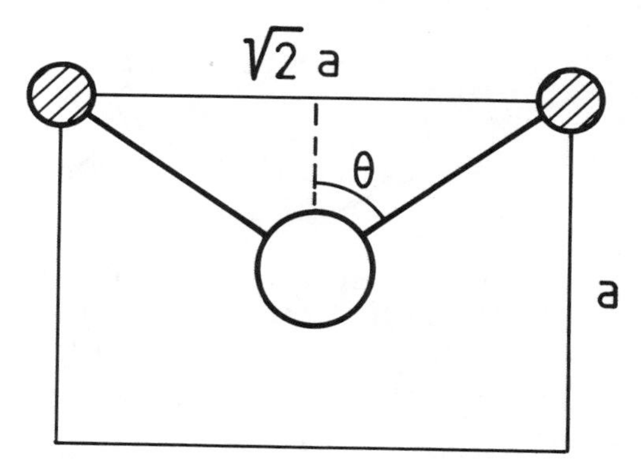

The angle indicated on this section through the cube is half the tetrahedral angle, $\theta_t = 2\theta$. The angle is given as an *inverse trigonometric function*. It is

$$\theta = \tan^{-1} \frac{\sqrt{2}\,a/2}{a/2} = \tan^{-1} \sqrt{2}$$

where $\tan^{-1}$ is known as the *arctangent*, sometimes written *arctan*. It means, in this case, the angle whose tangent is $\sqrt{2}$. (The inverse of each of the other trigonometric functions is similarly defined.) By use of a trigonometric table or a calculator having the arctan function, one finds that $\theta = 54.7356°$, so $\theta_t = 109.4712°$.

Example 1.4

The nature of the crystal structure for a given ionic compound is determined in part by the ratio of the ionic radii, known as the *radius ratio*. The ratio must fall within a certain range for a given coordination, so that the ions pack together in a reasonable fashion. The ideal ratio is that for which the cations and anions, viewed as hard spheres, just touch one another. Calculate the ideal radius ratio for cations that have a coordination number of 3.

Solution. For threefold coordination, the centers of the anions form an equilateral triangle. The spheres just touch one another and form a cavity in which the cation must be accommodated. The ideal radius ratio is that for which the cation just touches each of the anions forming the cavity:

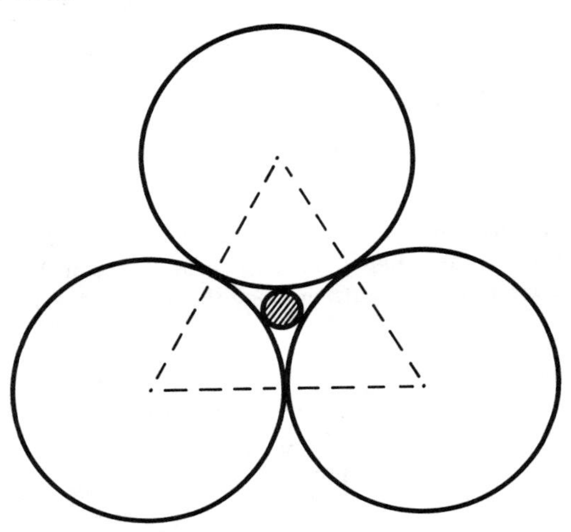

We know from elementary geometry, as can also be readily seen from the figure above, that the bisectors of each of the internal angles for an equilateral triangle intersect at two-thirds the distance between an apex and the opposite base. A right triangle that is half that of the equilateral triangle above is represented by

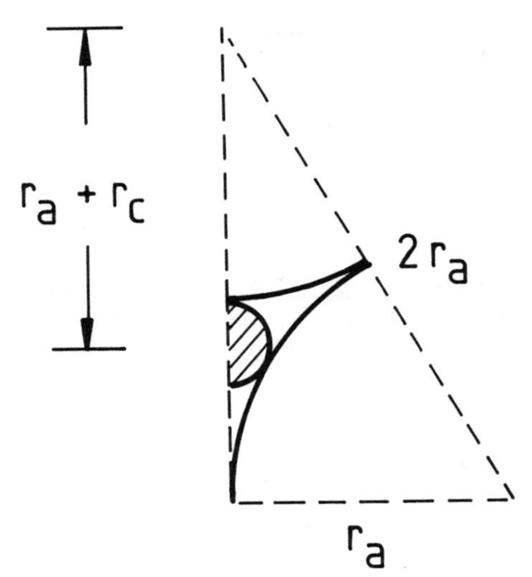

where the vertical line represents one of the bisectors. Its length is given by

$$\sqrt{(2r_a)^2 - (r_a)^2} = \sqrt{3}\, r_a$$

Since the cation center is at two-thirds this distance, we have

$$r_a + r_c = \frac{2}{3}\sqrt{3}\, r_a$$

$$r_c = \left(2\frac{\sqrt{3}}{3} - 1\right) r_a$$

or

$$\frac{r_c}{r_a} = 0.1547$$

1.1.4 Hyperbolic Functions

The so-called *hyperbolic functions* are related to the hyperbola in much the same way that the trigonometric functions are related to the circle (see Figure 1.4). We take P to be any arbitrary point with x-coordinate a and y-coordinate b, lying on a hyperbola whose vertex is at A. Then the hyperbolic sine (sinh) and hyperbolic cosine (cosh) are defined as follows:

$$\sinh z = \frac{MP}{OA} \qquad \cosh z = \frac{OM}{OA}$$

These can be expressed in terms of exponential functions by

$$\sinh z = \frac{e^z - e^{-z}}{2} \qquad \cosh z = \frac{e^z + e^{-z}}{2} \tag{1.24}$$

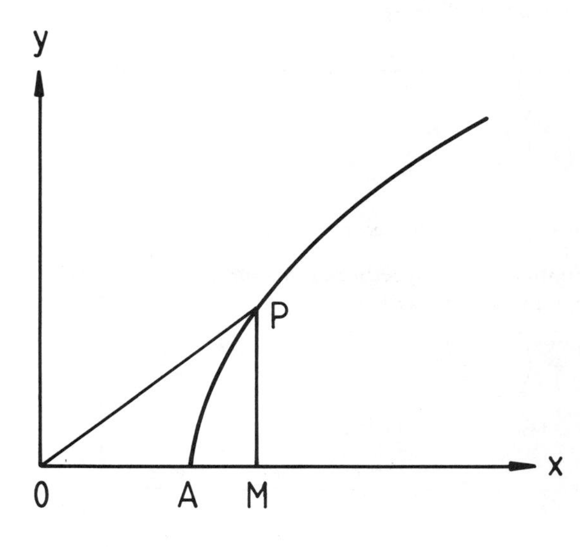

Figure 1.4 Relation of the parameters used to define the hyperbolic function to the coordinates of a hyperbola. The parameter z is actually given by $z = (2 \times \text{area } OAP)/(\text{length } OA)^2$.

and the other hyperbolic functions may be written by analogy to the trigonometric functions:

$$\tanh z = \frac{\sinh z}{\cosh z} \qquad \coth z = \frac{\cosh z}{\sinh z}$$
$$\operatorname{sech} z = \frac{1}{\cosh z} \qquad \operatorname{csch} z = \frac{1}{\sinh z} \tag{1.25}$$

From these definitions several important relationships among the hyperbolic functions may be derived, which include the following:

$$\begin{aligned} \sinh(-z) &= -\sinh z \\ \cosh(-z) &= \cosh z \\ \cosh^2 z - \sinh^2 z &= 1 \\ \operatorname{sech}^2 z - \tanh^2 z &= 1 \\ \coth^2 z - \operatorname{csch}^2 z &= 1 \end{aligned} \tag{1.26}$$

Example 1.5

Prove the following hyperbolic identity:

$$\sinh 2x = 2 \sinh x \cosh x$$

Solution. From the definitions of $\sinh x$ and $\cosh x$ given in Eq. 1.24, we have

$$2 \sinh x \cosh x = 2\left(\frac{e^x - e^{-x}}{2}\right)\frac{e^x + e^{-x}}{2}$$

Expansion using Eq. 1.4 gives the following:

$$\frac{e^{x+x} + e^{x-x} - e^{-x+x} - e^{-x-x}}{2} = \frac{e^{2x} - e^{-2x}}{2} = \sinh 2x$$

Example 1.6

Show that

$$\sinh(x + y) = \sinh x \cosh y + \cosh x \sinh y$$

Solution. Using the definitions of the hyperbolic functions given in Eq. 1.24, the right-hand side of the formula above may be written

$$\frac{e^x - e^{-x}}{2}\left(\frac{e^y + e^{-y}}{2}\right) + \frac{e^x + e^{-x}}{2}\left(\frac{e^y - e^{-y}}{2}\right)$$

From the law of exponents (Eq. 1.4) this becomes

$$\frac{e^{x+y} - e^{-x+y} + e^{x-y} - e^{-x-y}}{4} + \frac{e^{x+y} - e^{x-y} + e^{-x+y} - e^{-x-y}}{4}$$

$$= \frac{2e^{x+y} - 2e^{-x-y}}{4}$$

$$= \frac{e^{(x+y)} - e^{-(x+y)}}{2}$$

$$= \sinh(x + y)$$

Example 1.7

One of the interesting concepts to emerge from quantum mechanical calculations is the *tunnel effect*. We assume that a matter wave of energy E is traveling to the right and that it encounters a potential energy barrier of height V_0. We can visualize the process using the following diagram:

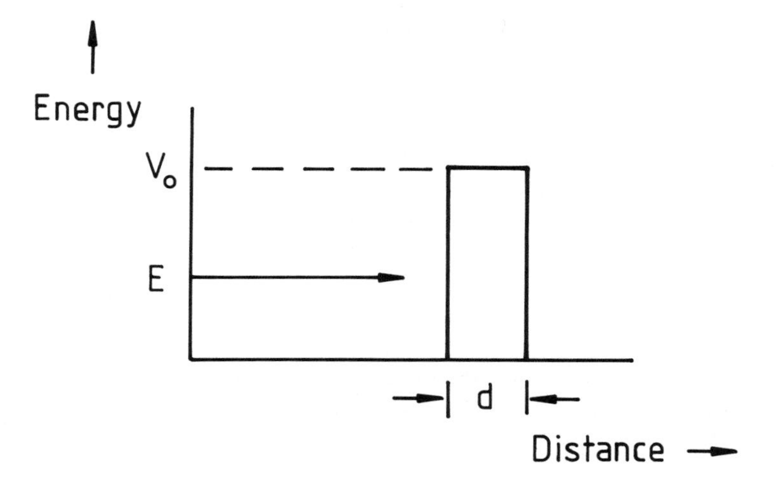

Let us assume that the energy E is only half that of the barrier. Classically, the wave would be reflected by the barrier. It is reflected by the barrier when treated quantum mechanically, too, but not completely. There is a certain probability that the wave will penetrate the barrier even though its energy is not sufficient to go over it. This probability is given by

$$P = \frac{1}{\cosh^2\beta d + \dfrac{(\alpha^2 - \beta^2)^2}{4\alpha^2\beta^2}\sinh^2\beta d}$$

where α and β are

$$\alpha = \frac{2\pi}{h}\sqrt{2mE} \qquad \beta = \frac{2\pi}{h}\sqrt{2m(V_0 - E)}$$

Here h is a small constant known as the quantum of action, or *Planck's constant*, and m is the mass. Show that as d becomes large, the probability of penetration of the barrier is given by

$$P \approx 4e^{-2\beta d}$$

Solution. Since we are given that $E = V_0/2$, we see that $\alpha = \beta$ and the second term in the denominator is zero. From Eq. 1.24 it is apparent that for large values of βd, $\exp(-\beta d)$ is small, so that

$$P = \frac{1}{\left(\dfrac{e^{\beta d} + e^{-\beta d}}{2}\right)^2} \approx \frac{1}{\left(\dfrac{e^{\beta d}}{2}\right)^2} = 4e^{-2\beta d}$$

In the extreme case where d becomes very large, the probability becomes zero. We can let Planck's constant become arbitrarily small in the above, in which case the classical result should be obtained in accordance with the *Bohr correspondence principle*. As h tends toward zero, we see that βd increases without bound, and $P \to 0$ as required classically.

1.2 SPHERICAL POLAR COORDINATES

It is essential for the simplification of mathematical manipulations that the coordinates be chosen in such a way as to conform as closely as possible to the natural symmetry of the problem. In Section 1.1.3 we defined the trigonometric functions in terms of a point that moves on a circle of radius r (see Figure 1.2). For a system having circular symmetry, we would probably use the coordinates r and θ rather than x and y, the relationship between them being (see Eq. 1.16)

$$r = \sqrt{x^2 + y^2} \qquad \theta = \tan^{-1}\frac{y}{x} \tag{1.27}$$

where $\tan^{-1}$ represents the arctangent as defined in Example 1.3. These are known as *polar coordinates*, and the reverse transformation to Cartesian coordinates is

$$x = r\cos\theta \qquad y = r\sin\theta \tag{1.28}$$

For three-dimensional problems the symmetry is often spherical, and the coordinate system of choice is then the *spherical polar coordinates*. The three coordinates used in this case to define the three-dimensional space are (1) the distance from the origin r, (2) the polar angle θ, and (3) the azimuthal angle ϕ. The relationship of these coordinates to Cartesian coordinates is illustrated in Figure 1.5. The transformations between them can be easily deduced using the trigonometric relationships defined above. From the figure it should be clear that the lengths a and b are given by

$$a = r\sin\theta \qquad b = r\cos\theta$$

and from these we see that

$$\begin{aligned} x &= a\cos\phi = r\sin\theta\cos\phi \\ y &= a\sin\phi = r\sin\theta\sin\phi \\ z &= b = r\cos\theta \end{aligned} \tag{1.29}$$

The inverse transformation is

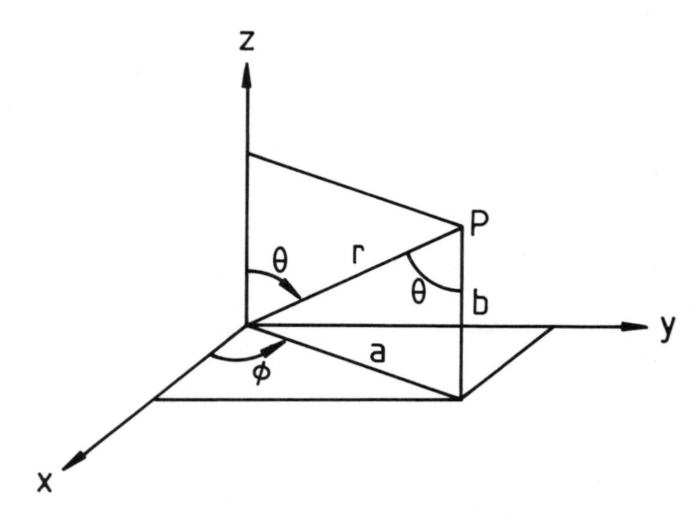

Figure 1.5 Relationship of the spherical polar coordinates r, θ, and ϕ to the Cartesian coordinates x, y, and z.

$$\begin{aligned} r &= \sqrt{x^2 + y^2 + z^2} \\ \theta &= \cos^{-1} \frac{z}{\sqrt{x^2 + y^2 + z^2}} \\ \phi &= \tan^{-1} \frac{y}{x} \end{aligned} \tag{1.30}$$

The range of the Cartesian variables that is required to cover three-dimensional space in its entirety is

$$-\infty \leq x \leq +\infty$$
$$-\infty \leq y \leq +\infty$$
$$-\infty \leq z \leq +\infty$$

We can correspondingly specify the location of any point in three-dimensional space using the spherical polar coordinates by defining their ranges as follows:

$$0 \leq r \leq \infty$$
$$0 \leq \theta \leq \pi$$
$$0 \leq \phi \leq 2\pi$$

There are other coordinate systems, such as toroidal coordinates, ellipsoidal coordinates, cylindrical coordinates, and so forth. Such alternatives are employed in certain special cases, but the spherical polar coordinate system defined above is most useful for problems of a chemical nature. We will find that all the problems introduced throughout the remainder of this book can be solved readily in Cartesian coordinates or spherical polar coordinates.

1.3 COMPLEX NUMBERS

The real numbers encountered thus far in our discussion are integers n, rational numbers n/m, and certain irrational numbers, such as $\sqrt{2}$, π, and e. These numbers can all be put into a one-to-one correspondence with points on a real number axis and ordered, so that we may always state unambiguously which one of any pair of real numbers is the larger.

It is required that we expand our definition to include complex numbers for certain applications. For example, in problems of atomic and molecular structure we find that the electrons and other elementary particles show certain wavelike characteristics (see Example 1.7), and the mathematical model used most successfully in such cases involves solution of a *wave equation*. The *wave function* that represents a solution to the wave equation is a complex quantity in many cases.

Within the real number system the square root of a negative number is not defined, since there is no number which when multiplied by itself can result in a negative quantity. The square root of a negative number can be defined in terms of imaginary numbers, however. We can factor any negative number into a -1 times a positive quantity, and formally we write $\sqrt{-n} = \sqrt{-1} \times \sqrt{n}$. The symbol i is used to represent the square root of -1:

$$i = \sqrt{-1} \qquad i^2 = -1 \tag{1.31}$$

and any number that can be written in the form bi, where b is real, is known as an *imaginary number*. If an imaginary number is combined with an arbitrary real number, which we will call a, the result is a *complex number* of the form

$$z = a + ib \tag{1.32}$$

We see that real and imaginary numbers are just special cases of more general complex numbers, where either b or a are equal to zero, respectively. The real part of a complex number is often written $\text{Re}(z)$, and the imaginary part $\text{Im}(z)$. A complex number is equal to zero only if the real and imaginary parts are both equal to zero, and two complex numbers are equal to one another only if their real parts are equal and their imaginary parts are equal.

The basic mathematical manipulations of real numbers, with which we are already familiar, can be extended to include complex numbers as well. *Complex addition* or *subtraction* is accomplished by separately adding or subtracting both the real and imaginary parts:

$$(a + bi) \pm (c + di) = (a \pm \text{c}) + (b \pm d)i \tag{1.33}$$

The law of *complex multiplication* is obtained by straightforward algebraic procedures using Eq. 1.31 to simplify the result:

$$\begin{aligned}(a + bi)(c + di) &= ac + adi + bci + bdi^2 \\ &= (ac - bd) + (ad + bc)i \end{aligned} \tag{1.34}$$

To obtain the law of *complex division*, we note that the complex number system is a closed system, so that if we divide two complex numbers, the result must be some other complex number. Thus we write

$$\frac{a + bi}{c + di} = e + fi \tag{1.35}$$

We multiply both sides of Eq. 1.35 by $c + di$ and make use of Eq. 1.34 to obtain

$$\begin{aligned} a + bi &= (c + di)(e + fi) \\ &= (ce - df) + (de + cf)i \end{aligned}$$

These can only be equivalent if

$$a = ce - df$$

and

$$b = de + cf$$

Solving these two equations for e and f gives

$$e = \frac{ac + bd}{c^2 + d^2} \qquad f = \frac{bc - ad}{c^2 + d^2} \tag{1.36}$$

It is noted that this same result could have been obtained by multiplication of both numerator and denominator of Eq. 1.35 by $c - di$. The factor $c - di$ is known as the *complex conjugate* of $c + di$. It is simply obtained for any complex number by changing the sign of the imaginary part. When a complex number is multiplied by its complex conjugate, the result is always real and equal to the square of the real part plus the square of the imaginary part. For the imaginary number $z = c + di$, the *modulus* or *absolute value* is defined as the positive square root of this real quantity:

$$|z| = \sqrt{c^2 + d^2} \tag{1.37}$$

The complex conjugate is usually represented by z^*, and it should be obvious using this notation that

$$\begin{aligned} \text{Re}(z) &= \frac{1}{2}(z + z^*) \\ \text{Im}(z) &= \frac{1}{2i}(z - z^*) \\ |z| &= (+)\sqrt{zz^*} \end{aligned} \tag{1.38}$$

A geometrical representation of complex numbers is given using the *Argand diagram*, as shown in Figure 1.6. A real number lies on the real x-axis, and a pure imaginary number lies on the imaginary y-axis. A complex number $z = a + bi$ is repre-

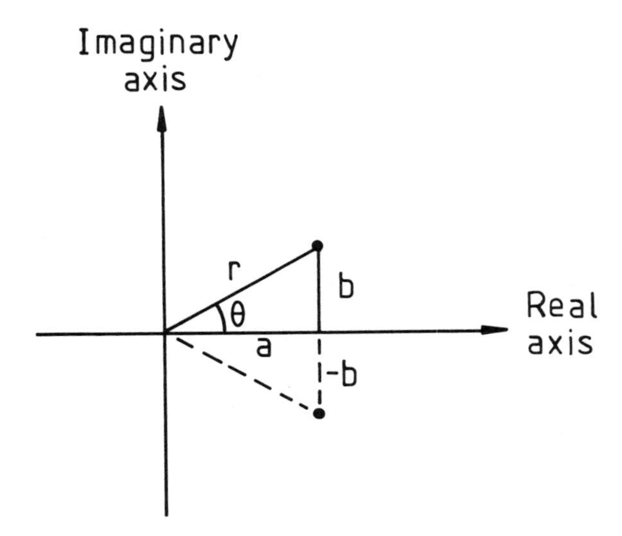

Figure 1.6 Argand diagram for the representation of a number in the complex plane.

sented as a point in the complex plane with components a and b as shown in the figure. Note that the complex conjugate, since it is obtained by changing the sign of the imaginary part, is a mirror image reflected through the x-axis. The point can also be described in polar form using the angle from the x-axis, and the distance from the origin r, as shown. Using the trigonometric functions defined previously, we have $a/r = \cos\theta$ and $b/r = \sin\theta$, or

$$z = r(\cos\theta + i\sin\theta) \tag{1.39}$$

Here r is the modulus of z as defined above, $r = |z|$, and $\theta = \tan^{-1} b/a$, with $-\pi \leq \theta \leq \pi$.

If we have two complex numbers of the form

$$z_1 = r_1(\cos\theta_1 + i\sin\theta_1)$$

$$z_2 = r_2(\cos\theta_2 + i\sin\theta_2)$$

their product may be written

$$\begin{aligned} z_1z_2 &= r_1r_2(\cos\theta_1 + i\sin\theta_1)(\cos\theta_2 + i\sin\theta_2) \\ &= r_1r_2[(\cos\theta_1\cos\theta_2 - \sin\theta_1\sin\theta_2) + i(\sin\theta_1\cos\theta_2 + \cos\theta_1\sin\theta_2)] \end{aligned}$$

Using Eqs. 1.21 and 1.22, this becomes

$$z_1z_2 = r_1r_2[\cos(\theta_1 + \theta_2) + i\sin(\theta_1 + \theta_2)] \tag{1.40}$$

Similarly, we find that

$$\frac{z_1}{z_2} = \frac{r_1}{r_2}[\cos(\theta_1 - \theta_2) + i\sin(\theta_1 - \theta_2)] \tag{1.41}$$

Let us now set $z_1 = z_2 = z$ in Eq. 1.40. We have

$$z^2 = r^2(\cos 2\theta + i \sin 2\theta)$$

We use this result and Eq. 1.40 to multiply by z again to get

$$z^3 = r^3(\cos 3\theta + i \sin 3\theta)$$

This may be repeated any number of times. The general result is

$$z^n = r^n(\cos n\theta + i \sin n\theta)$$

If we apply this formula to a complex number whose modulus is unity ($r = 1$), we have *De Moivre's theorem*:

$$(\cos \theta + i \sin \theta)^n = \cos n\theta + i \sin n\theta \tag{1.42}$$

Implicit in the development above is the assumption that n is a positive integer, but it is easily shown that the result is more general; n can be any rational number, positive or negative.

One important application of De Moivre's theorem is the derivation of trigonometric formulas for multiple angles. Let us, for example, take $n = 3$ in Eq. 1.42. We get

$$\begin{aligned}\cos 3\theta + i \sin 3\theta &= (\cos \theta + i \sin \theta)^3 \\ &= \cos^3\theta + 3i \cos^2\theta \sin \theta + 3i^2 \cos \theta \sin^2\theta + i^3 \sin^3\theta \\ &= [\cos^3\theta - 3 \cos \theta(1 - \cos^2\theta)] + i[3(1 - \sin^2\theta) \sin \theta - \sin^3\theta]\end{aligned}$$

By equating real and imaginary parts, we obtain two useful formulas,

$$\cos 3\theta = 4 \cos^3\theta - 3 \cos \theta$$

$$\sin 3\theta = 3 \sin \theta - 4 \sin^3\theta \tag{1.43}$$

Expansions of $\sin^n\theta$ and $\cos^n\theta$ are also obtained from De Moivre's theorem. We note that if $z = \cos \theta + i \sin \theta$, then

$$\frac{1}{z} = \frac{z^*}{zz^*} = \cos \theta - i \sin \theta$$

Thus

$$z + \frac{1}{z} = 2 \cos \theta \qquad z - \frac{1}{z} = 2i \sin \theta$$

and

$$z^n + \frac{1}{z^n} = 2 \cos n\theta \qquad z^n - \frac{1}{z^n} = 2i \sin n\theta \tag{1.44}$$

We calculate $\cos^4\theta$, for example, by setting $n = 4$ in the above. This gives

$$
\begin{aligned}
2^4 \cos^4\theta &= \left(z + \frac{1}{z}\right)^4 \\
&= z^4 + 4z^2 + 6 + \frac{4}{z^2} + \frac{1}{z^4} \\
&= \left(z^4 + \frac{1}{z^4}\right) + 4\left(z^2 + \frac{1}{z^2}\right) + 6 \\
&= 2 \cos 4\theta + 4 \cdot 2 \cos 2\theta + 6
\end{aligned}
$$

Therefore,

$$
\cos^4\theta = \frac{1}{8} (\cos 4\theta + 4 \cos 2\theta + 3) \tag{1.45}
$$

Complex numbers can be expressed in exponential form. The relationship is given by *Euler's formula* [L. Euler (1707–1783)]:

$$
e^{\pm i\theta} = \cos \theta \pm i \sin \theta \tag{1.46}
$$

This formula may be verified using series expansion methods, which we discuss in Section 1.8. From it we can see that

$$
\cos \theta = \frac{e^{i\theta} + e^{-i\theta}}{2}
$$

$$
\sin \theta = \frac{e^{i\theta} - e^{-i\theta}}{2i}
$$

Note that

$$
\begin{aligned}
\sin^2\theta + \cos^2\theta &= \left(\frac{e^{i\theta} - e^{-i\theta}}{2i}\right)^2 + \left(\frac{e^{i\theta} + e^{-i\theta}}{2}\right)^2 \\
&= \frac{e^{2i\theta} - 2 + e^{-2i\theta}}{-4} + \frac{e^{2i\theta} + 2 + e^{-2i\theta}}{4} \\
&= \frac{1}{2} + \frac{1}{2} = 1
\end{aligned}
$$

as required by Eq. 1.18.

Euler's formula can be used to perform useful manipulations and simplifications of the hyperbolic functions. For example, it is easily seen that

$$
\begin{aligned}
\sinh i\theta &= \frac{e^{i\theta} - e^{-i\theta}}{2} = i \sin \theta \\
\cosh i\theta &= \frac{e^{i\theta} + e^{-i\theta}}{2} = \cos \theta \qquad (1.47) \\
\tanh i\theta &= i \tan \theta
\end{aligned}
$$

Example 1.8

The solution of the wave equation for the hydrogen atom gives the following wave functions for the p-orbitals:

$$l = 1 \quad m = 0 \qquad \psi_0 = \sqrt{\frac{3}{4\pi}}\, f(r) \cos\theta$$

$$l = 1 \quad m = \pm 1 \qquad \psi_{\pm 1} = \sqrt{\frac{3}{8\pi}}\, f(r) \sin\theta e^{\pm i\phi}$$

Show that we can take combinations of these to get equivalent real orbitals that are directed along the Cartesian axes.

Solution. We use the Euler formula Eq. 1.46 to write

$$\sin\theta e^{+i\phi} = \sin\theta\cos\phi + i\sin\theta\sin\phi$$

$$\sin\theta e^{-i\phi} = \sin\theta\cos\phi - i\sin\theta\sin\phi$$

Thus

$$\sin\theta\cos\phi = \sin\theta\,\frac{e^{i\phi} + e^{-i\phi}}{2}$$

$$\sin\theta\sin\phi = \sin\theta\,\frac{e^{i\phi} - e^{-i\phi}}{2i}$$

and from Eq. 1.29 we see that

$$z \propto \cos\theta$$

$$x \propto \sin\theta\cos\phi$$

$$y \propto \sin\theta\sin\phi$$

Therefore, we make the following transformation:

$$P_z = \psi_0 = \sqrt{\frac{3}{4\pi}}\, f(r)\cos\theta$$

$$P_x = \frac{\psi_1 + \psi_{-1}}{\sqrt{2}} = \frac{\sqrt{\frac{3}{8\pi}}\, f(r)\sin\theta(e^{i\phi} + e^{-i\phi})}{\sqrt{2}}$$

$$= \sqrt{\frac{3}{4\pi}}\, f(r)\sin\theta\cos\phi$$

$$P_y = \frac{\psi_1 - \psi_{-1}}{\sqrt{2}i} = \frac{\sqrt{\frac{3}{8\pi}}\, f(r)\sin\theta(e^{i\phi} - e^{-i\phi})}{\sqrt{2}i}$$

$$= \sqrt{\frac{3}{8\pi}}\, f(r)\sin\theta\sin\phi$$

Example 1.9

The angular parts of the wave functions for the hydrogen atom, as obtained by solution of the wave equation, can all be expressed in terms of a set of functions called the *spherical harmonics*. One source lists the $l = 3$, $m = -3$ solution as

$$Y_3^{-3} = \sqrt{\frac{7}{4\pi}}\sqrt{\frac{5}{16}}\frac{(x - iy)^3}{r^3}$$

while another gives

$$Y_3^{-3} = \sqrt{\frac{7}{4\pi}}\sqrt{\frac{5}{16}}\sin^3\theta e^{-3i\phi}$$

for the same spherical harmonic. Show that the two are equivalent.

Solution. From Eq. 1.29 we have

$$\frac{x}{r} = \sin\theta\cos\phi \qquad \frac{y}{r} = \sin\theta\sin\phi$$

so

$$\left(\frac{x - iy}{r}\right)^3 = (\sin\theta\cos\phi - i\sin\theta\sin\phi)^3$$

$$= \sin^3\theta(\cos\phi - i\sin\phi)^3$$

With Eq. 1.46 we see that this may be written

$$\sin^3\theta(e^{-i\phi})^3$$

and using Eq. 1.6 gives

$$\sin^3\theta\, e^{-3i\phi}$$

1.4 VECTORS, MATRICES, AND DETERMINANTS

Many quantities require specification of their magnitude only in order that they be completely defined. Some examples are mass, electrical charge, and work. We call these *scalars*.

Other quantities, such as velocity or magnetic field strength, are completely defined only when we know both the magnitude and the direction. These are *vectors*, and they are generally represented in print by boldface type to distinguish them from scalars. A vector is most conveniently represented in terms of three unit vectors **i**, **j**, **k** directed along the three positive Cartesian axes x, y, and z (see Figure 1.7). We let a_1, a_2, and a_3 be the projections of the vector **A** along each of these three axes, and it may be written

$$\mathbf{A} = a_1\mathbf{i} + a_2\mathbf{j} + a_3\mathbf{k} \tag{1.48}$$

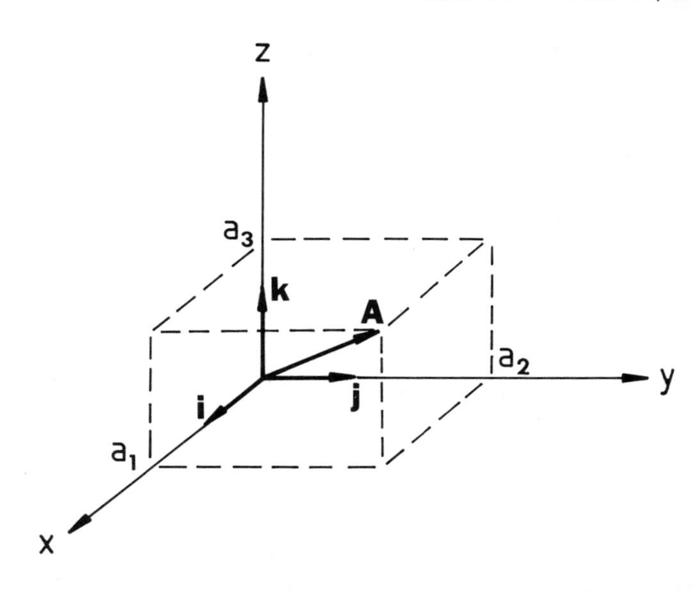

Figure 1.7 Representation of a vector **A** in terms of the unit vectors **i**, **j**, **k** directed along the Cartesian coordinate axes x, y, and z (see Eq. 1.48).

The magnitude of **A**, represented by $|\mathbf{A}|$, is

$$|\mathbf{A}| = \sqrt{a_1^2 + a_2^2 + a_3^2}$$

Any two vectors can be added or subtracted to get a resultant vector,

$$\mathbf{A} \pm \mathbf{B} = (a_1 \pm b_1)\mathbf{i} + (a_2 \pm b_2)\mathbf{j} + (a_3 \pm b_3)\mathbf{k}$$

and multiplication of a vector by a scalar is simply

$$q\mathbf{A} = qa_1\mathbf{i} + qa_2\mathbf{j} + qa_3\mathbf{k}$$

There are two ways in which a pair of vectors may be multiplied together. The first gives a scalar as the result and it is known as the *scalar product* or *dot product*. It may be written in the form

$$\mathbf{A} \cdot \mathbf{B} = a_1b_1 + a_2b_2 + a_3b_3 \tag{1.49}$$

Physically, the scalar product corresponds to multiplying the length of either one of the vectors by the projection of the other upon it, which is often written

$$\mathbf{A} \cdot \mathbf{B} = |\mathbf{A}| \cdot |\mathbf{B}| \cdot \cos\theta$$

where θ is the angle between the two vectors. Obviously, the scalar product is zero if the vectors are *orthogonal* (i.e., at an angle of 90 degrees).

Two vectors may also be multiplied together in such a way that the resultant is a vector. The *vector product* or *cross product* is

$$\mathbf{A} \times \mathbf{B} = (a_2b_3 - a_3b_2)\mathbf{i} + (a_3b_1 - a_1b_3)\mathbf{j} + (a_1b_2 - a_2b_1)\mathbf{k} \tag{1.50}$$

The product vector is perpendicular to the plane containing the vectors **A** and **B**, and its direction is given by the *right-hand rule*. That is, if the fingers of the right hand are used to represent the rotation from **A** toward **B**, the thumb points in the direction of the vector product. Its magnitude may be written

$$|\mathbf{A}| \cdot |\mathbf{B}| \cdot \sin \theta$$

where once again θ is the angle between the vectors **A** and **B**. We see that the vector product has its maximum value if the vectors **A** and **B** are orthogonal, and it vanishes if they are parallel.

We find it convenient in many cases to treat an entire rectangular array of quantities as a unit. Such an array is known as a *matrix*, and the individual quantities, designated by their position within the array,

$$\begin{pmatrix} a_{11} & a_{12} & \cdots & a_{1n} \\ a_{21} & a_{22} & \cdots & a_{2n} \\ \vdots & \vdots & & \vdots \\ a_{m1} & a_{m2} & \cdots & a_{mn} \end{pmatrix} \tag{1.51}$$

are called the *matrix elements*. The first subscript designates the row in which the element occurs, and the second subscript gives the column. If the first subscript were restricted to the value 1, we would have but a single row known as a *row matrix*, while if the second subscript were 1 only, we would have a *column matrix*. If $m = n$, we have as many rows as columns, which is a *square matrix*.

A square matrix whose elements are all equal to zero except $a_{11}, a_{22}, a_{33}, \ldots$ is called a *diagonal matrix*. A diagonal matrix whose diagonal elements are all equal to 1 is known as a *unit matrix*, and if all elements are equal to zero, including those on the diagonal, we have a *null matrix*.

A square matrix for which $a_{ij} = a_{ji}$ for all i and j is called *symmetric*, and if $a_{ij} = -a_{ji}$ for all i and j it is *skew symmetric*. A square matrix has a *determinant* associated with it, which may be computed according to certain rules of *expansion*. We represent the determinant by

$$|\mathbf{A}| = \begin{vmatrix} a_{11} & a_{12} & \cdots & a_{1n} \\ a_{21} & a_{22} & \cdots & a_{2n} \\ \vdots & \vdots & & \vdots \\ a_{n1} & a_{n2} & \cdots & a_{nn} \end{vmatrix} \tag{1.52}$$

One method of expansion is by successive reduction of the order of the determinant. We define the *minor* of an element a_{ij} of the determinant of order n to be the determinant of order $n - 1$, which is obtained by deletion of the ith row and jth column of the original determinant. The *cofactor* A_{ij} of the a_{ij} element is its minor times the factor $(-1)^{i+j}$. The determinant is obtained by summing the products of the elements of any given row or column times their corresponding cofactors.

Several characteristics of determinants are readily noted, including the following: (1) If all elements in a given row or column are zero, the determinant is zero. (2) If all elements of a given row or column are multiplied by a constant c, the result is the same as multiplying the determinant by c. (3) If the corresponding elements of any two rows or columns are proportional to one another, the determinant is zero. (4) Interchanging any two rows or columns changes the sign of the determinant. (5) The value of a determinant is unchanged by the addition of any multiple of a given row or column to the corresponding elements of another row or column. The latter property is particularly helpful for rapid simplification of determinant expansion, as shown in Example 1.10.

The addition or subtraction of two matrices is defined only if they contain the same number of rows and columns, in which case they are combined element by element as follows:

$$\mathbf{A} \pm \mathbf{B} = \begin{pmatrix} a_{11} \pm b_{11} & a_{12} \pm b_{12} & \cdots & a_{1n} \pm b_{1n} \\ a_{21} \pm b_{21} & a_{22} \pm b_{22} & \cdots & a_{2n} \pm b_{2n} \\ \vdots & \vdots & & \vdots \\ a_{m1} \pm b_{m1} & a_{m2} \pm b_{m2} & \cdots & a_{mn} \pm b_{mn} \end{pmatrix} \tag{1.53}$$

Scalar multiplication of a matrix is carried out by multiplication of each element by the scalar factor,

$$p\mathbf{A} = \begin{pmatrix} pa_{11} & pa_{12} & \cdots & pa_{1n} \\ pa_{21} & pa_{22} & \cdots & pa_{2n} \\ \vdots & \vdots & & \vdots \\ pa_{m1} & pa_{m2} & \cdots & pa_{mn} \end{pmatrix} \tag{1.54}$$

Matrix multiplication is defined for two matrices **A** and **B** only if the number of columns of **A** is the same as the number of rows of **B**. When this condition is met they are called *conformable matrices*. The ij element of the product matrix is obtained by multiplying across the ith row of **A** and down the jth column of **B** term by term and summing the products. For example, if

$$\mathbf{A} = \begin{pmatrix} 1 & 4 \\ 2 & 7 \\ 3 & 6 \end{pmatrix} \qquad \mathbf{B} = \begin{pmatrix} 5 & 3 & 9 \\ 4 & 8 & 2 \end{pmatrix}$$

then the product **AB** is

$$\mathbf{AB} = \begin{pmatrix} (1 \times 5) + (4 \times 4) & (1 \times 3) + (4 \times 8) & (1 \times 9) + (4 \times 2) \\ (2 \times 5) + (7 \times 4) & (2 \times 3) + (7 \times 8) & (2 \times 9) + (7 \times 2) \\ (3 \times 5) + (6 \times 4) & (3 \times 3) + (6 \times 8) & (3 \times 9) + (6 \times 2) \end{pmatrix}$$

$$= \begin{pmatrix} 21 & 35 & 17 \\ 38 & 62 & 32 \\ 39 & 57 & 39 \end{pmatrix} \tag{1.55}$$

Note that matrix multiplication in general is *not commutative*; that is, $\mathbf{AB} \neq \mathbf{BA}$.

If the determinant of a square matrix is equal to zero, the matrix is *singular*. For *nonsingular* matrices, an inverse exists such that

$$\mathbf{A}^{-1}\mathbf{A} = \mathbf{A}\mathbf{A}^{-1} = \mathbf{I} \tag{1.56}$$

where $\mathbf{I}$ is the unit matrix. This *inverse matrix* is calculated in a rather complicated way: We first construct the matrix in which each element is replaced by its cofactor and then write its *transpose*, which is obtained by interchanging rows and columns ($A_{ij} \rightarrow A_{ji}$). This transpose of the cofactor matrix is called the *adjoint matrix*. The inverse matrix $\mathbf{A}^{-1}$ is obtained by dividing the adjoint of $\mathbf{A}$ by its determinant $|\mathbf{A}|$. Fortunately, there are more efficient ways of carrying out the numerical computation so that one seldom resorts to the rather ghastly procedure suggested here.

Example 1.10

Evaluate the following determinant:

$$\begin{vmatrix} 4 & 2 & -2 & 2 \\ 1 & 7 & 0 & 6 \\ 3 & 8 & 5 & 1 \\ 9 & 3 & 1 & 3 \end{vmatrix}$$

Solution. The expansion is simplified by using property 5 above to place as many zeros as possible in a row or column. We add column 3 to column 2 and to column 4, and add twice column 3 to column 1. This gives

$$\begin{vmatrix} 0 & 0 & -2 & 0 \\ 1 & 7 & 0 & 6 \\ 13 & 13 & 5 & 6 \\ 11 & 4 & 1 & 4 \end{vmatrix}$$

Expanding by minors along the top row, we have

$$(-1)^{1+3}(-2) \begin{vmatrix} 1 & 7 & 6 \\ 13 & 13 & 6 \\ 11 & 4 & 4 \end{vmatrix}$$

$$= (-2)\left[(1)\begin{vmatrix} 13 & 6 \\ 4 & 4 \end{vmatrix} - 7\begin{vmatrix} 13 & 6 \\ 11 & 4 \end{vmatrix} + 6\begin{vmatrix} 13 & 13 \\ 11 & 4 \end{vmatrix} \right]$$

$$= (-2)[(1)(13 \cdot 4 - 6 \cdot 4) - 7(13 \cdot 4 - 6 \cdot 11) + 6(13 \cdot 4 - 13 \cdot 11)]$$

$$= 840$$

1.5 DIFFERENTIAL CALCULUS

A study of the variation of a function of one or more independent variables, in which use is made of the concept of infinitesimal change, is called *differential calculus*. The techniques of differential calculus allow us to determine slopes, maxima, minima, and other useful characteristics of mathematical functions in a very powerful way. They are essential tools for the construction of models that faithfully depict the behavior of the various chemical systems with which we will be concerned.

1.5.1 Slopes and Critical Points

Simple linear relationships occur often in the study of chemistry. They are usually written in the form

$$y = mx + b \tag{1.57}$$

In terms of two points x_1, y_1 and x_2, y_2 through which the line passes, the y-intercept b is

$$b = \frac{x_2 y_1 - x_1 y_2}{x_2 - x_1} \tag{1.58}$$

and the slope m of the line is given by

$$m = \frac{y_2 - y_1}{x_2 - x_1} \tag{1.59}$$

Other figures also have distinguishing slopes, but the value depends on where the points are located. If we think of two points on an arbitrary curve chosen such that there is but an infinitesimal separation between them, the line connecting them will be tangent to the curve, and it will be uniquely determined. We will take the slope of such a tangent to be our definition of the slope of the curve in the region where the points are taken.

If we were to draw such tangent lines and calculate their slopes at various places along the curve, we would find that in most cases the slope itself is a continuously varying function. Let us consider what a plot of the slope would look like for the schematic function represented in Figure 1.8. We first examine the curve in the region of the point A. Just to the left of A, the slope for any tangent drawn will be positive, while

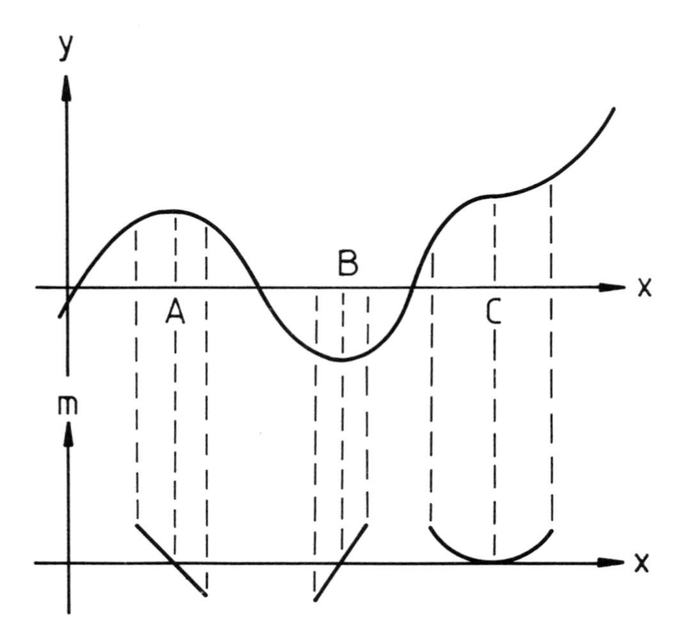

Figure 1.8 Schematic diagram showing how the slope of a function behaves in the vicinity of a maximum (A), a minimum (B), and a horizontal point of inflection (C).

that for any tangent drawn just to the right of A will be negative. At point A, the tangent is zero and the "slope of the slope" is also zero. Thus we see that if the slope passes from positive values to negative values as the independent variable increases, the curve has a *maximum* at the point where the slope is zero and the "slope of the slope" is negative at that point. Similarly, the slope in the vicinity of point B passes from negative to positive values as x increases, being equal to zero at the point B. In this case the "slope of the slope" is positive, and this is characteristic of a *minimum* in the curve. At point C, the slope has become zero but it has *not* changed sign. Any point such as C where the slope is a maximum or a minimum is called a *point of inflection*, and if the slope is equal to zero at that point, it is called a *horizontal point of inflection*. Maxima, minima, and horizontal points of inflection are different types of *critical points*.

We are now in a position to give a more precise definition of the slope of a curve. For some arbitrary function $y = f(x)$, two points on the curve $f(x)$ versus x may be written

$$x_1, f(x_1) \qquad x_2, f(x_2)$$

and the slope of the line connecting these points according to Eq. 1.59 is equal to

$$m = \frac{f(x_2) - f(x_1)}{x_2 - x_1}$$

Now let us take x_2 to be x_1 plus some small increment in x, say Δx:

$$x_2 = x_1 + \Delta x$$

The slope at the point x_1 is then given approximately by

$$m = \frac{f(x_1 + \Delta x) - f(x_1)}{x_1 + \Delta x - x_1} = \frac{f(x_1 + \Delta x) - f(x_1)}{\Delta x}$$

We can drop the subscript 1 since the argument is general and applies anywhere along the curve. The foregoing formula for the slope becomes exact in the limit of an infinitesimally small increment Δx. If we represent the true slope of the curve $f(x)$ at point x by $f'(x)$, then in terms of the above it is equal to

$$f'(x) = \lim_{\Delta x \to 0} \frac{f(x + \Delta x) - f(x)}{\Delta x} \tag{1.60}$$

where the notation

$$\lim_{\Delta x \to 0}$$

is read "in the limit where Δx becomes infinitesimally small."

For actual calculations we make note of the fact that if $y = f(x)$, then $f(x + \Delta x)$ may be written as $y + \Delta y$, and Eq. 1.60 becomes

$$f'(x) = \lim_{\Delta x \to 0} \frac{y + \Delta y - y}{\Delta x} = \lim_{\Delta x \to 0} \frac{\Delta y}{\Delta x} \tag{1.61}$$

The relationship between these quantities is illustrated in Figure 1.9. We represent $\Delta y/\Delta x$ in the limit $\Delta x \to 0$ by dy/dx, which is read "the derivative of y with respect to x". The process of calculating the derivative is known as *differentiation*. According to Eq. 1.61, the derivative thus defined is equivalent to the slope of the curve.

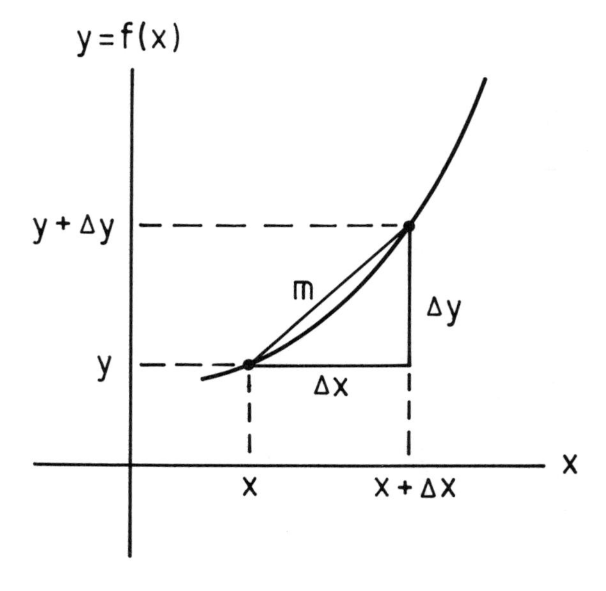

Figure 1.9 The slope of the curve $y = f(x)$ at the point x approximated by $m = \Delta y/\Delta x$.

In view of the discussion above, it is clear that if we have some function $y = f(x)$, we can locate its critical points by finding an appropriate expression for the derivative dy/dx and setting this equal to zero. Since the derivative of a curve is equal to its slope, the "slope of the slope" must be the equivalent of taking the derivative of the derivative. We call this the *second derivative of the function with respect to x*, and it is represented by

$$f''(x) = \frac{d\left(\frac{dy}{dx}\right)}{dx} = \frac{d}{dx}\frac{d}{dx}y = \frac{d^2y}{dx^2}$$

After the critical points of a curve have been found by setting the first derivative equal to zero, these points are then identified as maxima, minima, or horizontal points of inflection, depending on whether the second derivative is negative, positive, or zero at that point, respectively.

We will next consider the problem of finding the derivatives of various types of functions.

Example 1.11

Careful measurements show that at 1 atmosphere, the density of dry air is 0.001247 g mL^{-1} at 10°C and 0.001165 g mL^{-1} at 30°C. Use these data to estimate the temperature (in °C) of absolute zero.

Solution. The corresponding specific volumes (i.e., volume per gram) are

$$\text{at } 10°\text{C:} \qquad v = \frac{1}{0.001247 \text{ g mL}^{-1}} = 801.9 \text{ mL g}^{-1}$$

$$\text{at } 30°\text{C:} \qquad v = \frac{1}{0.001165 \text{ g mL}^{-1}} = 858.4 \text{ mL g}^{-1}$$

The slope is

$$m = \frac{v_2 - v_1}{t_2 - t_1} = \frac{858.4 - 801.9}{30 - 10} = 2.823 \text{ mL g}^{-1} \text{ deg}^{-1}$$

and the v-intercept is

$$b = \frac{t_2 v_1 - t_1 v_2}{t_2 - t_1} = \frac{(30)(801.9) - (10)(858.4)}{30 - 10} = 773.7 \text{ mL g}^{-1}$$

Thus the linear equation relating specific volume to temperature (in °C) may be written

$$v = (2.823 \text{ mL g}^{-1} \text{ deg}^{-1})t + 773.7 \text{ mL g}^{-1}$$

Solving for the value of t for which $v = 0$ gives

$$t = -\frac{773.7 \text{ mL g}^{-1}}{2.823 \text{ mL g}^{-1} \text{ deg}^{-1}} = -274.1 \text{ deg}$$

1.5.2 Derivatives of Algebraic Functions

Let us take y to be any general term from a polynomial in x which we represent by

$$y = ax^n$$

where a is a constant. Then we define

$$y + \Delta y = a(x + \Delta x)^n$$

and by a binomial expansion of the term $(x + \Delta x)^n$ we see that

$$y + \Delta y = a\left[x^n + nx^{n-1}(\Delta x) + \frac{n(n-1)}{2!}x^{n-2}(\Delta x)^2 + \cdots + (\Delta x)^n\right]$$

The first term of the expansion is just y, so we have

$$\Delta y = anx^{n-1}(\Delta x) + \frac{an(n-1)}{2!}x^{n-2}(\Delta x)^2 + \cdots + a(\Delta x)^n$$

or

$$\frac{\Delta y}{\Delta x} = anx^{n-1} + \frac{an(n-1)}{2!}x^{n-2}(\Delta x) + \cdots + a(\Delta x)^{n-1}$$

Now in the limit $\Delta x \to 0$, each of the terms except the first is equal to zero and we have

$$\frac{dy}{dx} = anx^{n-1} \tag{1.62}$$

We note that had we used $y = x^n$ in the derivation above, we would have gotten the result $dy/dx = nx^{n-1}$. Therefore we conclude that $d(ax^n)/dx = ad(x^n)/dx$. It should be obvious that $da/dx = 0$, since $y(x) = y(x + dx)$ if $y = a$, where a is a constant.

There are some other useful relationships that will help us calculate derivatives of algebraic functions. If we have a sum of polynomial terms, each may be differentiated independently according to Eq. 1.62. Summing the result, we would find that

$$\frac{d(f \pm g)}{dx} = \frac{df}{dx} \pm \frac{dg}{dx} \tag{1.63}$$

Sometimes we have a problem in which some function, say y, is a function of the variable u, but u itself is a function of the independent variable x. In that case the derivative of y with respect to x is given by the *chain rule*:

$$\frac{dy(u)}{dx} = \frac{du}{dx}\frac{dy}{du} \tag{1.64}$$

It may be worth noting that the derivative of y with respect to x is the reciprocal of the derivative of x with respect to y:

$$\frac{dy}{dx} = \frac{1}{\dfrac{dx}{dy}} \tag{1.65}$$

Two other differentiation formulas are often used. They deal with those cases where we have two functions that form a product or a quotient. In the first case we have $y = f \cdot g$. Then

$$\begin{aligned} y + \Delta y &= (f + \Delta f) \cdot (g + \Delta g) \\ &= f \cdot g + f\,\Delta g + g\,\Delta f + \Delta f\,\Delta g \end{aligned}$$

and

$$\Delta y = f\,\Delta g + g\,\Delta f + \Delta f\,\Delta g$$

We see that

$$\frac{\Delta y}{\Delta x} = f\frac{\Delta g}{\Delta x} + g\frac{\Delta f}{\Delta x} + \frac{\Delta f\,\Delta g}{\Delta x}$$

Now as Δx approaches zero, Δy, Δf, and Δg also approach zero, while $\Delta y/\Delta x$, $\Delta f/\Delta x$, and $\Delta g/\Delta x$ approach the corresponding derivatives. The last term is therefore zero in the limit $\Delta x \rightarrow 0$ and

$$\frac{d(fg)}{dx} = f\frac{dg}{dx} + g\frac{df}{dx} \tag{1.66}$$

If we use this formula for the differentiation of a product and let $y = f \cdot g^{-1}$, then

$$\frac{dy}{dx} = f\frac{dg^{-1}}{dx} + g^{-1}\frac{df}{dx}$$

Now from the chain rule (Eq. 1.64) and Eq. 1.62,

$$\frac{dg^{-1}}{dx} = \frac{dg}{dx}\frac{dg^{-1}}{dg} = (-1)g^{-2}\frac{dg}{dx}$$

so

$$\frac{dy}{dx} = -fg^{-2}\frac{dg}{dx} + g^{-1}\frac{df}{dx} = \frac{\dfrac{df}{dx}}{g} - \frac{f\dfrac{dg}{dx}}{g^2}$$

$$\frac{d\left(\dfrac{f}{g}\right)}{dx} = \frac{g\dfrac{df}{dx} - f\dfrac{dg}{dx}}{g^2} \tag{1.67}$$

The formulas that we have discussed here will allow any algebraic expression to be differentiated so that its slope may be determined. They are summarized in Appendix A.1 for ready reference.

Example 1.12

Find any critical points of the curve $y = 7 + 12x - 2x^2$ and identify them as maxima, minima, or horizontal points of inflection.

Solution. Taking the derivative with respect to x gives

$$\frac{dy}{dx} = 12 - 4x$$

This may be equated to zero and solved for x:

$$12 - 4x = 0$$

$$4x = 12$$

$$x = 3$$

There is only one critical point in this case; it occurs at $x = 3$, and the corresponding y value is

$$y = 7 + 12(3) - 2(3)^2 = 25$$

The second derivative is

$$\frac{d^2y}{dx^2} = -4$$

and since this is negative, the point (3,25) is a *maximum*.

Example 1.13

When silver chloride dissolves in HCl solutions, some of the silver is in the form of ions Ag^+, some is in the form of undissociated molecules $AgCl_{(aq)}$ (it is actually a *weak* electrolyte), and some is in the form of complex ions $AgCl_2^-$. Given the following data

$$AgCl_{(s)} \rightleftharpoons Ag^+ + Cl^- \qquad pK_{sp} = 9.75$$

$$AgCl_{(s)} \rightleftharpoons AgCl_{(aq)} \qquad pK_0 = 6.70$$

$$AgCl_{(s)} + Cl^- \rightleftharpoons AgCl_2^- \qquad pK_1 = 4.70$$

find the chloride concentration for which the solubility is a minimum.

Solution. We define S to be the molar solubility. Then the material balance condition can be written

$$S = [Ag^+] + [AgCl_{(aq)}] + [AgCl_2^-]$$

Using the three equilibrium constants

$$K_{sp} = [Ag^+][Cl^-]$$

$$K_0 = [AgCl_{(aq)}]$$

$$K_1 = \frac{[AgCl_2^-]}{[Cl^-]}$$

allows us to write the material balance condition in the form

$$S = \frac{K_{sp}}{[Cl^-]} + K_0 + K_1\,[Cl^-]$$

We find the minimum solubility as a function of chloride concentration by differentiation of this expression with respect to $[Cl^-]$, and setting the result equal to zero:

$$\frac{dS}{d[Cl^-]} = -\frac{K_{sp}}{[Cl^-]^2_{min}} + K_1 = 0$$

$$[Cl^-]_{min} = \sqrt{\frac{K_{sp}}{K_1}}$$

This leads to

$$\log [Cl^-]_{min} = \frac{1}{2}(\log K_{sp} - \log K_1)$$

$$= \frac{1}{2}(pK_1 - pK_{sp})$$

$$= -2.53$$

$$[Cl^-]_{min} = 3.0 \times 10^{-3}$$

Example 1.14

It is well known that water has a maximum density at about 4°C. Find the precise temperature of maximum density to the nearest 0.01°C, given that the volume of a gram of water is accurately represented at low temperatures (in °C) by

$$v = 1.0000 - 6.4270 \times 10^{-5}t + 8.5053 \times 10^{-6}t^2 - 6.790 \times 10^{-8}t^3$$

Solution. Differentiation with respect to t gives

$$\frac{dv}{dt} = -6.4270 \times 10^{-5} + 2(8.5053 \times 10^{-6})t - 3(6.790 \times 10^{-8})t^2$$

The minimum occurs where $dv/dt = 0$, or

$$2.037 \times 10^{-7}t^2 - 1.70106 \times 10^{-5}t + 6.4270 \times 10^{-5} = 0$$

By the quadratic formula,

$$t = \frac{1.70106 \times 10^{-5} \pm \sqrt{(1.70106 \times 10^{-5})^2 - 4(2.037 \times 10^{-7})(6.4270 \times 10^{-5})}}{2(2.037 \times 10^{-7})}$$

$$= \frac{1.70106 \times 10^{-5} \pm 1.53946 \times 10^{-5}}{4.074 \times 10^{-7}} = 79.54,\ 3.97$$

The root near 4°C is 3.97°C; the other has no meaning because the formula is accurate only in the range of small values of t.

1.5.3 Derivatives of Logarithmic Functions

We will let $y = \log_b u$, where u is any positive function of x that can be differentiated. Proceeding as in the preceding sections, we write

$$y + \Delta y = \log_b(u + \Delta u)$$

and

$$\begin{aligned}\Delta y &= \log_b(u + \Delta u) - \log_b u \\ &= \log_b \frac{u + \Delta u}{u} = \log_b\left(1 + \frac{\Delta u}{u}\right)\end{aligned}$$

Using Eq. 1.13 we can write this in the form

$$\Delta y = \frac{\Delta u}{u}\,\frac{u}{\Delta u}\log_b\left(1 + \frac{\Delta u}{u}\right) = \frac{\Delta u}{u}\log_b\left(1 + \frac{\Delta u}{u}\right)^{u/\Delta u}$$

so

$$\frac{\Delta y}{\Delta x} = \frac{1}{u}\,\frac{\Delta u}{\Delta x}\log_b\left(1 + \frac{\Delta u}{u}\right)^{u/\Delta u} \tag{1.68}$$

In the limit where Δx approaches zero, Δu and Δy also approach zero, and the argument of the log term in the limit of $\Delta x \to 0$ can be written

$$\lim_{t\to 0}\,(1 + t)^{1/t} \tag{1.69}$$

where we have made the substitution $t = \Delta u/u$. A plot of the function $(1 + t)^{1/t}$ for various small values of t is shown in Figure 1.10. The expression Eq. 1.69 defines a very important *transcendental number*, that is, one that cannot be obtained as a root of an algebraic equation with rational coefficients. This is the number $e = 2.71828$, which was mentioned in Section 1.1.2.

Equation 1.68 in the limit $\Delta x \to 0$ therefore becomes

$$\frac{d(\log_b u)}{dx} = \frac{1}{u}\log_b e\,\frac{du}{dx} \tag{1.70}$$

We can now see an obvious advantage to the use of natural rather than common logs. Since $\ln e = 1$, Eq. 1.70 simplifies to

$$\frac{d\ln u}{dx} = \frac{1}{u}\,\frac{du}{dx} \tag{1.71}$$

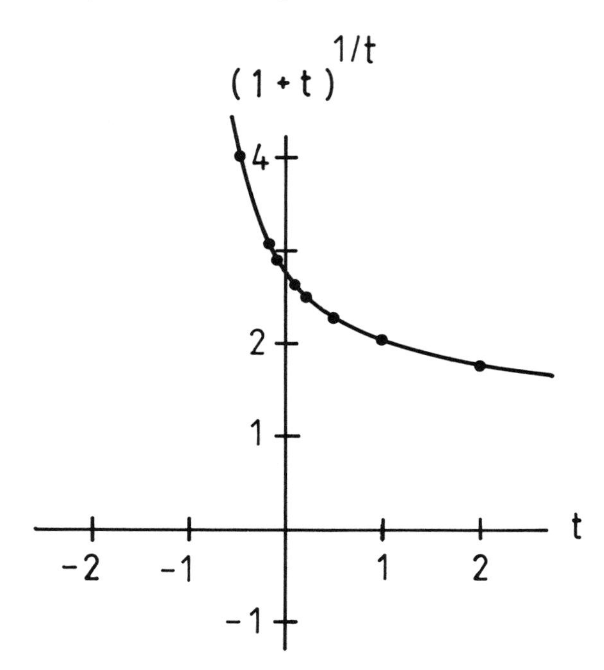

Figure 1.10 Graph of the function $(1 + t)^{1/t}$.

If u is simply equal to x, then

$$\frac{d \ln x}{dx} = \frac{1}{x} \tag{1.72}$$

Example 1.15

For a first-order reversible reaction it may be shown that the fraction of material reacted x and the time t are related by

$$t = \frac{c}{k} \ln \frac{c}{c - x}$$

where c and k are constants. Find an expression for dx/dt. What is the value of x at equilibrium?

Solution. Differentiation of the foregoing expression for t with respect to x may be written

$$\frac{dt}{dx} = \frac{c}{k} \frac{d \ln \dfrac{c}{c - x}}{dx}$$

If we let $u = c/(c - x)$ and use the chain rule Eq. 1.64, we get

$$\frac{d \ln u}{dx} = \frac{1}{u} \frac{du}{dx} = \frac{c - x}{c} [c(-1)(c - x)^{-2}(-1)]$$

$$= \frac{c - x}{c} \left[\frac{c}{(c - x)^2} \right] = \frac{1}{c - x}$$

so

$$\frac{dt}{dx} = \frac{c}{k}\left(\frac{1}{c - x}\right)$$

Using Eq. 1.65 gives the desired result:

$$\frac{dx}{dt} = \frac{k}{c}(c - x)$$

Now at equilibrium we require that

$$\frac{dx}{dt} = \frac{k}{c}(c - x) = 0$$

and since k and c are presumably not equal to zero, we have

$$x = c$$

1.5.4 Derivatives of Exponential Functions

Formulas for the differentiation of exponential functions are easily derived. Let $y = a^u$, where u is again some differentiable function of x. Then

$$\ln y = \ln a^u = u \ln a$$

Since $\ln a$ is a constant, we have

$$\frac{d \ln y}{dx} = \frac{1}{y}\frac{dy}{dx} = \ln a \frac{du}{dx}$$

Therefore,

$$\frac{da^u}{dx} = y \ln a \frac{du}{dx}$$

$$= a^u \ln a \frac{du}{dx} \qquad (1.73)$$

In the case where $a = e$, we have

$$\frac{de^u}{dx} = e^u \frac{du}{dx} \qquad (1.74)$$

and if $u = x$, then

$$\frac{de^x}{dx} = e^x \qquad (1.75)$$

Example 1.16

For the consecutive reaction scheme

$$\mathrm{A} \xrightarrow{k_1} \mathrm{B} \xrightarrow{k_2} \mathrm{C}$$

it may be shown that the concentration of B as a function of time is

$$b = a_0\left(\frac{k_1}{k_2 - k_1}\right)(e^{-k_1 t} - e^{-k_2 t})$$

It has been assumed that only A is present (at concentration a_0) at time $t = 0$. Derive an expression giving the time t for which b has its maximum value.

Solution. We differentiate with respect to t and set the result equal to zero:

$$\frac{db}{dt} = a_0\left(\frac{k_1}{k_2 - k_1}\right)(-k_1 e^{-k_1 t} + k_2 e^{-k_2 t}) = 0$$

This may be written in the form

$$k_1 e^{-k_1 t} = k_2 e^{-k_2 t}$$

Taking the natural logarithm of both sides of the equation gives

$$\ln k_1 - k_1 t = \ln k_2 - k_2 t$$

which may be solved for t to obtain

$$t = \frac{1}{k_1 - k_2} \ln \frac{k_1}{k_2}$$

1.5.5 Derivatives of Trigonometric Functions

We consider first the derivative of the sine function $y = \sin u$. We again note that if we increase u by a certain amount Δu, we must replace y by $y + \Delta y$:

$$y + \Delta y = \sin(u + \Delta u)$$

Solving this equation for Δy gives

$$\Delta y = \sin(u + \Delta u) - \sin u$$

We now apply Eq. 1.23 using $\gamma = u + \Delta u$ and $\delta = u$ to obtain

$$\Delta y = 2\cos\left(u + \frac{\Delta u}{2}\right)\sin\frac{\Delta u}{2}$$

If we now divide through by the change Δx in the independent variable x, we have

$$\frac{\Delta y}{\Delta x} = \frac{2\cos\left(u + \frac{\Delta u}{2}\right)\sin\frac{\Delta u}{2}}{\Delta u}\frac{\Delta u}{\Delta x}$$

$$= \cos\left(u + \frac{\Delta u}{2}\right) \frac{\sin \dfrac{\Delta u}{2}}{\dfrac{\Delta u}{2}} \frac{\Delta u}{\Delta x} \tag{1.76}$$

Now as $\Delta x \to 0$ we also have $\Delta u \to 0$, so

$$\frac{dy}{dx} = \cos u \left(\lim_{\Delta u \to 0} \frac{\sin \dfrac{\Delta u}{2}}{\dfrac{\Delta u}{2}} \right) \frac{du}{dx} \tag{1.77}$$

The term in parentheses may be evaluated as follows: From Figure 1.11 we see that if θ is a small angle measured in radians, then

$$\text{area } OAR = \frac{1}{2}(OR \sin \theta)(OR \cos \theta)$$

$$\text{area } OBR = \frac{1}{2}(OR)^2\theta$$

$$\text{area } OCR = \frac{1}{2}(OR)(OR \tan \theta)$$

It is obvious from the figure that

$$\text{area } OAR < \text{area } OBR < \text{area } OCR$$

or

$$\sin \theta \cos \theta < \theta < \frac{\sin \theta}{\cos \theta}$$

If we divide through by $\sin \theta$ and then take the reciprocal, we have

$$\frac{1}{\cos \theta} > \frac{\sin \theta}{\theta} > \cos \theta$$

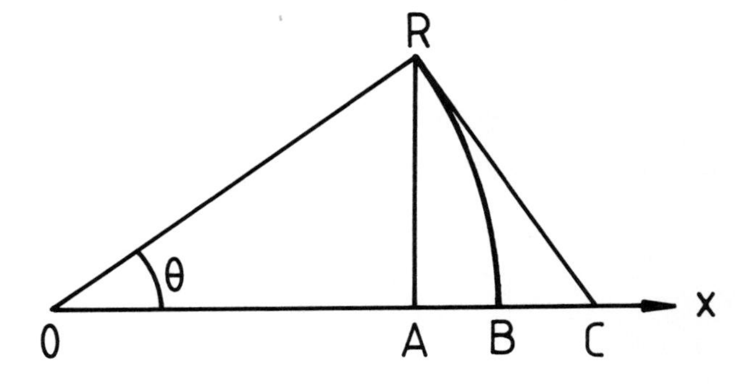

Figure 1.11 The line BR is the arc of a circle of radius OR, and CR is perpendicular to OR; that is, it is tangent to the circle at point R.

Now as $\theta \to 0$, $\cos\theta \to 1$, so we have

$$\lim_{\theta \to 0} \frac{\sin\theta}{\theta} = 1$$

Inserting this result in Eq. 1.77 gives

$$\frac{d \sin u}{dx} = \cos u \frac{du}{dx} \tag{1.78}$$

Since the cosine function is equivalent to the sine function shifted 90 degrees,

$$\cos u = \sin\left(\frac{\pi}{2} - u\right)$$

we can write

$$\frac{d \cos u}{dx} = \frac{d \sin\left(\frac{\pi}{2} - u\right)}{dx}$$

This may be differentiated using the chain rule to obtain

$$\frac{d \sin\left(\frac{\pi}{2} - u\right)}{d\left(\frac{\pi}{2} - u\right)} \frac{d\left(\frac{\pi}{2} - u\right)}{dx} = \cos\left(\frac{\pi}{2} - u\right)\left(-\frac{du}{dx}\right)$$

But once again $\cos(\pi/2 - u) = \sin u$, and we see that

$$\frac{d \cos u}{dx} = -\sin u \frac{du}{dx} \tag{1.79}$$

The formulas for differentiation of any of the other trigonometric functions may readily be derived using Eqs. 1.78 and 1.79. Formulas for the differentiation of the hyperbolic functions may also be derived with ease using the equations of Section 1.5.4, and they are not given here (see the Examples and the Exercises at the end of the chapter).

Example 1.17

Derive a formula for the differentiation of $\sinh x$.

Solution. Using the definition of $\sinh x$ given in Eq. 1.24, we can write

$$\frac{d \sinh x}{dx} = \frac{d\left(\frac{e^{x} - e^{-x}}{2}\right)}{dx}$$

Making use of Eq. 1.74, we get

$$\frac{d\left(\dfrac{e^x - e^{-x}}{2}\right)}{dx} = \frac{1}{2}\left(\frac{de^x}{dx} - \frac{de^{-x}}{dx}\right) = \frac{1}{2}(e^x + e^{-x})$$

But by definition (Eq. 1.24) this is just $\cosh x$, so

$$\frac{d \sinh x}{dx} = \cosh x$$

Example 1.18

Derive a formula for the differentiation of $\tan \theta$ with respect to θ.

Solution. Since by definition $\tan \theta = \sin \theta/\cos \theta$, we have

$$\frac{d \tan \theta}{d\theta} = \frac{d\left(\dfrac{\sin \theta}{\cos \theta}\right)}{d\theta}$$

Using Eq. 1.67, this can be written

$$\frac{\cos \theta\left(\dfrac{d \sin \theta}{d\theta}\right) - \sin \theta\left(\dfrac{d \cos \theta}{d\theta}\right)}{\cos^2\theta}$$

By Eqs. 1.18 and 1.19 this simplifies to

$$\frac{\cos^2\theta + \sin^2\theta}{\cos^2\theta} = 1 + \tan^2\theta = \sec^2\theta$$

1.5.6 Partial Derivatives

Suppose that we have a simple rectangular figure such as

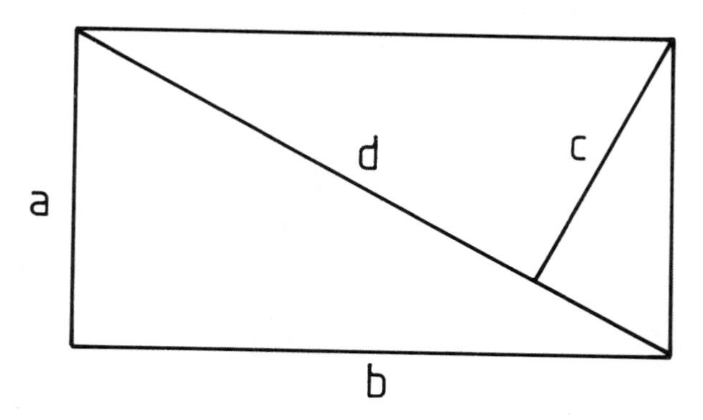

and we wish to evaluate its area. There are a number of ways in which this might be done; it is obvious that

$$A = ab \qquad \text{and} \qquad A = cd$$

are two ways to express it, and with a little algebraic manipulation we can show that

$$A = a(d^2 - a^2)^{1/2} \qquad \text{and} \qquad A = a^2c(a^2 - c^2)^{-1/2}$$

are also legitimate equations giving the area in terms of the parameters shown in the figure.

We now consider the question: How does the area A vary with respect to a change in the length a? This is given by the derivative dA/da, which we find by differentiation of the expression for A. However, we see that we will get a different result for each of the four equations given above. How can they all be correct?

The answer lies in the fact that the question as we have posed it is incomplete; some other condition must also be specified. If we wish to know how the area A varies with respect to a change in the length a, subject to the condition that the length b is held constant, we can use the first equation $A = ab$. Treating b as a constant, we get $dA/da = b$ as the answer. Similarly, we may ask how A varies with the length a if the diagonal length d is held constant. In this case we would use the expression $A = a(d^2 - a^2)^{1/2}$ and differentiate with respect to a treating d as a constant, and so forth. The various expressions that we obtain by differentiation with respect to a single variable, while treating the others as constants, are known as *partial derivatives*.

We will generalize the result by assuming some variable z which is a function of the independent variables x and y:

$$z = f(x, y)$$

The graphical representation of the relationship requires a three-dimensional plot such as that shown schematically in Figure 1.12. Also shown in Figure 1.12 are two

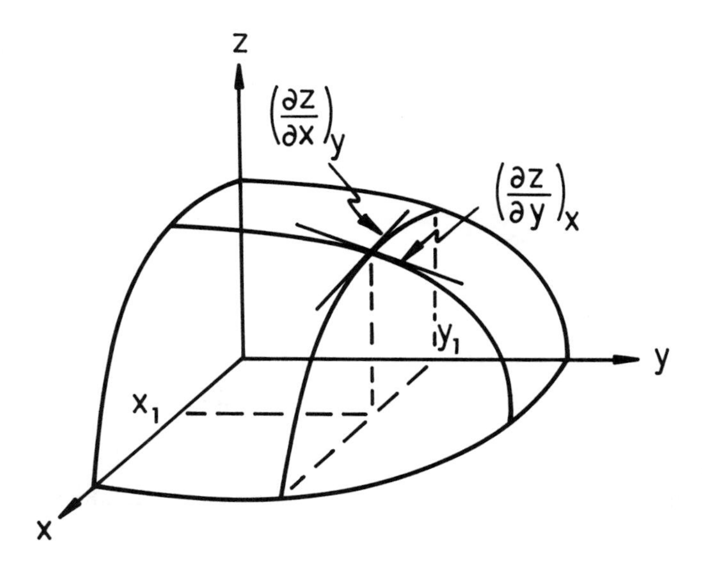

Figure 1.12 Three-dimensional schematic representation of the graph of $z = f(x, y)$. Two sections through the surface are shown, $f(x_1, y)$ and $f(x, y_1)$. The partial derivatives of z with respect to x and y at their point of intersection are also shown.

sections through the surface representing z as a function of x and y: one for which x is held constant at the value x_1, and one for which y is held constant at the value y_1. Holding one of the independent variables constant in essence reduces the problem to a two-dimensional one of the type already discussed in previous sections.

At the point $z_1 = f(x_1, y_1)$, the two curves $f(x, y_1)$ and $f(x_1, y)$ intersect. We can specify the slope at the point with respect to each of the independent variables. If y is held constant at the value y_1, then $f(x, y_1)$ involves only the one independent variable x, and we can evaluate the slope with respect to x. This is *the partial derivative of z with respect to x holding y constant*, and it is represented by

$$\left(\frac{\partial z}{\partial x}\right)_y = \lim_{\Delta x \to 0} \frac{f(x_1 + \Delta x, y_1) - f(x_1, y_1)}{\Delta x} \tag{1.80}$$

Similarly, *the partial derivative of z with respect to y holding x constant* is

$$\left(\frac{\partial z}{\partial y}\right)_x = \lim_{\Delta y \to 0} \frac{f(x_1, y_1 + \Delta y) - f(x_1, y_1)}{\Delta y} \tag{1.81}$$

The actual process of differentiation is the same as that already discussed. The only difference is that in the present case we are dealing with two independent variables, and differentiation is always carried out with respect to variation of a single variable, treating the other as a constant. There may be more than two independent variables. Then a three-dimensional visualization like that of Figure 1.12 is no longer possible, but the process is the same conceptually. We calculate partial derivatives for each independent variable in turn, holding all others constant, and thereby determine the slope with respect to each of the variables.

We do not restrict ourselves in practice to independent variations of a single variable, but must often consider the simultaneous variation of more than one of the independent variables. Let us assume again for the sake of simplicity that there are only two independent variables x and y as represented by

$$z = f(x, y)$$

If x and y are simultaneously varied, there will be a corresponding change in the variable z which may be written

$$\Delta z = f(x + \Delta x, y + \Delta y) - f(x, y)$$

We now add and subtract $f(x, y + \Delta y)$ to obtain

$$\Delta z = f(x + \Delta x, y + \Delta y) - f(x, y + \Delta y) + f(x, y + \Delta y) - f(x, y)$$

If the first two terms are multiplied by the factor $\Delta x/\Delta x$ and the second two by the factor $\Delta y/\Delta y$, the result may be written

$$\Delta z = \left[\frac{f(x + \Delta x, y + \Delta y) - f(x, y + \Delta y)}{\Delta x}\right]\Delta x + \left[\frac{f(x, y + \Delta y) - f(x, y)}{\Delta y}\right]\Delta y$$

In the limit as $\Delta x \to 0$ and $\Delta y \to 0$, this becomes

$$dz = \left[\lim_{\substack{\Delta x \to 0 \\ \Delta y \to 0}} \frac{f(x + \Delta x, y + \Delta y) - f(x, y + \Delta y)}{\Delta x} \right] dx + \left[\lim_{\Delta y \to 0} \frac{f(x, y + \Delta y) - f(x, y)}{\Delta y} \right] dy$$

For the first term in brackets y is constant, and the expression is just the equivalent of $(\partial z/\partial x)_y$, while x is constant in the second term in brackets, and it is simply $(\partial z/\partial y)_x$. Thus the expression becomes

$$dz = \left(\frac{\partial z}{\partial x}\right)_y dx + \left(\frac{\partial z}{\partial y}\right)_x dy \tag{1.82}$$

for infinitesimal changes in both x and y. Thus we see that the *total differential* of z is the rate of change of z with respect to a change in x, times the change in x, plus the rate of change of z with respect to a change in y, times the change in y. If there were more independent variables, they would be added on with the same format.

Many of the properties of simple derivatives apply to partial derivatives as well. For example, by analogy with Eq. 1.65, if a partial derivative is nonzero, its reciprocal exists and it is given by

$$\left(\frac{\partial z}{\partial x}\right)_y = \frac{1}{\left(\dfrac{\partial x}{\partial z}\right)_y} \tag{1.83}$$

Higher-order derivatives also exist and may be represented by

$$\left(\frac{\partial}{\partial x}\left(\frac{\partial z}{\partial x}\right)_y\right)_y \equiv \left(\frac{\partial^2 z}{\partial x^2}\right)_y$$

In this case the possibility also exists of mixed multiple derivatives, such as

$$\left(\frac{\partial}{\partial y}\left(\frac{\partial z}{\partial x}\right)_y\right)_x$$

It turns out that the order of differentiation is immaterial as long as we are dealing with functions that are single-valued and continuous, so we can write

$$\frac{\partial^2 z}{\partial y\, \partial x} = \frac{\partial^2 z}{\partial x\, \partial y} \tag{1.84}$$

Other important properties of partial derivatives may be readily derived from Eq. 1.82. We see from this equation that if we were to consider the variation of z with respect to some other variable u, treating y as a constant, the result would be

$$\left(\frac{\partial z}{\partial u}\right)_y = \left(\frac{\partial z}{\partial x}\right)_y \left(\frac{\partial x}{\partial u}\right)_y \tag{1.85}$$

Equation 1.85 may be regarded as a result of division of Eq. 1.82 by the quantity du, subject to the restriction that $dy = 0$. It should be noted that a differential is not a simple algebraic quantity, however, and its manipulation must be treated with some caution. Equation 1.85 is the equivalent of the chain rule of differentiation (Eq. 1.64) applied to partial derivatives.

From Eq. 1.82 it follows that for an infinitesimal process in which y is varied in such a way that z remains constant,

$$0 = \left(\frac{\partial z}{\partial x}\right)_y \left(\frac{\partial x}{\partial y}\right)_z + \left(\frac{\partial z}{\partial y}\right)_x$$

By use of Eq. 1.83 this may be written

$$\left(\frac{\partial z}{\partial x}\right)_y \left(\frac{\partial x}{\partial y}\right)_z = -\left(\frac{\partial z}{\partial y}\right)_x = -\frac{1}{\left(\frac{\partial y}{\partial z}\right)_x}$$

or

$$\left(\frac{\partial x}{\partial y}\right)_z \left(\frac{\partial y}{\partial z}\right)_x \left(\frac{\partial z}{\partial x}\right)_y = -1 \tag{1.86}$$

This is a very important result that is valid regardless of the functional relationship between x, y, and z. It is often used to simplify expressions involving their derivatives. The form is easily remembered, because the partial derivatives involve a *cyclic permutation* of the three variables: that is, $x \leftarrow y \leftarrow z$ (with z linked back to x).

Example 1.19

The equation of state obeyed by an ideal gas is $pV = nRT$. Evaluate the product

$$\left(\frac{\partial V}{\partial T}\right)_p \left(\frac{\partial p}{\partial V}\right)_T \left(\frac{\partial T}{\partial p}\right)_V$$

for an ideal gas.

Solution. Solving the gas law for V, we get $V = nRT/p$. Differentiation with respect to T, holding p constant, gives

$$\left(\frac{\partial V}{\partial T}\right)_p = \left[\frac{\partial\left(\frac{nRT}{p}\right)}{\partial T}\right]_p = \frac{nR}{p}$$

Similarly,

$$p = \frac{nRT}{V} \quad \text{and} \quad \left(\frac{\partial p}{\partial V}\right)_T = -\frac{nRT}{V^2}$$

$$T = \frac{pV}{nR} \quad \text{and} \quad \left(\frac{\partial T}{\partial p}\right)_V = \frac{V}{nR}$$

Therefore,

$$\left(\frac{\partial V}{\partial T}\right)_p\left(\frac{\partial p}{\partial V}\right)_T\left(\frac{\partial T}{\partial p}\right)_V = \frac{nR}{p}\left(-\frac{nRT}{V^2}\right)\frac{V}{nR}$$

$$= -\frac{nRT}{pV} = -1$$

as demanded by Eq. 1.86.

Example 1.20

The *isothermal compressibility* of a given material is the relative decrease in volume with pressure at constant temperature:

$$\kappa = -\frac{1}{V}\left(\frac{\partial V}{\partial p}\right)_T$$

Calculate κ for an ideal gas at $p = 5$ atm.

Solution. Since $V = nRT/p$, we have

$$\left(\frac{\partial V}{\partial p}\right)_T = -\frac{nRT}{p^2}$$

and

$$\kappa = -\frac{1}{V}\left(-\frac{nRT}{p^2}\right) = \frac{nRT}{pV}\left(\frac{1}{p}\right) = \frac{1}{p}$$

At $p = 5$ atm,

$$\kappa = \frac{1}{5\text{ atm}} = 0.2\text{ atm}^{-1}$$

Example 1.21

The *thermal expansivity* of a material is the relative change in volume with temperature,

$$\alpha = \frac{1}{V}\left(\frac{\partial V}{\partial T}\right)_p$$

Calculate the thermal expansivity of water at 20°C, given that its specific volume is given as a function of temperature as in Example 1.11.

Solution. An expression for the derivative of the specific volume v with respect to temperature was given in Example 1.11. Inserting $t = 20$°C gives

$$-6.4270 \times 10^{-5} + 1.70106 \times 10^{-5}\,(20) - 2.037 \times 10^{-7}\,(20)^2$$

$$= 1.9446 \times 10^{-4}\text{ cm}^3\text{ deg}^{-1}$$

Similarly, we find that the volume at $t = 20$°C is

$$1.0000 - 6.4270 \times 10^{-5}\,(20) + 8.5053 \times 10^{-6}\,(20)^2 - 6.790 \times 10^{-8}\,(20)^3$$

$$= 1.0016\text{ cm}^3$$

and therefore,

$$\alpha = \frac{1.9446 \times 10^{-4}\ \text{cm}^3\ \text{deg}^{-1}}{1.0016\ \text{cm}^3}$$
$$= 1.9416 \times 10^{-4}\ \text{deg}^{-1}$$

1.6 INTEGRAL CALCULUS

We have seen that the derivative of a function defines many of its very useful characteristics and we have learned how it can be calculated in a number of cases. In some applications, we may know the derivative and need the function itself. This may be termed antidifferentiation, but it is more usual to refer to the process as *integration*. This forms the basis for *integral calculus*, which is the subject matter dealt with in this section.

1.6.1 Indefinite Integrals

Let us assume that we are given a certain function $f(x)$ which we consider to be a derivative, and that we wish to determine the function $F(x)$ from which it is derived. In this context the function $F(x)$ is called the *integral*, and the function $f(x)$ that is to be integrated is known as the *integrand*. We represent the process by

$$F(x) = \int f(x)\, dx \tag{1.87}$$

First consider the case $f(x) = ax^n$. Referring back to Eq. 1.62, we see that the derivative of $ax^{n+1}/(n + 1)$ is equal to ax^n. However, since the derivative of a constant is zero, the derivative of $ax^{n+1}/(n + 1) + c$, where c is any constant, is also equal to ax^n. Therefore, we have

$$\int ax^n\, dx = a\frac{x^{n+1}}{n+1} + c \tag{1.88}$$

where c is called the *constant of integration*. This formula is valid for all values of n except $n = -1$.

For $n = -1$, the denominator of Eq. 1.88 is equal to zero, and this case must be dealt with separately. Inserting $n = -1$ in the integrand gives

$$\int ax^{-1}\, dx = \int a\frac{dx}{x}$$

From Eq. 1.72 we see that

$$d(a \ln x + c) = a\frac{dx}{x}$$

so

$$\int a \frac{dx}{x} = a \ln x + c \tag{1.89}$$

We have seen that $d(af(x))/dx = a\, df(x)/dx$, so we would expect that

$$\int af(x)\, dx = a \int f(x)\, dx$$

Also, just as we can differentiate term by term (Eq. 1.63), we can likewise integrate term by term:

$$\int [f(x) \pm g(x)]\, dx = \int f(x)\, dx \pm \int g(x)\, dx \tag{1.90}$$

Exponential integrals are easily handled. We see by reference to Eq. 1.73 that

$$\int a^u\, du = \frac{a^u}{\ln a} + c \tag{1.91}$$

and from Eq. 1.74 that

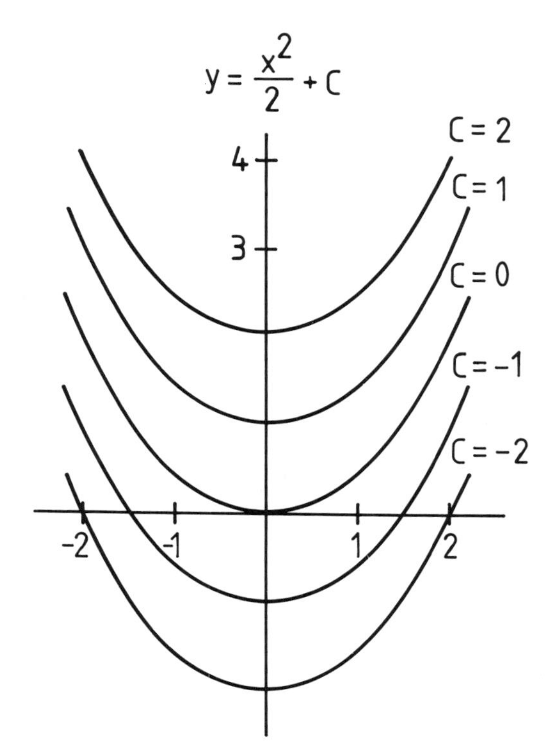

Figure 1.13 Family of curves whose slope $m = x$ (see Eq. 1.93).

$$\int e^u \, du = e^u + c \tag{1.92}$$

Some of the more commonly used integrals are tabulated in Appendix A.2 for quick reference.

The presence of a constant of integration in each of the integral formulas indicates that an indefinite integral is not a unique curve, but rather an entire *family of curves*. For example, let us assume that we wish to know the equation of a curve whose slope m is equal to x, $dy/dx = m = x$. By integration of $dy = x\,dx$ we find that

$$y = \int dy = \int x \, dx = \frac{x^2}{2} + c \tag{1.93}$$

We see that the solution in this case is a family of *parabolas*, all of which have the slope $m = x$, as shown in Figure 1.13. If we wish to have a unique solution, the constant of integration must be determined. For example, we might require that the solution be equal to $+1$ at $x = 0$. Then we have

$$+1 = \frac{(0)^2}{2} + c = c$$

or $c = +1$ and the curve is completely specified.

1.6.2 Definite Integrals

Let us assume we have some function $y = f(x)$ whose graph is shown schematically in Figure 1.14. A is defined as the area under the curve between the limits $x = a$ and $x = b$, which is represented by the shaded region. We can calculate A by summing all the various small contributions ΔA_i. Taking ΔA_3 as an example, we let $x_3 - x_2 = \Delta x$, in

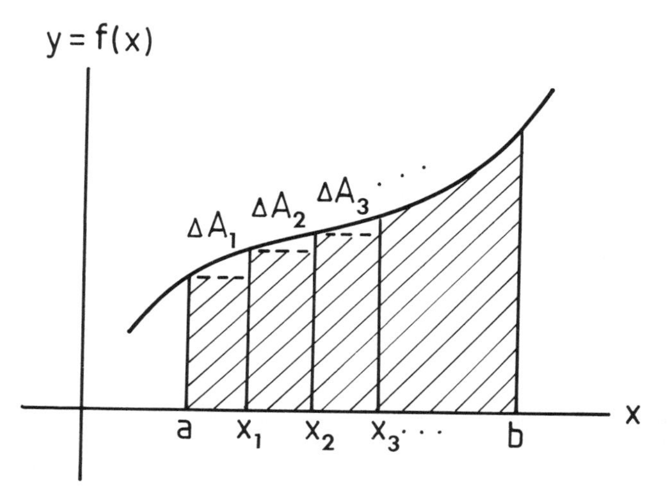

Figure 1.14 Graph of the function $f(x)$. A is the shaded area under the curve within the interval $x = a$ to $x = b$.

which case the area ΔA_3 is equal to $f(x_2)\,\Delta x$ plus a small nearly triangular area approximated by $[f(x_3) - f(x_2)]\,\Delta x/2$. We can write this as

$$\frac{\Delta A_3}{\Delta x} = f(x_2) + \frac{f(x_3) - f(x_2)}{2}$$

Now in the limit of an infinitesimally small Δx, the quantity $f(x_3) - f(x_2)$ is vanishingly small, so

$$\lim_{\Delta x \to 0} \frac{\Delta A_3}{\Delta x} = \frac{dA_3}{dx} = f(x_2)$$

By generalization of this result we see that at any point along the curve

$$dA = f(x)\,dx \tag{1.94}$$

or in other words, *the differential of the area dA is equal to f(x) dx*.

We conclude that the area A is the function of x whose derivative with respect to x is $f(x)$. Integration of the differential area dA with respect to x gives A, the total area under the curve. Let us write the integral as

$$A = \int f(x)\,dx = F(x) + c$$

where $F(x)$ can be found for various functions $f(x)$ by consulting Appendix A.2.

We can evaluate the constant of integration by noting that the area must be zero if we let $x = a$:

$$0 = F(a) + c$$

or

$$c = -F(a)$$

Then the area for arbitrary values of x is given by

$$A = F(x) - F(a)$$

The total area under the curve bounded by $x = a$ and $x = b$ is found by letting $x = b$ in the equation above, which leads to

$$A = F(b) - F(a) \tag{1.95}$$

We call this area the *definite integral* of $f(x)$ in the interval of integration a, b. Here a is called the lower limit and b the upper limit. Symbolically, we write

$$\int_a^b f(x)\,dx = [F(x)]_a^b = F(b) - F(a) \tag{1.96}$$

Example 1.22

The molar heat capacity $\bar{C}_p$ is the amount of heat required to raise the temperature of 1 mol of a material by 1 degree at constant pressure. We can assume that the heat capacity follows a T-cubed law $\bar{C}_p = aT^3$ at very low temperatures. Given that the heat capacity of silver at 20 K is 0.3995 cal mol^{-1} K^{-1}, how much heat would be required to raise the temperature of 2 mol of Ag from 5 K to 10 K?

Solution. First we solve for the value of a:

$$a = \frac{\bar{C}_p}{T^3} = \frac{0.3995 \text{ cal mol}^{-1}\text{ K}^{-1}}{(20 \text{ K})^3}$$

$$= 4.99 \times 10^{-5} \text{ cal mol}^{-1}\text{ K}^{-1}$$

The heat required is then given by integration as follows:

$$\int_5^{10} \bar{C}_p \, dT = a \int_5^{10} T^3 \, dT = a \left[\frac{T^4}{4}\right]_5^{10}$$

$$= \frac{a}{4}(10{,}000 - 625) \text{ K}^4$$

$$= \frac{4.99 \times 10^{-5} \text{ cal mol}^{-1}\text{ K}^{-4}}{4}(9375) \text{ K}^4$$

$$= 0.117 \text{ cal mol}^{-1}$$

$$Q = (0.117 \text{ cal mol}^{-1})(2 \text{ mol}) = 0.234 \text{ cal}$$

1.6.3 Some Special Integration Techniques

Although there are extensive tables of integrals in the *CRC Handbook of Chemistry and Physics* and numerous other sources, we still need to be able to carry out integrations ourselves to meet special needs. No general procedure can be outlined, since a wide variety of methods are used depending on the nature of the integrand. A few of the more important of these are outlined here.

Substitution Methods

The substitution method of integral evaluation makes use of an appropriate definition of variables within the integrand so that it is brought into a standard form whose integral is known. Suppose, for example, that we need to evaluate the integral

$$\int e^{-ax} \, dx$$

We note that this is similar, although not exactly equal to integral number 7 of Appendix A.2. However, if we insert $u = -ax$, which in differential form becomes $du = -a\,dx$, we have

$$\int e^{-ax} \, dx = \int e^u \left(-\frac{du}{a}\right) = -\frac{1}{a}\int e^u \, du$$

$$= -\frac{1}{a}[e^u + c]$$

Since c is just an integration constant whose value must be determined, and a is also a constant, we need not write $-c/a$ but can simply define a new constant C so the result is

$$\int e^{-ax}\,dx = -\frac{1}{a}e^{-ax} + C \tag{1.97}$$

As a second example let us consider the following integral:

$$\int \frac{x\,dx}{1+x^2}$$

If we let $u = 1 + x^2$, then $du = 2x\,dx$ and the result is

$$\int \frac{x\,dx}{1+x^2} = \int \frac{\frac{du}{2}}{u} = \frac{1}{2}\int \frac{du}{u} = \frac{1}{2}[\ln u + c]$$

$$= \frac{1}{2}\ln(1+x^2) + C \tag{1.98}$$

Numerous other possibilities exist for simplification by substitution, many of which involve trigonometric identities. Take the following integral as an example:

$$\int \frac{\sin x \cos x}{1 - \cos x}\,dx$$

As a general rule we look for a way of making the denominator as simple as possible. Thus we take $u = 1 - \cos x$, and then we have $\cos x = 1 - u$ and $-\sin x\,dx = -du$. Inserting these in the integral above gives

$$\int \frac{\cos x \sin x}{1-\cos x}\,dx = \int \frac{(1-u)\,du}{u} = \int \frac{du}{u} - \int du$$

$$= \ln u - u + c = \ln(1 - \cos x) - (1 - \cos x) + c$$

$$= \ln(1 - \cos x) + \cos x + C \tag{1.99}$$

Note that in this case we define the integration constant as $C = c - 1$.

Another example of integration using trigonometric identities is provided by the integral

$$\int \sin^3 x\,dx$$

This can be written in the form

$$\int \sin x(1 - \cos^2 x)\,\mathrm{dx} = \int \sin x\,dx - \int \cos^2 x \sin x\,dx$$

If in the second integral we make the substitution $u = \cos x$, then $du = -\sin x \, dx$ and the result is

$$\int \sin^3 x \, dx = \int \sin x \, dx + \int u^2 \, du$$

$$= -\cos x + \frac{1}{3} \cos^3 x + C \tag{1.100}$$

In dealing with definite integrals, it is customary to change the limits of integration when we make a substitution and then it is not necessary to change back to the original variable after the integration is carried out. Suppose, for example, that we wish to integrate

$$\int_0^2 \frac{dx}{1 + x}$$

We make the obvious substitution $u = 1 + x$ with $du = dx$. We note, however, that at $x = 0$, $u = 1$ and at $x = 2$, $u = 3$. The integral is therefore

$$\int_0^2 \frac{dx}{1 + x} = \int_1^3 \frac{du}{u} = [\ln u]_1^3 = \ln 3$$

Of course, the same result would have been obtained from

$$[\ln(1 + x)]_0^2$$

Partial Fractions

If the integrand involves a polynomial in the denominator, it may be very difficult to handle unless it can be broken up into simpler terms. This may be done by the method of partial fractions if the polynomial is factorable. Consider the following:

$$\int \frac{5x + 3}{x^2 - 6x - 7} \, dx$$

The denominator in this case is factorable, since we note that

$$x^2 - 6x - 7 = (x - 7)(x + 1)$$

The integrand may therefore be written as the sum of two partial fractions,

$$\frac{5x + 3}{x^2 - 6x - 7} = \frac{A}{x - 7} + \frac{B}{x + 1}$$

if the terms A and B are properly chosen. They can be determined through clearing this equation of fractions by putting the terms on the right over the common denominator $(x - 7)(x + 1)$:

$$5x + 3 = A(x + 1) + B(x - 7)$$

If we let $x = -1$ in this equation, we see that $-2 = -8B$ or $B = 1/4$. Insertion of the value $x = +7$ gives $38 = 8A$, or $A = 19/4$. With these values of A and B the integral can easily be evaluated:

$$\int \frac{5x + 3}{x^2 - 6x - 7}\,dx = \frac{1}{4}\int \frac{dx}{x + 1} + \frac{19}{4}\int \frac{dx}{x - 7}$$

$$= \frac{1}{4}\ln(x + 1) + \frac{19}{4}\ln(x - 7) + C \qquad (1.101)$$

When the method of partial fractions is applied to a polynomial that contains a repeated factor, a term must be included for each possible power. For example, if we wish to evaluate the integral

$$\int \frac{2x + 1}{x^3 - 4x^2 + 4x}\,dx$$

we note that the denominator in this case is factorable into $x(x - 2)^2$. Here $(x - 2)$ is a repeating factor that occurs to the second power. Therefore, the integrand is decomposed as follows:

$$\frac{2x + 1}{x^3 - 4x^2 + 4x} = \frac{A}{x} + \frac{B}{x - 2} + \frac{C}{(x - 2)^2}$$

We clear this equation of fractions to obtain

$$2x + 1 = A(x - 2)^2 + Bx(x - 2) + Cx$$

The technique for determining the constants that we used in the previous example will not work here, since there are only two values of x (-2 and 0) which zero the coefficients on the right, but there are three constants to be determined. Since the equation must be satisfied for all values of x, we could use these two plus any other arbitrary value. A second method is to equate the coefficients of like powers of x on the two sides of the equation. We find by equating the coefficients of x^0 that $1 = 4A$, or $A = 1/4$. Similarly, for the x^2 terms we get $0 = A + B$ or $B = -A = -1/4$. Finally, for x^1 the result is $2 = -4A - 2B + C$, or $C = 2 + 4(1/4) + 2(-1/4) = 5/2$. Having determined the coefficients, we can readily complete the integration; the result is

$$\int \frac{2x + 1}{x^3 - 4x^2 + 4x}\,dx = \frac{1}{4}\int \frac{dx}{x} - \frac{1}{4}\int \frac{dx}{x - 2} + \frac{5}{2}\int \frac{dx}{(x - 2)^2}$$

$$= \frac{1}{4}\ln x - \frac{1}{4}\ln(x - 2) + \frac{5}{2}\left(\frac{(x - 2)^{-1}}{-1}\right) + C$$

$$= \frac{1}{4}\ln\frac{x}{x - 2} - \frac{5}{2(x - 2)} + C \qquad (1.102)$$

Integration by Parts

For the differential of a product of two terms we can write (see Eq. 1.66)

$$d(uv) = u\,dv + v\,du$$

or

$$u\,dv = d(uv) - v\,du$$

Integration of both sides leads to the very important equation

$$\int u\,dv = uv - \int v\,du \tag{1.103}$$

This is the basic formula for integration by parts and we use it, often repeatedly, when the integrand involves products of dissimilar factors.

We take as an example the integral

$$\int x^2 \cos x\,dx$$

If we let

$$u = x^2 \qquad dv = \cos x\,dx$$

$$du = 2x\,dx \qquad v = \sin x$$

we can use the formula for integration by parts to get

$$\int x^2 \cos x\,dx = x^2 \sin x - 2\int x \sin x\,dx$$

This has simplified the original integral, but it involves a second which we must now evaluate. In this case we take

$$u = x \qquad dv = \sin x\,dx$$

$$du = dx \qquad v = -\cos x$$

and by a second application of the formula we obtain, finally,

$$\int x^2 \cos x\,dx = x^2 \sin x - 2\left[-x \cos x + \int \cos x\,dx\right]$$

$$= x^2 \sin x + 2x \cos x - 2 \sin x + C \tag{1.104}$$

As a second example let us consider the evaluation of the definite integral

$$\int_0^\infty x^2 e^{-x}\,dx$$

We let

$$u = x^2 \qquad dv = e^{-x}\,dx$$

$$du = 2x\,dx \qquad v = -e^{-x}$$

and thereby obtain

$$\int_0^\infty x^2 e^{-x}\,dx = [-x^2 e^{-x}]_0^\infty + 2\int_0^\infty x e^{-x}\,dx$$

The first term is clearly zero at the lower limit, but the result at the upper limit is less obvious. Setting $x = \infty$ in $x^2 e^{-x}$ actually gives $\infty \cdot 0$, which is called an *indeterminate form*, because its actual value is uncertain. Other expressions that are indeterminate forms are $0/0$ and ∞/∞. In such cases we have to evaluate the limit by use of *l'Hospital's rule*, which states that if $f_1(x)/f_2(x)$ is indeterminate, the proper limit is given by $f_1'(x)/f_2'(x)$. In our case we can write $f_1(x) = x^2$ and $f_2(x) = e^x$, so we have

$$\lim_{x\to\infty} \frac{x^2}{e^x} = \lim_{x\to\infty} \frac{2x}{e^x} = \frac{\infty}{\infty}$$

We see that the result is still indeterminate. However, a second application of l'Hospital's rule can be made and in this case we obtain the result

$$\lim_{x\to\infty} \frac{f_1''(x)}{f_2''(x)} = \frac{2}{e^x} = \frac{2}{\infty} = 0$$

so the integral simplifies to

$$\int_0^\infty x^2 e^{-x}\,dx = 2\int_0^\infty x e^{-x}\,dx$$

A second integration by parts can now be made using

$$u = x \qquad dv = e^{-x}\,dx$$

$$du = dx \qquad v = -e^{-x}$$

This gives

$$\int_0^\infty x e^{-x}\,dx = [-x e^{-x}]_0^\infty + \int_0^\infty e^{-x}\,dx$$

The first term is again equal to zero, but the latter is readily integrated and the final result is

$$\int_0^\infty x^2 e^{-x}\,dx = 2\int_0^\infty e^{-x}\,dx = 2[-e^{-x}]_0^\infty$$

$$= 2(-e^{-\infty} + e^{-0}) = 2 \qquad (1.105)$$

1.6.4 Multiple Integrals

In Section 1.5.6 we learned that a function of several variables may be differentiated by each in turn to obtain the various partial derivatives. By the same token, a function of several variables may be subjected to a *partial integration*, in which the integration process is applied to a single variable with all others treated as constants. The formulas used to carry out such partial integrations are the same as those developed in the preceding sections.

We may also carry out a *multiple integration*, which implies that we integrate with respect to more than one variable. The nature of the result depends on the integrand and on the way in which the integration is carried out.

As a simple example, let us consider the graph shown in Figure 1.15, where two arbitrary functions of x, $y_1 = f_1(x)$ and $y_2 = f_2(x)$ are plotted. Assume that we desire to determine the area bounded by these two curves and the lines $x = a$ and $x = b$. This may be done by dividing the total area into a number of very small squares as shown, and summing all those squares that lie within the area or touch the boundaries. The result is

$$A = \sum_{x=a}^{b} \sum_{y=y_1}^{y_2} \Delta y \, \Delta x \tag{1.106}$$

and the formula should become exact in the limit where the mesh size becomes infinitesimally small. This limit may be written

$$\lim_{\Delta x \to 0} \sum_{x=a}^{b} \left[\lim_{\Delta y \to 0} \sum_{y=y_1}^{y_2} \Delta y \right] \Delta x$$

We note that the term in brackets is a straight-line segment that extends vertically from y_1 to y_2, and it is equivalent to the integral

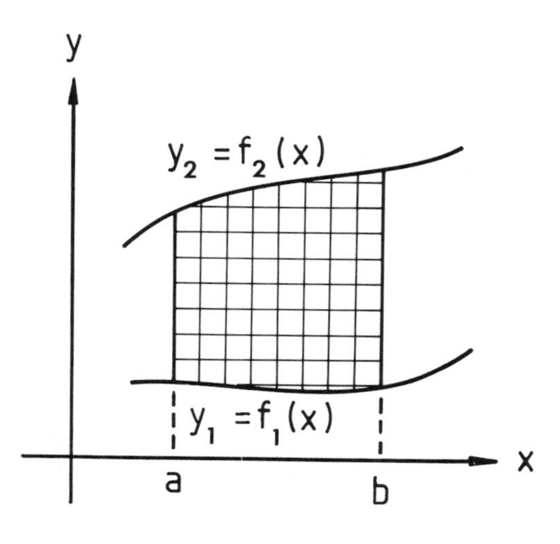

Figure 1.15 The total area enclosed by the curves y_1, y_2 and the lines $x = a$ and $x = b$ may be found by constructing a grid and adding the areas of all the small squares lying within.

$$\int_{f_1(x)}^{f_2(x)} dy = f_2(x) - f_1(x)$$

Each of these lines is multiplied by its width Δx, which is equal to the area of the narrow strip, and the sum of these gives the total area:

$$A = \lim_{\Delta x \to 0} \sum_{x=a}^{b} \int_{f_1(x)}^{f_2(x)} dy\, \Delta x = \int_a^b \int_{f_1(x)}^{f_2(x)} dy\, dx \tag{1.107}$$

Equation 1.107 is called an *iterated integral*, although we frequently use the less descriptive term *double integral*. In this special case the integrand is simply unity, but it may be some arbitrary function of x and y in the general case. We understand that the integration is always done from inside out. That is, we carry out the integration over y, treating x as a constant, and then carry out the integration over x, treating y as a constant. There are instances where the order of integration is interchangeable, but this is not always the case.

Let us now examine the situation where our integrand, say z, is some arbitrary function of x and y: $z = f(x, y)$. We note by reference to Figure 1.16 that this case is analogous to that considered above, except that now each square on our grid is multiplied by a third dimension z, whose magnitude depends on both independent variables x and y. Each of the small elements $z\ \Delta y\ \Delta x$ is a volume in this case, and the sum represents the total volume of the figure as shown:

$$V = \sum_{x=a}^{b} \sum_{y=y_1}^{y_2} z(x, y)\, \Delta y\, \Delta x$$

Again the formula becomes exact in the limit and we have

$$V = \int_a^b \int_{f_1(x)}^{f_2(x)} f(x, y)\, dy\, dx \tag{1.108}$$

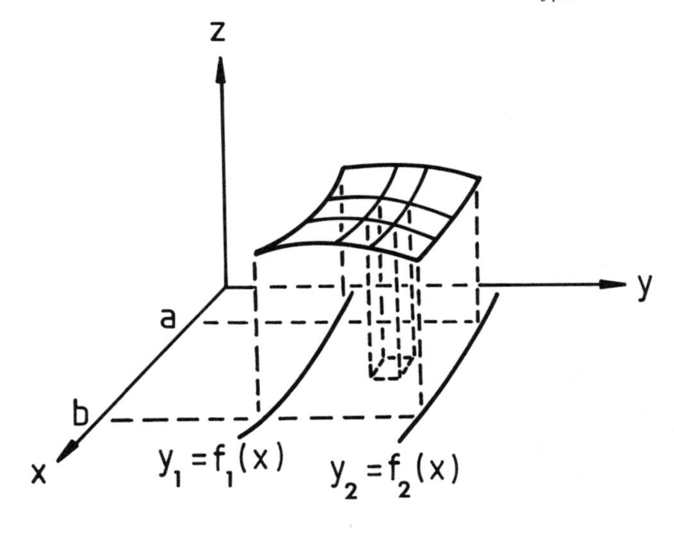

Figure 1.16 Each area of the grid in the x-y plane is multiplied by a height z, making it a volume element of the solid figure shown.

We often carry out multiple integrations in chemistry to determine average properties for an entire atom, molecule, solution, crystal, or any other unit under consideration. It is required in such cases that we carry out the multiple integration over the entire unit with an appropriate weighting function ρ_A, which describes the variation in the property A as a function of the coordinates. This is represented by

$$\bar{A} = \int_{\text{vol}} \rho_A \, d\tau \tag{1.109}$$

where $\bar{A}$ represents the average value of the property A, and $d\tau$ is a differential element of volume. In the case of Cartesian coordinates, the volume element $d\tau$ is $dx\ dy\ dz$. However, in general we make our coordinates conform to the symmetry of the problem to avoid undue complexity, as indicated in Section 1.2.

For the spherical polar coordinates defined in Section 1.2, the volume element $d\tau$ is shown in Figure 1.17. The lengths of the sides of an infinitesimal volume element resulting from differential increases in each of the coordinates dr, $d\theta$, and $d\phi$ are shown; the product of the three gives

$$d\tau = r^2\, dr \sin\theta\, d\theta\, d\phi \tag{1.110}$$

Example 1.23

If a molecule with dipole moment $\boldsymbol{\mu}$ is oriented at an angle θ with respect to an electric field of strength $\mathbf{E}$, then the potential energy of interaction may be written

$$-\boldsymbol{\mu} \cdot \mathbf{E} = \mu E \cos\theta$$

where μ and E are the respective magnitudes of the $\boldsymbol{\mu}$ and $\mathbf{E}$ vectors. The molecules will experience a preferential orientation in the field which is described by the *Boltzmann equation*,

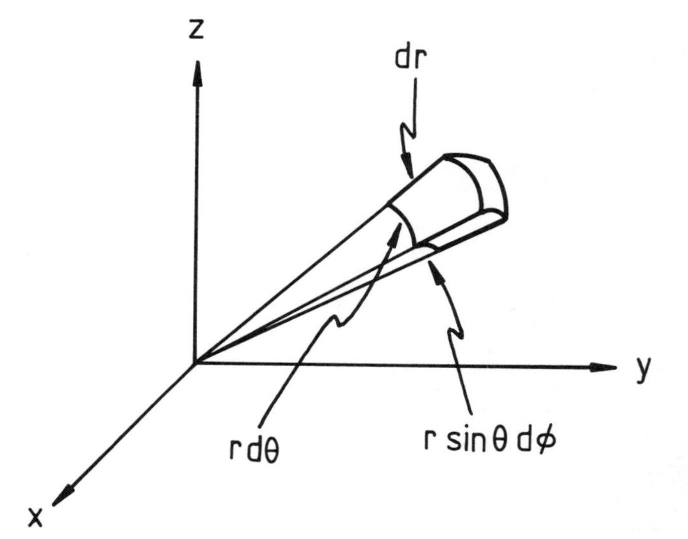

Figure 1.17 The volume element $d\tau$ in terms of the spherical polar coordinates r, θ, and ϕ.

$$N = N_0 e^{-\epsilon/kT}$$

where ϵ is the potential energy. Derive an equation relating the average value of the dipole moment in the direction of the field $\bar{\mu}$ to the total dipole moment μ.

Solution. The potential energy in this case is a function of orientation only, and we can write the Boltzmann equation in the differential form

$$dN = N_0 \exp\left[\frac{\mu E \cos\theta}{kT}\right] \sin\theta \, d\theta \, d\phi$$

The projection of a given dipole moment in the direction of the field is $\mu \cos\theta$. The average value is obtained through weighting this by the Boltzmann factor and integrating over the entire range of coordinates, and dividing by the total number of molecules:

$$\bar{\mu} = \frac{\int \mu \cos\theta \, dN}{\int dN} = \frac{\int_0^{2\pi}\int_0^{\pi} N_0\mu \exp\left[\frac{\mu E \cos\theta}{kT}\right] \cos\theta \sin\theta \, d\theta \, d\phi}{\int_0^{2\pi}\int_0^{\pi} N_0 \exp\left[\frac{\mu E \cos\theta}{kT}\right] \sin\theta \, d\theta \, d\phi}$$

$$= \frac{N_0\mu \int_0^{2\pi} d\phi \int_0^{\pi} \exp\left[\frac{\mu E \cos\theta}{kT}\right] \cos\theta \sin\theta \, d\theta}{N_0 \int_0^{2\pi} d\phi \int_0^{\pi} \exp\left[\frac{\mu E \cos\theta}{kT}\right] \sin\theta \, d\theta}$$

$$= \frac{\mu \int_0^{\pi} \exp\left[\frac{\mu E \cos\theta}{kT}\right] \cos\theta \sin\theta \, d\theta}{\int_0^{\pi} \exp\left[\frac{\mu E \cos\theta}{kT}\right] \sin\theta \, d\theta}$$

If we let $\mu E/kT = x$ and $\cos\theta = y$, then at $\theta = 0$ we have $y = 1$, while at $\theta = \pi$ we have $y = -1$. The expression becomes

$$\frac{\bar{\mu}}{\mu} = \frac{-\int_{+1}^{-1} e^{xy} y \, dy}{-\int_{+1}^{-1} e^{xy} \, dy}$$

Interchanging the limits of integration changes the sign of the integral and we have

$$\frac{\bar{\mu}}{\mu} = \frac{\int_{-1}^{+1} e^{xy} y \, dy}{\int_{-1}^{+1} e^{xy} \, dy}$$

The integral in the denominator by Eq. 1.97 is

$$\int_{-1}^{+1} e^{xy}\, dy = \frac{1}{x}\,[e^{xy}]_{-1}^{+1} = \frac{e^x - e^{-x}}{x}$$

and that in the numerator is evaluated through integration by parts to give (you should verify this)

$$\int_{-1}^{+1} e^{xy} y\, dy = \frac{e^x + e^{-x}}{x} - \frac{e^x - e^{-x}}{x^2}$$

The result may be written

$$\frac{\bar{\mu}}{\mu} = \frac{\dfrac{e^x + e^{-x}}{x} - \dfrac{e^x - e^{-x}}{x^2}}{\dfrac{e^x - e^{-x}}{x}} = \frac{e^x + e^{-x}}{e^x - e^{-x}} - \frac{1}{x} = \coth x - \frac{1}{x} \equiv \mathcal{L}(x)$$

where $\mathcal{L}(x)$ is known as the *Langevin function*.

1.7 DIFFERENTIAL EQUATIONS

An ordinary differential equation is a function of two variables, containing derivatives of one of the variables with respect to the other. There are differential equations of many different types, but we will review only a few which are of special interest to the chemist.

The following is an example of a simple ordinary differential equation:

$$y\frac{dy}{dx} + 2x = 0$$

Here the highest-order derivative involved is first order, so this is called a first-order equation. When dealing with first-order equations it is common to write them in a differential form,

$$y\, dy = -2x\, dx$$

In this case the variables are *separable*. Whenever we can separate the variables and obtain an equation of the form

$$P(x)\, dx + Q(y)\, dy = 0 \tag{1.111}$$

we can obtain a solution by direct integration.

First-order differential equations are not always separable, but if they can be put in the general form

$$\frac{dy}{dx} + P(x)y = Q(x) \tag{1.112}$$

then a solution may still be quite readily obtained. Equation 1.112 is *linear* because it is of first degree in both y and dy/dx. If $Q(x) = 0$ as well, it is also *homogeneous*, meaning that all terms contain the dependent variable y. If we multiply Eq. 1.112 by the factor

$$e^{\int P dx}$$

we obtain

$$e^{\int P dx} \frac{dy}{dx} + yPe^{\int P dx} = Qe^{\int P dx}$$

It will be noted that the left-hand side of this equation is just the derivative of

$$ye^{\int P dx}$$

Therefore, integration of both sides of the equation is equivalent to

$$ye^{\int P dx} = \int Qe^{\int P dx} dx + C$$

or

$$y = e^{-\int P dx}\left[\int Qe^{\int P dx} dx + C\right] \tag{1.113}$$

Because the equation becomes integrable when multiplied by

$$e^{\int P dx}$$

this is known as an *integrating factor*. In practice, we do not generally use Eq. 1.113 directly, but rather determine the integrating factor and then integrate the resulting equation by inspection.

In dealing with linear differential equations of higher order, it is useful to note that if $y(x)$ is a solution, then $c \cdot y(x)$ is also a solution, where c is a constant. Furthermore, if there are two functions $y_1(x)$ and $y_2(x)$ which are solutions to the differential equation, then a linear combination of the form

$$y_3(x) = c_1 y_1(x) + c_2 y_2(x)$$

is also a solution, where once again c_1 and c_2 are constants.

The usual procedure is to guess a solution of the form

$$y = e^{mx}$$

This will generally lead to an *auxiliary equation* from which we can determine the values of m that satisfy the equation. A general solution can then be constructed of the form

$$y = c_1 e^{m_1 x} + c_2 e^{m_2 x} + \cdots \tag{1.114}$$

If the auxiliary equation has two identical roots, then $c_1 e^{mx}$ and $c_2 x e^{mx}$ are used.

For example, consider the following second-order linear homogeneous differential equation with *constant coefficients*:

$$\frac{d^2y}{dx^2} - 2\frac{dy}{dx} - 3y = 0$$

If we assume a solution of the form e^{mx}, we get

$$\frac{d^2e^{mx}}{dx^2} - 2\frac{de^{mx}}{dx} - 3e^{mx} = 0$$

This reduces to

$$m^2e^{mx} - 2me^{mx} - 3e^{mx} = 0$$

or

$$m^2 - 2m - 3 = 0$$

as the auxiliary equation. This equation has the roots $m_1 = +3$ and $m_2 = -1$, so the general solution is of the form

$$y = c_1e^{3x} + c_2e^{-x}$$

The constants c_1 and c_2 can be determined by specification of certain *boundary conditions*.

Now let us consider a still more general method which will be applicable to those cases where the methods described above fail. We assume that a *power series* solution is always possible,

$$y = a_0 + a_1x + a_2x^2 + \cdots = \sum_{n=0}^{\infty} a_nx^n \tag{1.115}$$

provided that enough terms are taken and that the coefficients a_n are properly chosen. In some instances the solution will be a finite number of terms, but often the solution is an *infinite series*.

As an example, suppose that we have the equation

$$(1 + x^2)\frac{d^2y}{dx^2} + 2x\frac{dy}{dx} - 2y = 0$$

and we desire a solution subject to the boundary conditions $y(0) = 1$ and $y'(0) = 0$. If we assume a solution of the form Eq. 1.115, then

$$y' = \sum_{n=0}^{\infty} na_nx^{n-1}$$

$$y'' = \sum_{n=0}^{\infty} n(n-1)a_nx^{n-2}$$

Putting these expressions for y, y', and y'' in the differential equation gives

$$\sum_{n=0}^{\infty} n(n-1)a_nx^{n-2} + \sum_{n=0}^{\infty} n(n-1)a_nx^n + 2\sum_{n=0}^{\infty} na_nx^n - 2\sum_{n=0}^{\infty} a_nx^n = 0$$

Since this must be satisfied for all values of the variable x, it follows that the coefficients of each power of x must be individually equal to zero. For the x^{n-2} term we get

$$n(n-1)a_n + (n-2)(n-3)a_{n-2} + 2(n-2)a_{n-2} - 2a_{n-2} = 0$$

$$n(n-1)a_n + n(n-3)a_{n-2} = 0$$

$$a_n = -\frac{n-3}{n-1}a_{n-2}$$

This is known as a *recursion formula*, as it tells us the value of each coefficient in the power series in terms of previous coefficients. In our case all the odd terms are referred back to the coefficient a_1, while the even coefficients are all determined once the value of a_0 is known. We note, however, that no matter what the value of a_1, a_3 will be equal to zero, and so will $a_5, a_7, \ldots$. According to the boundary condition $y(0) = 1$, we require that $a_0 = 1$, while we must have $a_1 = 0$ in order to satisfy the second condition $y'(0) = 0$. We find that

$$a_0 = 1 \qquad a_1 = 0$$

$$a_2 = -\frac{-1}{1}a_0 = 1 \qquad a_3 = 0$$

$$a_4 = -\frac{1}{3}a_2 = -\frac{1}{3} \qquad a_5 = 0$$

$$\vdots \qquad\qquad \vdots$$

and the solution is therefore

$$y = 1 + x^2 - \frac{1}{3}x^4 + \frac{1}{5}x^6 - \cdots$$

A differential equation of another type is often encountered by the chemist. It may be written in the form

$$dz = M(x,y)\,dx + N(x,y)\,dy \tag{1.116}$$

An integration of Eq. 1.116 may be possible if we are given a functional relationship between x and y. Then we can eliminate one or the other of these variables from the integrand on the right and it becomes possible to carry out the integration along a certain specified path from (x_1,y_1) to (x_2,y_2). The result depends on the path of integration and is known as a *line integral*.

If, on the other hand, Eq. 1.116 is a special case of Eq. 1.82, the situation is quite different. Then we know that some function z exists whose total differential is Eq. 1.116, and we can make the identification

$$M = \left(\frac{\partial z}{\partial x}\right)_y \qquad \text{and} \qquad N = \left(\frac{\partial z}{\partial y}\right)_x$$

In such a case Eq. 1.116 is called an *exact differential*, and we can test for exactness by differentiating each of the coefficients as follows:

$$\left(\frac{\partial M}{\partial y}\right)_x = \left[\frac{\partial}{\partial y}\left(\frac{\partial z}{\partial x}\right)_y\right]_x \qquad \left(\frac{\partial N}{\partial x}\right)_y = \left[\frac{\partial}{\partial x}\left(\frac{\partial z}{\partial y}\right)_x\right]_y$$

By Eq. 1.84 these must be equivalent if the function z exists, and we have

$$\left(\frac{\partial M}{\partial y}\right)_x = \left(\frac{\partial N}{\partial x}\right)_y \tag{1.117}$$

This is known as the *Euler reciprocity relation*. If it is satisfied, the differential equation 1.116 is exact and the integral depends only on the initial and final states. Otherwise, no simple function z exists that is a solution to Eq. 1.116 and the integral will depend on the path of integration, as mentioned above.

This is of special importance in the study of thermodynamics, where one deals with *state functions*, such as internal energy, enthalpy, entropy, and free energy, whose differentials are exact. As such they may simply be evaluated and related as long as the initial and final states are known, regardless of the path followed by the actual system.

Equation 1.116 involves more than two variables and it is a type of *partial differential equation*. Other partial differential equations are encountered in chemistry. One well-known example is the *Schrödinger wave equation*,

$$\nabla^2\psi + \frac{8\pi^2 m}{h^2}(E - V)\psi = 0 \tag{1.118}$$

where V is the potential energy, E is the total energy, and

$$\nabla^2 \equiv \frac{\partial^2}{\partial x^2} + \frac{\partial^2}{\partial y^2} + \frac{\partial^2}{\partial z^2} \tag{1.119}$$

is a three-dimensional differential operator known as the *Laplacian operator*.

Depending on the form of the potential energy V, other coordinate systems may be more appropriate for ∇^2. The general approach used to solve partial differential equations is to (we hope) find a coordinate system such that the variables are separable, and then the equations can be solved as ordinary differential equations.

As a very simple example, let us assume that we have a free particle, that is, that there is no potential ($V = 0$). Then we assume a solution of the form

$$\psi = X(x)\,Y(y)\,Z(z)$$

where, as the notation implies, $X(x)$ is a function of the independent variable x only, $Y(y)$ is a function only of y, and $Z(z)$ is a function only of z. Insertion of this solution in Eq. 1.118 with $V = 0$ leads to

$$\frac{h^2}{8\pi^2 m}\left(YZ\frac{d^2X}{dx^2} + XZ\frac{d^2Y}{dy^2} + XY\frac{d^2Z}{dz^2}\right) + EXYZ = 0$$

We divide through by XYZ and rearrange to get

$$\frac{h^2}{8\pi^2 m}\frac{1}{X}\frac{d^2X}{dx^2} = -\frac{h^2}{8\pi^2 m}\frac{1}{Y}\frac{d^2Y}{dy^2} - \frac{h^2}{8\pi^2 m}\frac{1}{Z}\frac{d^2Z}{dz^2} - E$$

If x, y, and z are independent variables, as we have assumed, this equation must be satisfied for all possible values for each of them. The only way that this equation can be valid for independent arbitrary variations of the variables is for each side to be a constant. Let us call the *separation constant* $-E_x$. From the left-hand side of the equation we obtain

$$\frac{h^2}{8\pi^2 m}\frac{1}{X}\frac{d^2X}{dx^2} = -E_x$$

The same argument may be used to separate the Y and Z parts, and we find that the three identical equations,

$$\frac{h^2}{8\pi^2 m}\frac{1}{X}\frac{d^2X}{dx^2} + E_x = 0 \qquad \frac{h^2}{8\pi^2 m}\frac{1}{Y}\frac{d^2Y}{dy^2} + E_y = 0 \qquad \frac{h^2}{8\pi^2 m}\frac{1}{Z}\frac{d^2Z}{dz^2} + E_z = 0$$

are obtained with the total energy E being equal to $E_x + E_y + E_z$. These ordinary differential equations can be solved using techniques such as those described above.

Example 1.24

Solve the differential equation

$$x\frac{dy}{dx} - y - x^2 = 0$$

Solution. Division by x allows us to put the equation in the form of Eq. 1.112. This equation then has $e^{\int P dx}$ as an integrating factor, where $P = -1/x$. This simplifies to

$$e^{\int -(1/x)dx} = e^{-\ln x} = e^{\ln(1/x)} = \frac{1}{x}$$

and the equation becomes

$$\frac{1}{x}\frac{dy}{dx} - \frac{1}{x^2}y = 1$$

We recognize the left-hand side of this equation as the differential of y/x:

$$\frac{d\left(\frac{y}{x}\right)}{dx} = \left(\frac{1}{x}\right)\frac{dy}{dx} + y\left(-\frac{1}{x^2}\right)$$

Thus we can write

$$\frac{d\left(\frac{y}{x}\right)}{dx} = 1$$

and the solution is

$$d\left(\frac{y}{x}\right) = dx$$

$$\frac{y}{x} = x + c$$

$$y = x^2 + Cx$$

Example 1.25

Show that for the electron moving in the field of the nucleus of a hydrogen atom, where the potential energy is a spherically symmetric term of the form $V = -e^2/r$, the ϕ part of the Schrödinger wave equation may readily be separated if the Laplacian operator is written in terms of the spherical polar coordinates:

$$\nabla^2 = \frac{1}{r^2}\frac{\partial}{\partial r}r^2\frac{\partial}{\partial r} + \frac{1}{r^2 \sin\theta}\frac{\partial}{\partial\theta}\sin\theta\frac{\partial}{\partial\theta} + \frac{1}{r^2\sin^2\theta}\frac{\partial^2}{\partial\phi^2}$$

Solution. We assume that the wave function is a product of three terms, each of which is a function of only one of the independent variables:

$$\psi = R(r)\Theta(\theta)\Phi(\phi)$$

Putting this in the Schrödinger wave equation (Eq. 1.118) gives

$$\Theta\Phi\frac{1}{r^2}\frac{d}{dr}r^2\frac{d}{dr}R + \frac{R\Phi}{r^2\sin\theta}\frac{d}{d\theta}\left(\sin\theta\frac{d\Theta}{d\theta}\right) + \frac{R\Theta}{r^2\sin^2\theta}\frac{d^2\Phi}{d\phi^2}$$
$$+ \frac{8\pi^2 m}{h^2}\left(E + \frac{e^2}{r}\right)R\Theta\Phi = 0$$

We multiply through by the factor $r^2\sin^2\theta/R\Theta\Phi$ and separate the terms in ϕ to one side of the equation:

$$-\frac{1}{\Phi}\frac{d^2\Phi}{d\phi^2} = \frac{\sin^2\theta}{R}\frac{d}{dr}r^2\frac{d}{dr}R + \frac{\sin\theta}{\Theta}\frac{d}{d\theta}\left(\sin\theta\frac{d\Theta}{d\theta}\right)$$
$$+ \frac{8\pi^2 m r^2\sin^2\theta}{h^2}\left(E + \frac{e^2}{r}\right)$$

Since the variables are independent, the two sides of the equation can be equal only if they are equal to a constant; we will call the separation constant c, and the equation in ϕ that must be solved as an ordinary differential equation may be written

$$\frac{d^2\Phi}{d\phi^2} + c\Phi = 0$$

1.8 INFINITE SERIES

In the preceding section we saw one example of how an infinite series may arise. Such series are used in many other contexts as well. As with other functions, they can be manipulated in numerous ways to obtain a desired result. For example, term-by-term addition, subtraction, differentiation, or integration is possible.

When a series is summed, it may approach some finite value as the number of terms becomes infinitely large. It is then called *convergent*. In other cases the sum does not approach a definite value and the series is *divergent*. Our principal interest, of course, is in convergent series.

It is obvious that an infinite series cannot approach a finite limit unless the term a_n tends toward zero as $n \to \infty$. For example, the series

$$\frac{1}{4} + \frac{2}{7} + \frac{3}{10} + \cdots + \frac{n}{3n + 1} + \cdots$$

is divergent since

$$\lim_{n \to \infty} \frac{n}{3n + 1} = \frac{1}{3}$$

An alternating series such as

$$1 - 1 + 1 - 1 + \cdots$$

is also divergent. Although the sum does not increase indefinitely as in the previous example, it continues to oscillate between the values 1 and 0 and fails to approach a limit.

It will not always be so obvious whether a given series converges or not. There are, however, several *convergence tests* that can be used to determine the behavior of a given series. For a general series that may be represented as follows:

$$a_0 + a_1 + a_2 + \cdots + a_n + \cdots \tag{1.120}$$

we may state the following:

a. *Integral test*. If a general formula $f(n) = a_n$ can be written to represent the nth term of the series, and if $f(x)$ is a continuous function of x that never increases with x for $x \geq c$, the series is convergent if the integral

$$\int_c^\infty f(x)\, dx$$

exists.

b. *Comparison test*. If the series

$$b_0 + b_1 + b_2 + \cdots + b_n + \cdots$$

is known to be convergent and $a_n < b_n$ for all values of n, then the series 1.120 is convergent.

c. *Ratio test*. For a series of positive constants we consider the limit

$$\lim_{n\to\infty} \frac{a_{n+1}}{a_n} = r$$

If $r < 1$, the series is convergent; if $r > 1$, the series is divergent; and if $r = 1$, the test fails and another must be used.

d. *Alternating series test*. An alternating series is convergent if each term is less in magnitude than that which precedes it and if the limit of the nth term as $n \to \infty$ is zero.

These same tests may be applied to a series of constant terms or to those involving variables. An example of the latter is the power series 1.115 introduced in Section 1.7. It will generally be found that such a series is convergent only within a certain range of values for the variable x. We call this the *interval of convergence*, and it is important that it be determined and specified in each case.

We often find that the simple power series 1.115 is divergent or at least is slow to converge. In such cases we can generally accelerate the convergence by replacing x by $x - c$ in the series to obtain

$$a_0 + a_1(x - c) + a_2(x - c)^2 + \cdots + a_n(x - c)^n + \cdots \tag{1.121}$$

In general, we can express any arbitrary function $f(x)$ in terms of such a power series. If the value of c is chosen such that the first few terms in Eq. 1.121 are a close representation of the function $f(x)$, the higher-order terms should be correspondingly small and the series should be rapidly convergent. The series 1.115 is obviously just a special case of Eq. 1.121 where $c = 0$.

Expressing the function $f(x)$ in terms of Eq. 1.121 is called a *power series expansion*, and we determine the expansion coefficients for a given case as follows: We write

$$f(x) = a_0 + a_1(x - c) + a_2(x - c)^2 + a_3(x - c)^3 + \cdots$$

and note that if we set $x = c$ in this equation we have

$$f(c) = a_0$$

Thus the first term in the expansion is just the function itself evaluated at $x = c$. Next we differentiate term by term to obtain

$$f'(x) = a_1 + 2a_2(x - c) + 3a_3(x - c)^2 + \cdots + (n - 1)a_{n-1}(x - c)^{n-2} + \cdots$$

Evaluation at $x = c$ in this case gives

$$f'(c) = a_1$$

Now a second differentiation gives

$$f''(x) = 2a_2 + 2 \cdot 3a_3(x - c) + \cdots + (n - 2)(n - 1)a_{n-1}(x - c)^{n-3} + \cdots$$

which is evaluated at $x = c$ to obtain

$$f''(c) = 2a_2$$

Similarly,

$$f'''(c) = 2 \cdot 3a_3$$

and so forth. Collecting these results, we see that

$$f(x) = f(c) + \frac{f'(c)}{1!}(x - c) + \frac{f''(c)}{2!}(x - c)^2 + \frac{f'''(c)}{3!}(x - c)^3 + \cdots + \frac{f^{n-1}(c)}{(n-1)!}(x - c)^{n-1} + \cdots \tag{1.122}$$

This is known as a *Taylor series expansion of the function f(x) about the point x = c.* We use the Taylor series expansion a great deal, often in a truncated form, to carry out mathematical manipulations and numerical evaluations. Once again we note that such an expansion is a useful representation of the function *only* in the vicinity of the point about which the expansion is carried out. Outside this range the series will be slow to converge, if it converges at all.

If the expansion is made about the point $x = 0$, we are dealing with the simple power series 1.115. It follows from Eq. 1.122 that the coefficients are then given by

$$f(x) = f(0) + \frac{f'(0)}{1!}x + \frac{f''(0)}{2!}x^2 + \cdots + \frac{f^{n-1}(0)}{(n-1)!}x^{n-1} + \cdots \tag{1.123}$$

This special case of a Taylor series expansion about the point $x = 0$ is known as a *Maclaurin series expansion.*

Example 1.26

Expand $\ln(1 + x)$ in a power series about the point $x = 0$.

Solution. The function and its various derivatives evaluated at the point $x = 0$ are

$$f = \ln(1 + x) \qquad f(0) = 0$$

$$f' = \frac{1}{1 + x} \qquad f'(0) = 1$$

$$f'' = \frac{-1}{(1 + x)^2} \qquad f''(0) = -1$$

$$f''' = \frac{2}{(1 + x)^3} \qquad f'''(0) = 2!$$

$$\vdots \qquad \qquad \vdots$$

Putting these in Eq. 1.123 gives

$$\ln(1 + x) = 0 + \frac{1}{1!}x + \frac{-1}{2!}x^2 + \frac{2!}{3!}x^3 + \cdots$$

$$= x - \frac{x^2}{2} + \frac{x^3}{3} - \cdots$$

Example 1.27

Expand the function $(1 - x)^{1/2}$ in series about the point $x = 0$ (i.e., in a Maclaurin expansion), and use the expansion to evaluate the square root of 0.9.

Solution. The function and its various derivatives evaluated at the point $x = 0$ may be written as follows:

$$f = (1 - x)^{1/2} \qquad f(0) = 1$$

$$f' = -\frac{1}{2}(1 - x)^{-1/2} \qquad f'(0) = -\frac{1}{2}$$

$$f'' = -\frac{1}{4}(1 - x)^{-3/2} \qquad f''(0) = -\frac{1}{4}$$

$$f''' = -\frac{3}{8}(1 - x)^{-5/2} \qquad f'''(0) = -\frac{3}{8}$$

$$\vdots \qquad \vdots$$

Putting these in Eq. 1.123, we have

$$f(x) = 1 - \frac{1}{2}x - \frac{1/4}{2}x^2 - \frac{3/8}{2 \cdot 3}x^3 - \cdots$$

$$= 1 - \frac{1}{2}x - \frac{1}{8}x^2 - \frac{1}{16}x^3 - \frac{5}{128}x^4 - \cdots$$

We put in the value $x = 0.1$ to obtain

$$f(0.1) = 1 - \frac{0.1}{2} - \frac{(0.1)^2}{8} - \frac{(0.1)^3}{16} - \frac{5(0.1)^4}{128} - \cdots$$

$$= 0.948684$$

The actual value correct to six decimal places is 0.948683.

SUGGESTED ADDITIONAL READING

1. James R. Barrante, *Applied Mathematics for Physical Chemistry*, Prentice-Hall, Inc., Englewood Cliffs, N.J., 1974.
2. Joseph B. Dence, *Mathematical Techniques in Chemistry*, Wiley-Interscience, New York, 1975.
3. P. G. Francis, *Mathematics for Chemists*, Chapman & Hall, New York, 1984.

4. D. M. Hirst, *Mathematics for Chemists*, Chemical Publishing Company, Inc., New York, 1979.
5. Robert G. Mortimer, *Mathematics for Physical Chemistry*, Macmillan Publishing Co., Inc., New York, 1981.
6. Charles L. Perrin, *Mathematics for Chemists*, Wiley-Interscience, New York, 1970.

EXERCISES

1.1. Plot each of the following functions, and determine the slope of the curve graphically at the point $x = 0.5$ by drawing the tangent. Compare the result with the precise value obtained by differentiation.
(a) $3x^2 - 5x + 8$
(b) $x^2 + x + 2$
(c) x^3
(d) $-2x^2 + 3x$

1.2. Find the value of x for which a maximum or a minimum occurs for each of the following by plotting the function, and compare with the exact value determined by differentiation.
(a) $-x^2 + 6x + 3$
(b) $2x^2 + 2x + 2$
(c) $x(x + 7) - 6$
(d) $-3x^2 + 20x + 4$

1.3. Find the critical points of the following functions and identify each as a maximum, minimum, or horizontal point of inflection.
(a) $y = x^4 + 4x^3 + 6x^2$
(b) $y = 5 + 6x + 3x^2$
(c) $y = 7 + 12x - 2x^2$
(d) $y = 2x^3 + 6$

1.4. Evaluate df/dx for each of the following functions.
(a) $f = \sqrt{a^2 - x^2}$
(b) $f = (1 - x^2)^2$
(c) $f = 1/\sqrt{2x - 7}$
(d) $f = 6y^2 + 4y$ where $y = 2x + 2$

1.5. Differentiate each of the following with respect to x.
(a) $(x + 1)(x + 2)(x + 4)$
(b) $x/\sqrt{1 + x}$
(c) $\ln(x^2 - x)$
(d) $e^{x^4 + x}$

1.6. Evaluate the partial derivatives of the function $f = \sqrt{xy} + 3xyz + z^3$ as follows:
(a) $\left(\frac{\partial f}{\partial x}\right)_{y,z}$ at (2,2,3)
(b) $\left(\frac{\partial f}{\partial x}\right)_{y,z}$ at (5,5,0)
(c) $\left(\frac{\partial f}{\partial y}\right)_{x,z}$ at (2,2,3)

(d) $\left(\dfrac{\partial f}{\partial z}\right)_{x,y}$ at (0,0,4)

1.7. Expand the function $\log x$ in a Taylor's series about the point $x = 10$.

1.8. Use the series expansion of Exercise 1.7 to evaluate the common logarithms of 9.99, 9.9, and 9. How many terms are required in each case for four-place accuracy?

1.9. Expand each of the following functions in a Maclaurin series and evaluate it at the point $x = 0.1$.

(a) e^{2x}
(b) e^{-x}
(c) 5^x
(d) $\ln x$

1.10. Give the indefinite integral for each of the following functions.

(a) x^4
(b) $3x^3 + 7x$
(c) $14/x$
(d) 3^x

1.11. What is the area under each of the following curves?

(a) $f = x^3 - x^2$ from $x = 5$ to $x = 10$
(b) $f = 3y^2 - 2y + 10$ from $y = 1$ to $y = 2$
(c) $f = e^{-x}$ from $x = 0$ to $x = \infty$
(d) $f = 2/r$ from $r = 1.5$ to $r = 2.5$

1.12. The molar heat capacity $\bar{C}_p$ is a measure of the amount of heat required to raise the temperature of 1 mol of a substance by 1 degree. For gases, the value of $\bar{C}_p$ is not constant, and it is usually expressed in the form of a power series

$$\bar{C}_p = a + bT + cT^2 \qquad (\text{cal K}^{-1}\ \text{mol}^{-1})$$

Plot the heat capacity for each of the following gases in the range 300 to 1500 K, and determine the amount of heat required to raise the temperature of each from 300 to 1500 K by estimating the area under the curve.

(a) H_2: $a = 6.943$, $b = -1.999 \times 10^{-4}$, $c = 4.808 \times 10^{-7}$
(b) CO_2: $a = 6.369$, $b = 1.010 \times 10^{-2}$, $c = -3.405 \times 10^{-6}$
(c) CH_4: $a = 3.381$, $b = 1.804 \times 10^{-2}$, $c = -4.300 \times 10^{-6}$
(d) C_6H_6: $a = -0.409$, $b = 7.762 \times 10^{-2}$, $c = -2.643 \times 10^{-5}$

1.13. Compare the heats obtained for each of the gases in Exercise 1.12 with the exact values obtained by integration of

$$\int_{300}^{1500} \bar{C}_p\, dT$$

1.14. The heat capacities of solids cannot be measured at extremely low temperatures, but it has been shown theoretically by Debye that the heat capacity should be proportional to T^3 near absolute zero:

$$\bar{C}_p = aT^3$$

Given that $\bar{C}_p = 0.83$ cal K^{-1} mol^{-1} at 15 K for SO_2, plot the molar heat capacity of SO_2 from 0 to 15 K, and estimate the area under the curve.

1.15. Compare the precise amount of heat required to raise the temperature of 1 mol of SO_2 from 0 to 15 K obtained by integration with the value estimated in Exercise 1.14.

1.16. The electron density for a 2*s* orbital of carbon can be represented approximately by

$$\frac{c^5}{96\pi} r^2 e^{-cr}$$

with $c = 3.25$ and r measured in units of the Bohr radius a_0 (0.529×10^{-8} cm).

(a) Plot the electron density as a function of r, and determine the value of r for which it is a maximum.

(b) Find an analytic expression for the most probable value of the radius, and compare the result calculated from it with the value obtained in part (a).

1.17. The ideal gas law $pV = nRT$ is a generalization that includes both Boyle's law and Charles' law as special cases. Here n is the number of moles of gas, and $R = 82.06$ cm^3 atm K^{-1} mol^{-1} is the molar gas constant.

(a) How does V vary with respect to p if T is held constant?

(b) Plot $(\partial V/\partial p)_T$ as a function of p for 1 mol of an ideal gas at $T = 300$ K.

(c) Discuss the physical significance of the *isothermal compressibility curve* you have drawn (e.g., why is it negative, how compressible is the gas for a given pressure change at low pressures compared with high pressures, etc.).

1.18. The difference in heat capacity at constant pressure C_p and the heat capacity at constant volume C_v can be shown to be

$$C_p - C_v = TV\frac{\alpha^2}{\kappa}$$

where α and κ are the thermal expansivity and isothermal compressibility defined in Examples 1.20 and 1.21. Evaluate $C_p - C_v$ for 1 mol of an ideal gas.

1.19. Evaluate

$$\left(\frac{\partial x}{\partial y}\right)_z \left(\frac{\partial y}{\partial z}\right)_x \left(\frac{\partial z}{\partial x}\right)_y$$

for the function $xy^2/z = c$, where c is a constant.

1.20. Integrate each of the following using a substitution method.

(a) $$\int e^{\sqrt{x}}\, dx$$

(b) $$\int \frac{dx}{1 - \sqrt{x + 1}}$$

1.21. Integrate each of the following using the method of partial fractions.

(a) $$\int \frac{dx}{x^2 - 1}$$

(b) $$\int \frac{x - 1}{(x - 2)^2}\, dx$$

1.22. Integrate each of the following by parts.

(a) $\int xe^x\,dx$

(b) $\int x \ln x\,dx$

1.23. The work required to expand a gas at constant temperature is

$$w = \int_{V_1}^{V_2} p\,dV$$

If an ideal gas is expanded adiabatically (i.e., with no exchange of heat with the surroundings), the temperature drops, and the pressure and volume are related by

$$pV^\gamma = p_0 V_0^\gamma$$

where γ is a constant and p_0 and V_0 are the initial pressure and volume. Derive an expression for the work of adiabatic expansion for an ideal gas from volume V_0 to V.

1.24. Derive a formula for the volume of a sphere of radius R by multiple integration of the volume element given in Eq. 1.110.

1.25. Assume that the density of a sphere of radius R varies as the distance from its center, $\rho(r) = r$. Calculate the mass of the sphere.

1.26. Thermodynamic calculations show that the effective pressure of any fluid is the sum of the applied pressure and the internal pressure, which is due to the intermolecular forces. The internal pressure may be written

$$p_i = T\left(\frac{\partial p}{\partial T}\right)_V - p$$

Assume that you have 1 mol of CO_2 (MW = 44.01) and that it obeys the van der Waals equation of state ($\bar{V}$ is the molar volume),

$$\left(p + \frac{a}{\bar{V}^2}\right)(\bar{V} - b) = RT$$

with $a = 3.592\ \text{L}^2\ \text{atm mol}^{-2}$, $b = 0.04267\ \text{L mol}^{-1}$, and $R = 0.08205\ \text{L atm K}^{-1}\ \text{mol}^{-1}$. What is the internal pressure p_i if the gas is compressed to a density of $0.516\ \text{g mL}^{-1}$ at $t = -16.7°\text{C}$?

1.27. Assume a function $u = e^{xy^2}$ with $x = r^2 s$ and $y = rs^3$. Evaluate the partial derivatives $(\partial u/\partial r)_s$ and $(\partial u/\partial s)_r$.

1.28. Derive formulas for each of the following.

(a) $\dfrac{d \csc \theta}{d\theta}$

(b) $\dfrac{d \cosh x}{dx}$

(c) $\dfrac{d \tanh x}{dx}$

1.29. Solve the following first-order differential equations.

(a) $x \dfrac{dy}{dx} - 4y = x^5$

(b) $y\,dx + x\,dy = \sin x\,dx$

1.30. Find general solutions to the following second-order differential equations by substitution of $y = e^{mx}$.

(a) $\dfrac{d^2y}{dx^2} + 4\dfrac{dy}{dx} + 4y = 0$

(b) $y'' = 9y$

(c) $y'' - 4y' + 4y = 0$ Evaluate the constants for the initial values $y = 3$ and $y' = 4$ at $x = 0$.

1.31. Find a series solution of the differential equation

$$\frac{d^2y}{dx^2} + x\frac{dy}{dx} - 2y = 0$$

with the initial values $y = 1$ and $dy/dx = 0$ for $x = 0$.

1.32. Test the following differential equations for exactness using the Euler reciprocity relation.

(a) $dz = y\,dx + x\,dy$

(b) $dz = (x + y)\,dx + (x + y)\,dy$

(c) $dz = 6x^2y\,dx + 2x^3\,dy$

(d) $dz = -\dfrac{y}{x^2}\,dx + \dfrac{1}{x}\,dy$

(e) $dz = (3x + 2y^2)\,dx + 2xy\,dy$

1.33. Verify Eq. 1.46 by expanding $\sin\theta$, $\cos\theta$, and $e^{\pm i\theta}$ in Taylor's series and comparing terms.

1.34. Show that the vector product given by Eq. 1.50 is equivalent to the following expression:

$$\mathbf{A} \times \mathbf{B} = \begin{vmatrix} \mathbf{i} & \mathbf{j} & \mathbf{k} \\ a_1 & a_2 & a_3 \\ b_1 & b_2 & b_3 \end{vmatrix}$$

1.35. Plot the Langevin function $\mathcal{L}(x)$ given in Example 1.23 as a function of x, and discuss its high-temperature and low-temperature limits.

1.36. Complete the separation of variables begun in Example 1.25 to obtain the ordinary differential equations in θ and r.

1.37. Compute the radius ratio for tetrahedral coordination (see Example 1.4).

1.38. Derive Eq. 1.41.

1.39. Show that negative integers satisfy De Moivre's theorem (Eq. 1.42).

1.40. Show that the fourth-order *Legendre polynomial*

$$P_4 = \frac{1}{64}(35\cos 4\theta + 20\cos 2\theta + 9)$$

may also be written

$$P_4 = \frac{1}{8}(35\cos^4\theta - 30\cos^2\theta + 3)$$

1.41. If a beam of amplitude A_i is perpendicularly incident on a shiny surface, the amplitude of a reflected beam A_r may be written

$$A_r = A_i \frac{N - 1}{N + 1}$$

Here $N = n(1 - i\kappa)$, where n is the relative refractive index and κ is the absorption index. Show that the reflectivity defined by

$$R = \left(\frac{A_r}{A_i}\right)\left(\frac{A_r}{A_i}\right)^*$$

is equal to

$$\frac{(n - 1)^2 + n^2\kappa^2}{(n + 1)^2 + n^2\kappa^2}$$

chapter two

COMPUTER PROGRAMMING

2.1 INTRODUCTION

Computers have become so commonplace in chemistry that it is highly unlikely that a person now beginning a career in chemistry or in any of its closely related fields would not soon become faced with the use of them in one way or another. They occur in many sizes and shapes. At times we rely on a large central computer facility that may be controlled and run directly at the site, or through the use of remote terminals that allow concurrent usage by many persons through time sharing. Small personal computers can be used to accomplish many of the same purposes, or they may be interfaced with laboratory instruments to exercise direct control in recording and reduction of data without the intervention of an operator. A computer that is interfaced directly to a given laboratory instrument, making it unavailable for other purposes, is described as a *dedicated* computer. Our particular concern is with the use of the computer in numerical problem solving related to chemistry rather than in instrument control.

A straightforward numerical calculation (often referred to these days as "number crunching") may be accomplished either to achieve greater accuracy or to remove the tedium associated with hand calculations. It should be recognized that we must have a thorough grasp of the problem so that we can communicate to the computer the manner in which the calculation is to be carried out. If we are faced with a "one-shot" calculation, it may be preferable to do it by hand. It may be that no labor would be saved by the time we develop a computer code and test it sufficiently to be assured that it is entirely free of errors (the latter process is often called "debugging" the program). The computer is only of real value for repetitive calculations, which will be carried out with great speed once the program or code functions properly. We can rest assured that the computer result using a carefully debugged program will be very reliable, since the internal checks preclude the possibility of arithmetic errors for all practical purposes.

Computers can also be used for information storage and retrieval. A considerable amount of information can be stored internally, but the use of very large quantities necessitates the use of peripheral devices, such as magnetic disks. Both data and programs may be stored permanently in this way; they can then be read into the computer and processed as required.

Computers can be used to generate predicted results making use of known relationships as an aid to understanding chemical behavior. We call this *computer simulation*. In principle we can always do the actual experiment, but there may be cases where time limitations, safety, or accessibility of equipment or supplies might render a simulation more desirable. Such simulations are often used in conjunction with *display graphics,* so that a more rapid assessment of the results can be made.

There are numerous other ways in which computers can be used to solve problems. They have found increasing educational use in recent years in drill and autotutorial applications. These uses are generally grouped under the heading CAI (computer-assisted instruction).

"Number crunching" applications often involve the input of a large quantity of data, or extensive arithmetic operations, or both. Then computation by a *batch process* is preferred. In such cases the program together with the necessary data may be submitted directly to a computer center. It may also be input from a remote terminal, using stored data files which may be edited. In either case, the user has no further control once the job has been submitted.

Many other applications can best be carried out making use of an *interactive process*. Here the user takes advantage of an appropriate display device (usually, a CRT tube), so that back-and-forth communication with the computer is possible.

Many different computer languages have been developed, each with its own set of "dialects." Some are more useful than others for the sciences, but the choice of language must be governed in the end by the particular application.

Ultimately, the computer performs all its work by manipulation of magnetized spots, called *bits*. A bit can have but one of two values. If it is magnetized, it represents the value 1; if not, it represents a 0. Both the instructions of the computer program and the data that it manipulates are therefore represented internally in a binary system. A small segment of a machine-language code may then have an appearance such as the following:

000111000100100110100010111100001

Fortunately, it is seldom necessary for us to communicate with the computer on this level.

In modern computers the bits are often grouped together in fours, called the 1-bit, 2-bit, 4-bit, and 8-bit. With no spots magnetized we represent the digit 0, with the 1-bit magnetized the digit 1, the 2-bit the digit 2, the 1-bit and 2-bit the digit 3, and so forth. It will be observed that with various combinations of these four bits we can construct any number between 0 and 15, or 16 in all. We represent these *hexadecimal numbers* by the digits 0 through 9, followed by the letters A through F.

A number of bits may be associated together to construct a *byte*. The actual number of bits per byte depends on the computer architecture, but one very common system uses 8 bits per byte for the internal representation of numbers and characters. There are then $16 \times 16 = 256$ values, and these are assigned according to one of the various codes in common use, such as the BCD (binary-coded decimal) or ASCII (American Standard Code for Information Interchange) systems. The latter is used for most microcomputers. It has the advantage of assigning character codes for both letters and numbers in ascending order, which facilitates alphabetizing as well as numerical sorting. For example, the digit 6 is represented by 36, and the digit 7 by 37. The letter M is represented by 4D, N by 4E, and O by 4F. The lowercase letters have their own distinct representations (6D for m, 6E for n, etc.), and special symbols are also included (e.g., 5C for \, 22 for ", 3D for =).

The storage of numerical quantities involves varying numbers of bits, depending on their complexity. For example, integers are normally represented using 2 bytes, or 16 bits. One of these bits is reserved for the designation of the sign of the number, leaving the other 15 to specify its magnitude. Any number in the range 0 to 32,767 can be represented using various combinations of the 15 bits, and so integers specified by this system are restricted to the range −32,768 to +32,767. Larger numbers and decimal fractions are usually stored and manipulated in floating-point form, as either single-precision or double-precision numbers, which will be described later. As might be anticipated, they require a larger bit pattern for their representation.

Various *high-level programming languages* have been designed to remove the extremely tedious exercise of constructing machine-language computer codes. The high-level languages replace the machine-language code by instructions that are more familiar and easily recognized by the programmer. Each must be used in conjunction with its own unique set of instructions which converts the statements made by the programmer into the machine-language instructions recognized by the computer. The codes that accomplish such tasks are generally referred to as part of the *software* of the computer, as opposed to the real nuts and bolts (more accurately the resistors, capacitors, ICs, magnetic disk drives, etc.), which constitute the computer *hardware*.

The software codes may be read into the computer from a magnetic disk, or they may be permanently entered into a portion of the internal storage system of the computer known as *ROM* (read-only memory). The working internal storage system where data or program steps can be both stored and retrieved is known as *RAM* (random access memory). The size of a computer's memory is usually given in kilobytes or megabytes. For example, we may state that a certain microcomputer has 128K of RAM. The actual size of the RAM memory is slightly larger than the 128,000 bytes implied by this statement. It will be recalled that we are dealing with a binary system here. "Kilo" in this case is not actually 1000, but is a close approximation to the $2^{10} = 1024$ bytes of memory actually specified by this number.

FORTRAN, an acronym for FORmula TRANslation, was the first high-level language to be developed for scientific work. It is still in common use today, but in a much more refined form (most systems with FORTRAN capability use a FORTRAN IV or a FORTRAN 77 version at the present time). The *FORTRAN source program* is processed by use of a *compiler*, which analyzes the entire program as a unit, checking it for errors in syntax and logic, and reducing it to an elementary machine-compatible form. Because the program is treated as a whole, the speed of operation is more rapid than with some other high-level languages, and it is generally the language of choice for batch processing, particularly if a great deal of looping and branching may be involved.

A simpler language has been developed more recently called BASIC, which stands for Beginner's All-Purpose Symbolic Instruction Code. It has gained great popularity not only because it is easier for the beginner to master, but also because it uses an *interpreter* rather than a compiler. This means that the program steps are individually analyzed and executed. Although some efficiency is sacrificed in the process, it readily allows for the use of the computer in an interactive mode. This has some obvious advantages. One can monitor the progress of a calculation and input information from the keyboard which may be dependent on intermediate results at that point. Calculations of the type we consider in this book can all be readily carried out in good time on a personal computer. Our discussion will be limited to the BASIC language, but we note in passing that it has many similarities to FORTRAN. The serious computer programmer will ultimately want to learn other high-level languages, such as FORTRAN or Pascal, but BASIC is quite a powerful language in its own right and serves as a convenient starting point. BASIC, in one of its manifold versions, is available on virtually all computers.

The storage and retrieval of files on floppy disks or hard disks is a common practice, which makes possible the saving of both programs and data for later use. The commands for such manipulations are carried out through operating systems which vary from computer to computer, and we will not attempt to describe them here. For the specifics you should consult the operating system manual supplied with the computer you are using.

2.2 THE BASIC LANGUAGE

Many books and manuals have been published that contain descriptions of the BASIC computer language. We will not go into all the intricacies, but will present sufficient information to write programs that will effectively solve the simple numerical problems we will encounter in later chapters. There are some small differences in the language depending on the computer system being used. We will confine our discussion for the most part to those general statements that are executable on most computer systems with BASIC software. Some minor changes may be required in a few cases, however.

We begin in Section 2.2.1 with instructions for entering and executing programs. It is recommended that you study the subsequent sections by sitting down at a computer keyboard and running the short illustrative programs as they are discussed.

2.2.1 The Command Mode

The *COMMAND mode* is also sometimes called the immediate mode, because the instructions, entered directly from the keyboard, are interpreted and executed immediately under user control. For example, we might type and enter the following:

PRINT 36*36

The command would be immediately executed and the result, 1296, would be displayed on the screen. For instructions of this type the computer can be used in much the same way as a calculator.

If a program has been stored in the computer memory, it can be executed in the *RUN mode,* which is initiated by entering the command RUN. The program statements are each given a number, and the computer processes them beginning with the lowest number and proceeding in order throughout the program, taking note of appropriate branch points that may be included. It returns to the COMMAND mode only upon normal termination (through an END or STOP command) or by intervention from the keyboard through pressing the BREAK key. The running of a program can be resumed by entering CONT, which stands for continue.

If we have more than one program in memory, or if we wish to enter the program at some point other than the beginning, we can enter the command

RUN m

and execution begins at the line number m. This can be useful, for example, if the raw data might be given in alternative forms.

In order to enter a program into the computer memory, we type the various line numbers followed by the corresponding computer statements, depressing the ENTER key in each case. The lines need not be given in order; they are sorted out internally and always listed and executed in order of ascending line numbers no matter in what order they are given. We can therefore insert additional lines by assigning appropriate numbers that will place them in the proper sequence in the program. To replace an existing line by another, we type the line number followed by the new BASIC statement, and the replacement will occur when the ENTER key is depressed. If we simply type a line number and enter it, the statement with that particular line number will be deleted from the program if it exists.

We can facilitate the entering of a long program by use of the *AUTO command*. The AUTO command causes line numbers to be supplied automatically. The command is of the form

AUTO n,m

where n is the initial line number and m is the increment to be used in assigning each successive line number. If n and m are not specified, $n = 10$ and $m = 10$ are assigned.

We often have statements that allow for certain options but automatically set values for them if they are omitted. We call these the *default values*. After the last line of the program has been entered, the BREAK key may be depressed to exit the AUTO mode.

There are several commands that may be helpful in editing an existing program. If several consecutive lines are to be deleted, we can use the command

DELETE n—m

and all lines in the range n to m inclusive will be eliminated. We may have made quite a large number of insertions in a program, so that additional statements cannot be included in the proper sequence. It is then useful to execute the command

RENUM i,j,k

This renumbers the program beginning at existing line number j, assigns it the new number i, and uses an increment k in the renumbering. If i,j,k are omitted, the default values are i = 10, k = 10, and j is equal to the first statement of the program. Any references to statement numbers that may occur in the program are automatically changed to the new numbers when a program is renumbered using the RENUM command. If we execute the command

EDIT m

then line m of our program is displayed and we are in an editing mode. Changes can then be made anywhere in the line by moving the curser to the appropriate point and depressing certain keys. The specifics of line editing vary from computer to computer; you should consult the manual supplied if you wish to make use of this feature.

After we have entered a program, we will generally want to examine it to be sure that it has been entered properly. It can be displayed on the screen using the command

LIST n—m

This causes a listing of the program in memory from line number n to line number m, inclusive. If n and m are omitted, the entire program is listed. The listing will take place very quickly, and if the program is long, it will scroll by on the screen too rapidly to be read. It may then be necessary to limit the range by specification of m and n. Either one or the other, or both, may be given. For example, the commands LIST 100—200, LIST 100—, LIST —200 cause listing of the program from statements 100 to 200, from 100 to the end of the program, and from the beginning to line 200, respectively. The command

LIST n

will display the single line n. We frequently need a permanent printed record of our computer output, which is generally called a *hard copy*. If we desire a hard copy of the program listing, we can use LLIST in place of LIST, and the listing will be routed to the line printer rather than to the CRT display.

Finally, we may wish to erase all the current program lines and variables from the computer memory so that we can make a fresh start with a new program. This can be done by turning the computer off, but it can be accomplished more simply without the necessity of going through the turn-on procedure (often called *booting* the system) by entering the command NEW.

A summary of these BASIC commands is given in Appendix A.3. Others specific to the manipulation of files on your computer may be found in the reference manual supplied.

2.2.2 Computer Statements

Most scientific applications make use of the *arithmetic replacement statement*. The following is an example:

```
50 LET X=A+B
```

The use of the term LET is optional, so this can as well be written

```
50 X=A+B
```

This has the appearance of a simple algebraic equation, but there are some important differences that should be noted. The above is an instruction to the computer to take whatever values are currently assigned to the two variables A and B, add them together, and assign this value to the variable X. Only a single new variable can be defined by such a statement. Thus an expression such as

```
30 X+Y=A+B
```

has meaning as an algebraic equation, but it is *not allowed* in BASIC. Statements such as

```
70 N=N+1
```

would be of dubious value as algebraic equalities, but they are frequently used in computer programs. In this simple case the current value of the variable N is augmented by 1, and this then becomes the new value of N. Note that the statement 70 N=N+1 contains a constant (1) and a variable (N), both of which may occur in various combinations in arithmetic replacement statements.

The arithmetic operations available in BASIC are

Negation	−X
Addition	X+Y
Subtraction	X−Y
Multiplication	X*Y
Division	X/Y
Exponentiation	X^Y

There are no implied operations; the symbol must always be included. For example, if we wish to multiply A by B, we must write A*B, not AB. The latter would be interpreted as a new variable whose name is AB.

A certain *hierarchy of operations* is observed in processing arithmetic replacement statements. There may be some uncertainty when we see Y=X/A+B as an algebraic statement. Should this be interpreted as $\frac{X}{A} + B$ or $\frac{X}{A + B}$? There is no ambiguity where the computer is concerned. The evaluation proceeds from left to right doing (1) exponentiations, (2) negations, (3) multiplications and divisions, and (4) additions and subtractions. In the example Y=X/A+B, X/A would be evaluated first and then this result would be added to B. If we want $\frac{X}{A + B}$, we must write this as X/(A+B). Parentheses can always be used to remove any ambiguity, but this slows the execution, so they should only be used when necessary. As a further example, the expression

```
5*7^2+3*7
```

would be evaluated by first computing 7 squared, then multiplying this by 5. Next, 3 would be multiplied by 7, and finally, the two terms 245 and 21 would be summed to get 266. Note that two arithmetic operators can never occur in succession. For example, if we want to raise 5 to the −2 power, we cannot write 5^−2; this must be written 5^(−2).

As stated above, both constants and variables are used in arithmetic replacement statements. Constants may for convenience be entered using the *floating-point form*, which is similar to scientific notation. For example, Avogadro's number, 6.022×10^{23} mol^{-1}, is written in floating point form as 6.022E23. Similarly, Planck's constant, 6.63×10^{-34} J · s, is written 6.63E−34. The Faraday constant, 96,485 coulombs, could be entered in the *fixed-point form* 96485., or in any of the floating-point forms .96485E5, 9648.5E1, 9.6485E4, 96485E0, and so forth. There is no difference in how these numbers would be stored internally. Note that we never set off digits with commas when entering a number.

Variables may be used so that a given expression can be evaluated for arbitrary values. For example, we may wish to evaluate the quadratic expression $y = 5x^2 + 2x + 3$ for various values of x, so we write

```
120 Y=5*X^2+2*X+3
```

Before this statement is encountered, there must be other program statements which define the current value of X for which the expression is to be evaluated.

The set of all letters of the alphabet together with the numerical digits 0 through 9 constitute the *alphanumeric character set*. A variable name may be made up of some suitable combination of alphanumeric characters, as long as the first character is alphabetic. The alphabetic characters used in BASIC statements (unless part of a literal character string to be discussed below) are always the uppercase letters.

The number of characters that can be used to define a variable is sometimes rather limited. One should determine the maximum number recognized by the version of BASIC being used. More characters are generally allowed, but they will simply be ignored and this can lead to some confusion. For example, there may be two variables called T1A and T1B in a particular program. If the version of BASIC software in use only recognizes two-character variable names, these two variables which we may regard as distinct will be treated as identical by the computer, and this could lead to erroneous results.

Some combinations of characters are not allowed as variable names, because they serve other special purposes. Again there are some differences, but the list of *reserved words* shown in Appendix A.4 is typical of the BASIC software used on IBM-PC-compatible computers and contains many reserved words which are common to other computers as well. One generally tries to use some mnemonic technique in selecting variable names, as this helps avoid confusion in the construction and later use of more complex programs.

If a given constant is to be used repeatedly, it is preferable to treat it as a variable since this speeds execution. For example, if we were to calculate areas for circles of many different radii, we might want to first write PI=3.14159, and then π can be treated as a variable: A=PI*R^2.

It is often convenient to place more than a single statement on a program line. This may be done to group several operations that occur as a sequence, to conserve storage space, or simply to make a listing of the program more readable. Each statement on a given line of the program must be separated from others by a colon (:), and as many statements may be included as desired as long as the total number of characters does not exceed 255.

For example, the two statements

```
80 R=D/2
90 V=4*PI*R^3/3
```

could be placed on a single line as follows:

```
80 R=D/2 : V=4*PI*R^3/3
```

Blank spaces may be inserted anywhere in a program line to improve readability, but they are (generally) ignored by the interpreter.

As already noted, each program line is numbered. Branching to various parts of the program is possible through reference to these line numbers (see Section 2.2.6). If we need to branch to a particular program statement, it must be on a line by itself, or be the first statement of a line that contains multiple statements.

Example 2.1

For each of the following, determine the value of Z if A = 2, B = 3, and C = 7.

(a) Z=A^2*B^2
(b) Z=A/B*C
(c) Z=A+B/C+B

(d) Z=C^B^A
(e) Z=A/B^A*B

Solution. (a) Exponentiation takes precedence over multiplication, so the result is equivalent to

$$(2^2) \times (3^2) = 4 \times 9 = 36$$

(b) Multiplication and division are of equivalent level, so execution proceeds from left to right to give

$$(2/3) \times 7 = 0.6666667 \times 7 = 4.666667$$

(c) Multiplication takes precedence over addition, so

$$2 + 3/7 + 3 = 2 + 0.4285714 + 3 = 5.428571$$

(d) Evaluation proceeds from left to right to give

$$(7^3)^2 = (243)^2 = 117649$$

(e) Exponentiation takes precedence over multiplication and division and the result is

$$(2/3^2) \times 3 = (2/9) \times 3 = 0.2222222 \times 3 = 0.6666667$$

Example 2.2

Indicate which of the following statements are in error and state the reason why.
(a) X=A^2*300/70E−7
(b) X=((A+C)/(D*E)/(A+B))*(B+C)(A+D)
(c) X=30,000*(3+Y)
(d) R=A+B−2*R
(e) A=R−3T+D4

Solution. (a) OK.
(b) Statement is in error because there is no operator indicated between the terms (B+C) and (A+D).
(c) Statement is in error because there can be no comma in a numerical constant.
(d) OK.
(e) Statement is in error. If 3T is a variable name, it is incorrect because it cannot begin with a numerical digit; if 3 and T are separate entities, there must be an operator between them.

2.2.3 Types of Variables

Sometimes we might wish to restrict a particular variable to integral values. For example, maybe the variable N will be used as a simple index to indicate the number of times we have cycled through a given part of the program. Each time through we may have a statement such as N=N+1 and then a test to see if we have gone through the requisite number of times. If we treat N as an *integer variable,* less computer space is required for storage (typically, 2 bytes), and less time is required for arithmetic operations involving it.

If we begin our program with the *type declaration*

```
10 DEFINT N
```

any variable beginning with the letter N, whenever encountered, is treated as an integer unless a redefinition is given by a second type declaration later in the program. The letters DEFINT are easily remembered as an acronym for DEFine INTeger. Note that if N were recomputed anywhere in the program, the result, if a fraction, would be converted to an integer. Thus if N is 3, N=N/2 would result in the value N = 1 or N = 2, depending on whether the version of BASIC being used carries out integer conversion by truncation or by rounding. Obviously, one must be very cautious with such integer arithmetic.

An implicit definition can be given within the type declaration. This can be very convenient to use if the type for many variables is to be specified. The implicit type statement is of the form

```
20 DEFINT I—N
```

This causes any variable beginning with the letters I, J, K, L, M, or N to be treated as integer variables.

There is an alternative method of defining integer variables. We can omit the type declaration statement altogether, and simply end any variable name we wish to treat as an integer with %. Thus the variables I%, NIT%, and MM% would all be treated as integer variables. It should be noted that if the latter method is used to designate integer variables, it must be used consistently throughout the program wherever the name occurs. For example, MM% and MM would be treated as two different variables, only one of which is restricted to integral values.

Computer calculations are generally carried out for *single-precision variables* with about seven-digit accuracy. All variables, even if given an integral value, are treated as single-precision variables unless otherwise specified. This is the default option. Single-precision variables generally require 4 bytes for storage, and it is not necessary to give a type declaration for them unless they have been used in another way in an earlier part of the program, and we desire to redefine them to serve another purpose. Then the type declaration

```
30 DEFSNG A,B−E
```

may be used (DEFSNG = DEFine SiNGle). The symbol for the second method in this case is !, and variable names ending in !, such as A!, DAF!, or SS!, will be treated as single-precision variables.

At times the approximate seven-digit accuracy of single-precision arithmetic may be insufficient. We may be dealing with small differences of large numbers, or simply have repetitive calculations in which there may be excessive accumulation of round-off errors. We may then make use of *double-precision variables,* whose type may be declared with the statement (DEFDBL = DEFine DouBLe)

```
10 DEFDBL A,B,F-H
```

This causes any variable beginning with A, B, F, G, or H to be treated as a double-precision variable. Double-precision variables require twice as many bytes for storage (i.e., typically 8 bytes), and the execution times are approximately four times longer than for comparable operations involving single-precision variables. The character used to specify double precision for individual variables without using the type declaration is #; DF#, C#, and H1P# are examples of acceptable double-precision variable names.

Single- or double-precision constants can be indicated using the floating-point form. If the exponent is represented by the usual E (the E-format), the constant is treated as a single-precision number, but if the exponent is indicated with a D (the D-format), the constant is treated as a double-precision number. The designation can also be made by specifying seven or fewer digits for a single-precision constant, or eight or more digits for a double-precision constant in a fixed-point form. Finally, the ! and # signs may be used as suffixes on numbers as well as variables to indicate their precision.

If both single- and double-precision variables and/or constants are used in an arithmetic replacement statement, care must be exercised to ensure that the benefits of defining some to be of double precision are not nullified by round-off errors. For example,

```
40 A#=2/3
```

is correct to only about seven digits even though A# is a double-precision variable, because 2 and 3 are single-precision constants and the calculation is carried out using single-precision arithmetic, and then the result is converted and stored as a double-precision variable. If we want the full double-precision accuracy in this case, we might write

```
40 A#=2#/3#
```

The type declaration

```
10 DEFSTR R,U-Z
```

defines any variable beginning with the letters R, U, V, W, X, Y, or Z to be *string variables* (DEFSTR = DEFine STRing). Alternatively, we can designate a string variable by ending the variable name with the dollar sign $; for example, C$ and F$ are acceptable string variable names. As with other variable designations, if the latter method is used, the dollar sign becomes an essential part of the variable name and must be used consistently throughout the program.

As the name suggests, a string variable is a string of alphanumeric or special characters that can be stored and manipulated. The character string is defined by enclosing it in quotation marks:

```
80 Y$="This is "
```

The + sign when used with string variables produces *concatenation*, or a joining of the two strings. For example, with Y$ as defined above and Z$="dumb", the statement

```
100 X$=Y$+Z$
```

would cause the string "This is dumb" to be assigned to the variable X$.

It can be very tedious, if not impossible, to assign different variable names to every variable used in some cases. Suppose that we have 150 individual data points which need to be subjected to a particular analysis. These may be stored in a one-dimensional array. The index by which we identify a particular value within the array is enclosed in parentheses, and the variable is said to be a *subscripted* variable. If the numbers are 1.5, 2.7, 3.8, 2.8, . . . and the variable name is EE, the array, which is a column matrix in this case, can be constructed as follows:

```
20 EE(1)=1.5
30 EE(2)=2.7
40 EE(3)=3.8
50 EE(4)=2.8
.
.
.
```

The variables within the array can be referred to by an index known as the subscript. For example, if the subscript I is equal to 3, then EE(I) = 3.8, but if I = 1, then EE(I) = 1.5, and so forth.

The elements of an array may depend on more than one subscript, in which case we use a two-dimensional, a three-dimensional, or in general an n-dimensional array. A two-dimensional array has double subscripts, and it represents a matrix. For example, the array XC might contain the numbers

$$\begin{matrix} 8 & 6 & 2 & 4 \\ 3 & 1 & 0 & 9 \\ 4 & 4 & 5 & 3 \\ 7 & 7 & 2 & 9 \\ 4 & 8 & 4 & 6 \end{matrix}$$

In this example the array has five rows and four columns, and it is a 5×4 matrix. A particular matrix element is referred to by specification of two subscripts, the first representing the row and the second the column. Thus we might write XC(K,L) to represent an element of the matrix above, in which case K is the row index and L is the column index. If K = 3 and L = 4, then XC(K,L) = 3, while XC(4,3) = 2 and XC(2,3) = 0, and so forth.

Note that we may also wish to define arrays of string variables, the individual members of which can be addressed by specification of their indices. This can be done in the same way as for numerical arrays as long as the variable name carries the proper designation or type declaration. For example, V$(I) and W$(I,J,K) are proper designations of one-dimensional and three-dimensional string variable arrays.

2.2.4 Some Initialization Statements

Data can be entered from the keyboard during execution of a program (see Section 2.2.5), but it is often more convenient to supply the data in advance by means of a *DATA statement*. Numerical constants or tables of reference data, for example, can be made a part of the program and need not be entered each time the program is run. If the three numbers 1.75, 3.32, and 7.56 will be needed during the course of the calculation, we may write

```
20 DATA 1.75,3.32,7.56
```

Prior to execution, the program is scanned for DATA statements. The numbers are placed in a library where they can be read later as required. The DATA statements can be placed anywhere throughout the program, but the order in which the data are called must be preserved. Most programmers find it convenient to group the DATA statements together, either at the beginning or at the end of the program, so that they can be located readily.

Variables are assigned storage locations within a given block, and the computer must know how many spaces will be required for the subscripted variables. Sufficient space is automatically reserved for subscripted variables whose indices do not exceed 10, but extra space must be reserved for larger arrays by means of a *DIMENSION statement*. The abbreviation DIM is used for dimension. For example, we might need three arrays in a given program, the first (XA) being a one-dimensional array with 20 elements, and the other two (XB and XC) being 30 × 30 matrices. The DIMENSION statement would then read

```
10 DIM XA(20),XB(30,30),XC(30,30)
```

The DIMENSION statements can appear anywhere in the program prior to the first use of the variables involved, but most programmers prefer to place them at the beginning of the program, where they can readily be located. It is good programming practice to dimension subscripted variables to the actual dimensions required even if the indices do not exceed 10, as it makes the programming process more systematic and leads to more efficient use of the available computer memory.

In complex computer programs, it is often convenient to include statements that identify the variables and/or program operations. The *REMARK statement*, abbreviated REM, is ignored by the computer during execution of a program; it is included solely for the convenience of the programmer. Suppose, for example, that somewhere in the middle of a long, complex program we perform the numerical integration of a function. We might include the following statements in the program:

```
800 REM——STATEMENTS 820—920 PERFORM A NUMERICAL
810 REM——INTEGRATION OF THE FUNCTION FA.
820 .
830 .
840 .
```

As with other program statements, the REMARK statement can be placed on a line with other statements by use of the colon. For example,

```
70 XS=0 : REM INITIALIZE XS
```

A single quotation mark (') is a shorthand notation used to represent REM. When a REMARK statement is added to the end of a program line as above, the single quotation mark can be used to represent the colon as well, and the statement reads

```
70 XS=0 'INITIALIZE XS
```

Note that if a REMARK statement is placed on a line with other BASIC statements, it must be the last statement on the line.

2.2.5 Input-Output Statements

Values of some of the variables used in a program will be generated by arithmetic replacement statements, but others must be supplied by the user. As mentioned above, we may define certain numerical constants or tabular data with the DATA statement. These numbers are assigned to variables within the program using the *READ statement*. For example, we may need to make repeated use of the number π in our calculations. We can then use the DATA statement

```
10 DATA 3.14159
```

in conjunction with the READ statement

```
20 READ PI
```

to assign the value 3.14159 to the variable PI.

Multiple READ statements and DATA statements can be spread throughout the program, but proper sequencing must be observed. The data will be read in order unless a *RESTORE statement* is encountered, in which case the reading of data starts again at the beginning of the data library. For example, the statements

```
10 DATA 1,2,3
20 READ A,B,C
30 READ D,E,F
40 DATA 4,5,6
```

upon execution lead to A = 1, B = 2, C = 3, D = 4, E = 5, and F = 6. However the statements

```
10 DATA 1,2,3
20 READ A,B,C
30 RESTORE
40 DATA 4,5,6
50 READ D,E,F
```

upon execution give A = 1, B = 2, C = 3, D = 1, E = 2, F = 3.

Numerical values can be assigned to variables during the execution of an interactive program using the *INPUT statement*. If the program contains the statement

```
120 INPUT PI
```

a question mark (?) will be displayed on the screen and the program will then wait for the user to enter the number to be assigned to this variable, say 3.14159.

If we wish to input a number of variables, it may be difficult to keep track of the order in which they are to be given. In this case it is convenient to use an INPUT statement of the form

```
120 INPUT"What is the value of PI";PI
```

Upon execution of this statement the screen displays

```
What is the value of PI?
```

and the computer waits for the number to be entered. Whatever is placed in quotes (the literal character string) is printed exactly and the semicolon is a control feature that causes the pause with a ? to occur on the same line.

We can also enter several values at once. For example,

```
180 INPUT"INPUT VALUES OF A,B,C";A,B,C
```

would result in display of the line

```
INPUT VALUES OF A,B,C?
```

If we respond with 1,2,3 then A = 1, B = 2, and C = 3 are assigned. If we entered only part of the numbers required to satisfy the INPUT statement, we would be told to enter the numbers again and this would continue until the entire INPUT statement is satisfied.

Suppose that we have made a calculation and we wish to display the result. We use a *PRINT statement*:

```
330 PRINT PH
```

As with the input, we may want to define the output more clearly using a modified PRINT statement:

```
330 PRINT"The pH of the solution is";PH
```

The literal character string is printed followed by the current value of the variable PH. If the value of PH were 7.543, for example, the display would read

```
The pH of the solution is 7.543
```

Once again the semicolon in the PRINT statement is a control character that causes the variable PH to be printed on the same line as the literal character string without any intervening space.

The statements

```
440 PRINT A
450 PRINT B
460 PRINT C
```

would result in the printing of the three variables A, B, and C on three successive lines. Several numbers can be printed on the same line using a single PRINT statement, such as

```
440 PRINT A,B,C
```

or

```
440 PRINT A;B;C
```

In the first case the numbers are spaced across the display at print intervals (of about 14 to 16 spaces, depending on the version of BASIC), whereas the numbers are juxtaposed to one another on the line if the semicolons are used. Printed numbers are all followed by one trailing space. Positive numbers are also preceded by a space because they carry a + sign, which does not print. Negative numbers are of course preceded by a − sign. In either case, if many numbers are printed with a single PRINT statement, they spill over if necessary to subsequent lines until all are displayed. If we wish to have a blank line in order to make our output more readable, this can be done with a simple PRINT statement with no variables given:

```
630 PRINT
```

An ordinary PRINT statement includes what is generally known as a carriage return at the end of the line, so that the next print will occur on a new line. On occasion we may wish to prevent the carriage return so that the next print will also occur on the same line. This may be done by ending the statement with a semicolon as in the following example:

```
320 PRINT X;
```

Numbers or string variables may be spaced to arbitrary positions on the display line using the *PRINT TAB statement*. For example,

```
820 PRINT TAB(20) A TAB(40) B
```

causes the value of A to be printed beginning at position 20 and B to be printed beginning at position 40. For multiple printing the subsequent tabs must be larger than the preceding ones (i.e., a backspace is not allowed). Furthermore, if the number in parentheses is larger than the screen width, a wraparound to the next line occurs.

The question mark (?) may be used as a shorthand symbol for the PRINT command. For example, the statement

```
80 ?L
```

is equivalent to

```
80 PRINT L
```

If our computer is equipped with a line printer, we may obtain a hard-copy output. If we replace PRINT with LPRINT, the output will be routed to the line printer rather than to the screen. For convenience, we often use both PRINT and LPRINT statements in the same program so that we can generate hard-copy output and still operate the computer in an interactive mode, monitoring the results with a CRT display. Note that the shorthand notation ? applies only to the PRINT command; there is no corresponding notation for LPRINT.

Example 2.3

What value of X will be printed by the following BASIC program?

```
10 READ C,D
20 DATA 2,5,6
30 READ A,B
40 RESTORE
50 READ E,F
60 DATA 3,14,2
70 X=C*D^E/A^B+F*A
80 PRINT X
```

Solution. The data library is set up initially as 2,5,6,3,14,2. Reading in order C = 2 and D = 5 from statement 10, and A = 6 and B = 3 from statement 30. Because of the RESTORE command, E and F are read starting again at the top of the library, so E = 2 and F = 5. Statement 70 is interpreted as

$$X = (C)(D^E/A^B) + (F)(A) = (2)(5^2/6^3) + (5)(6)$$

$$= 0.2314815 + 30 = 30.23148$$

2.2.6 Normal Program Flow and Branching

The statements in a BASIC computer program are processed in ascending order in normal execution. Any numbering sequence may be used as long as the order is preserved, but it is a common practice to begin numbering at 10 with each succeeding statement numbered in increments of 10, according to the default options given in

```
10 REM***PROGR TO CALC MOL WT AND PERCENT COMPOSITION
20 REM***FOR ANY COMPD OF C, H, AND O
30 READ AC,AH,AO
40 INPUT"HOW MANY CARBON ATOMS ARE THERE";NC
50 INPUT"HOW MANY HYDROGEN ATOMS ARE THERE";NH
60 INPUT"HOW MANY OXYGEN ATOMS ARE THERE";NO
70 MW=NC*AC+NH*AH+NO*AO
80 PRINT"THE MOLECULAR WEIGHT IS";MW
90 PRINT"THE COMPOUND IS";NC*AC/MW*100;"% CARBON,"
100 PRINT NH*AH/MW*100;"% HYDROGEN, AND";NO*AO/MW*100;"% OXYGEN."
110 DATA 12.011,1.0079,15.9994
120 END
```

Figure 2.1 BASIC program for the calculation of the molecular weight and percentage composition of any arbitrary compound of C, H, and O.

Section 2.2.1. This allows for later insertion of a limited number of additional statements without having to renumber the remainder of the program in order to accommodate the change. In most cases the final statement of the program is the END statement, but it is optional for many computers.

We are now ready to examine the structure of an actual BASIC program. Figure 2.1 gives a simple example. Statements 10 and 20 are REMARK statements that simply tell what the program does. Statement 30 serves to define the atomic weights of carbon, hydrogen, and oxygen (AC, AH, and AO, respectively), using the data supplied in statement 110. Statements 40 to 60 are used for the interactive input of the three numbers NC, NH, and NO, which represent the number of carbon, hydrogen, and oxygen atoms in the molecule, respectively. The molecular weight is calculated in statement 70. This value is printed in statement 80, and then used to calculate and print the percentage composition of the molecule in statements 90 and 100. Note that we obviate the naming and storing of these percentages by defining them in the PRINT statement itself rather than generating them with arithmetic replacement statements. Execution is halted by the END statement, number 120.

The normal program flow described above can be modified by various branching and decision-making commands, and it is through the use of these that we begin to realize the full potential of the computer. The simplest branching is by the *unconditional GOTO statement,* which, when encountered, transfers control to the statement number given. For example, we can insert a GOTO statement after line 100 in the program of Figure 2.1 to obtain

```
10 REM***PROGR TO CALC MOL WT AND PERCENT COMPOSITION
20 REM***FOR ANY COMPD OF C, H, AND O
30 READ AC,AH,AO
40 INPUT"HOW MANY CARBON ATOMS ARE THERE";NC
50 INPUT"HOW MANY HYDROGEN ATOMS ARE THERE";NH
60 INPUT"HOW MANY OXYGEN ATOMS ARE THERE";NO
70 MW=NC*AC+NH*AH+NO*AO
```

```
80 PRINT"THE MOLECULAR WEIGHT IS";MW
90 PRINT"THE COMPOUND IS";NC*AC/MW*100;"% CARBON,"
100 PRINT NH*AH/MW*100;"% HYDROGEN, AND";NO*AO/MW*100;"% OXYGEN."
105 GOTO 40
110 DATA 12.011,1.0079,15.9994
120 END
```

Then control is transferred immediately back to statement 40 after the results of a given calculation are printed. This allows the program to be run any number of times without halting and beginning execution again. In this simple case the cycling back to statement 40 occurs indefinitely unless we regain control of the computer (return to the command mode) by use of the BREAK key.

More sophisticated branching control is provided by the *computed GOTO statement,* whose general form is

$$\text{ON (exp) GOTO } n_1, n_2, n_3, \ldots$$

In this case the expression following the word ON is evaluated, converted to an integer if necessary, and then control is passed to any of a number of different statements n_1, n_2, n_3, . . . , depending on the value of the integer. The statement

80 ON K+5 GOTO 90,150,200,40

serves as an example. In this case control is passed to statement 90, 150, 200, or 40, depending on whether the current value of the expression K+5 is 1, 2, 3, or 4, respectively. If K+5 = 0 or an integer greater than 4, control passes to the next statement in the program (i.e., branching is bypassed).

If we wish to make repetitive calculations for a range of values, we make use of a *FOR-NEXT loop*. The structure of the FOR-NEXT loop is as follows:

```
FOR X=A TO B STEP C
   :
NEXT X
```

The sequence of steps between the FOR and NEXT statements is executed first with X=A, then with X=A+C, X=A+2C, X=A+3C, ... Control is passed back until the value of X is equal to or greater than the upper limit B, at which point the loop is satisfied and control passes on to the statement following NEXT X. If STEP C is omitted, the increment is unity by default.

Suppose, for example, that we modify the program of Figure 2.1 as follows:

```
10 REM***PROGR TO CALC MOL WT AND PERCENT COMPOSITION
20 REM***FOR ANY COMPD OF C, H, AND O
30 READ AC,AH,AO
33 INPUT"HOW MANY COMPOUNDS ARE THERE";N
35 FOR I-1 TO N
```

```
40 INPUT"HOW MANY CARBON ATOMS ARE THERE";NC
50 INPUT"HOW MANY HYDROGEN ATOMS ARE THERE";NH
60 INPUT"HOW MANY OXYGEN ATOMS ARE THERE";NO
70 MW=NC*AC+NH*AH+NO*AO
80 PRINT"THE MOLECULAR WEIGHT IS";MW
90 PRINT"THE COMPOUND IS";NC*AC/MW*100;"% CARBON,"
100 PRINT NH*AH/MW*100;"% HYDROGEN, AND";NO*AO/MW*100;"% OXYGEN."
105 NEXT I
110 DATA 12.011,1.0079,15.9994
120 END
```

We would then input N, the number of calculations to be made, at statement 33 and control would loop through statements 35 to 105 just N times, following which a normal termination would be made with control passed back to the command mode at statement 120. Here I is used as a simple index to control the number of times looping occurs. In more complex programs the index is often used in various calculations within the loop, but care must be exercised to avoid recalculation of it, or of the lower limit, upper limit, or step size, within the loop. By the same token, control should not be transferred into a loop as, for example, by a GOTO statement, since the index is usually undefined.

We can have a FOR-NEXT loop within a FOR-NEXT loop, which is called a *nesting* of loops:

```
FOR R=1.5 TO 15 STEP .5
FOR S=5.1 TO 8.1 STEP .1
:
NEXT S
NEXT R
```

One must be careful to terminate the inner before the outer loop when writing programs involving nested loops.

As an example of how nested FOR-NEXT loops might be used, assume we have two $N \times N$ matrices **A** and **B**, and we desire the product matrix **C** = **AB**. Matrix multiplication was described in Section 1.4; it may be summarized by the formula

$$C_{ij} = \sum_k A_{ik} B_{kj}$$

The first index represents the row and the second the column. From the formula we see that the C_{ij} element of the product matrix is obtained by multiplying the ith row of **A** and the jth column of **B** term by term, and summing these products. Note how the following program sequence carries out this matrix multiplication process:

```
100 FOR I=1 TO N
110 FOR J=1 TO N
120 C(I,J)=0
130 FOR K=1 TO N
140 C(I,J)=C(I,J)+A(I,K)*B(K,J)
150 NEXT K
160 NEXT J
170 NEXT I
```

Some versions of BASIC have simple statements for matrix manipulations such as this [e.g., MATC=A*B for multiplication, MATD=INV(A) for inversion, etc.], but many do not and then programming a sequence such as the above is required.

Program branching can also be controlled using the *relational operators*

$<$	is less than
$>$	is greater than
$=$	is equal to
$<=$ or $=<$	is less than or equal to
$>=$ or $=>$	is greater than or equal to
$<>$ or $><$	is not equal to

together with the *IF statement,* which is of the form

IF (exp) THEN (clause_1) ELSE (clause_2)

(exp) is a relational expression to be evaluated, and (clause_1) and (clause_2) are executable program steps. If the condition required by (exp) is satisfied, (clause_1) is executed and control passes to the next statement, but if the condition required by (exp) is not satisfied, (clause_2) is executed before control passes to the next statement. In many cases the ELSE part is omitted, and control passes directly to the next statement if the condition is not satisfied. (clause_1) and/or (clause_2) are often GOTO statements.

The relational operators in various forms together with the IF statement provide a set of very powerful techniques for decision making and branching. A simple illustration will be given by again modifying the simple program of Figure 2.1:

```
10 REM***PROGR TO CALC MOL WT AND PERCENT COMPOSITION
20 REM***FOR ANY COMPD OF C, H, AND O
30 READ AC,AH,AO
33 INPUT"NUMBER OF CALCULATIONS";N
35 M=0
40 INPUT"HOW MANY CARBON ATOMS ARE THERE";NC
50 INPUT"HOW MANY HYDROGEN ATOMS ARE THERE";NH
60 INPUT"HOW MANY OXYGEN ATOMS ARE THERE";NO
70 MW=NC*AC+NH*AH+NO*AO
```

```
80 PRINT"THE MOLECULAR WEIGHT IS";MW
90 PRINT"THE COMPOUND IS";NC*AC/MW*100;"% CARBON,"
100 PRINT NH*AH/MW*100;"% HYDROGEN, AND";NO*AO/MW*100;"% OXYGEN."
101 M=M+1
102 IF M<N THEN GOTO 40
110 DATA 12.011,1.0079,15.9994
120 END
```

Here we set the index M to zero (statement 35), and we increase it by 1 each time we make a calculation (statement 101). Statement 102 passes control back through the calculation loop until the index equals the number of calculations desired, following which control passes to the normal termination in statement 120.

The *logical operators* AND, OR, and NOT are also used to make program branching decisions. The expression

$$N<>0 \text{ AND } M>=N$$

is true (not equal to 0) if both conditions, N not equal to zero and M greater than or equal to N, are satisfied. Similarly,

$$N<>0 \text{ OR } M>=N$$

is true if either one or both of the conditions are satisfied; otherwise, it is false (equal to 0). The logical NOT

$$\text{NOT } N>5$$

returns a value -1 if the expression is false, or 0 if it is true.

A variable can be defined using these logical operators and its value then used in a decision-making process. For example, we might again modify the program listed in Figure 2.1, to obtain

```
10 REM***PROGR TO CALC MOL WT AND PERCENT COMPOSITION
20 REM***FOR ANY COMPD OF C, H, AND O
30 READ AC,AH,AO
33 INPUT"NUMBER OF CALCULATIONS";N
35 M=0
40 INPUT"HOW MANY CARBON ATOMS ARE THERE";NC
50 INPUT"HOW MANY HYDROGEN ATOMS ARE THERE";NH
60 INPUT"HOW MANY OXYGEN ATOMS ARE THERE";NO
70 MW=NC*AC+NH*AH+NO*AO
80 PRINT"THE MOLECULAR WEIGHT IS";MW
90 PRINT"THE COMPOUND IS";NC*AC/MW*100;"% CARBON,"
100 PRINT NH*AH/MW*100;"% HYDROGEN, AND";NO*AO/MW*100;"% OXYGEN."
101 M=M+1
102 B=NOT M=N
103 ON B+2 GOTO 40,120
110 DATA 12.011,1.0079,15.9994
120 END
```

We see that B will be -1 (true) as long as $M < N$, but when $M = N$, B will be 0 (false). This is used in a computed GOTO statement (103), and control passes to statement 40 until the desired number of calculations have been made.

The logical operators can be used directly in the IF statement and two or more conditions can be tested simultaneously to improve the efficiency of the program. For example, we might write our example program as follows:

```
10 REM***PROGR TO CALC MOL WT AND PERCENT COMPOSITION
20 REM***FOR ANY COMPD OF C, H, AND O
30 READ AC,AH,AO
40 INPUT"HOW MANY CARBON ATOMS ARE THERE";NC
50 INPUT"HOW MANY HYDROGEN ATOMS ARE THERE";NH
60 INPUT"HOW MANY OXYGEN ATOMS ARE THERE";NO
65 IF NC=0 OR NH=0 AND NO=0 THEN GOTO 120
70 MW=NC*AC+NH*AH+NO*AO
80 PRINT"THE MOLECULAR WEIGHT IS";MW
90 PRINT"THE COMPOUND IS";NC*AC/MW*100;"% CARBON,"
100 PRINT NH*AH/MW*100;"% HYDROGEN, AND";NO*AO/MW*100;"% OXYGEN."
105 GOTO 40
110 DATA 12.011,1.0079,15.9994
120 END
```

This would prevent a calculation being made if the number of carbon atoms were zero, or if there were neither hydrogen nor oxygen in the molecule. Statements of this kind can be very useful; we may not know at the outset how many calculations are desired, but a normal termination can still be made by entering zeros for NC and/or NH and NO.

One other branching technique is the use of the WHILE-WEND loop, whose general structure is of the form

```
WHILE (exp)
   .
   .
   .
WEND
```

Here (exp) is a numeric, relational, or logical expression. The set of statements between WHILE and WEND will continue to be executed as long as the condition represented by (exp) is not zero. If it becomes zero, control passes to the next statement beyond WEND. As an example, suppose that we wish to calculate the molecular weights and percent compositions of all the alkanes up to and including that with 10 carbon atoms. Recalling that the general formula for the homologous series of saturated hydrocarbons is C_nH_{2n+2}, we again modify our example program, this time as follows:

```
10 REM***PROGR TO CALC MOL WT AND PERCENT COMPOSITION
20 REM***FOR ALL ALKANES UP TO DECANE
30 READ AC,AH
```

```
40 NC=1
45 WHILE NC<11
50 NH=2*NC+2
70 MW=NC*AC+NH*AH
80 PRINT"THE MOLECULAR WEIGHT IS";MW
90 PRINT"THE COMPOUND IS";NC*AC/MW*100;"% CARBON,"
100 PRINT "AND ";NH*AH/MW*100;"% HYDROGEN."
105 NC=NC+1
107 WEND
110 DATA 12.011,1.0079
120 END
```

Each time the loop is executed, the value of NC is increased by 1 in statement 105. When it becomes equal to 11, NC $<$ 11 becomes false and control passes to the END statement.

Example 2.4

If you input the values 1,2,3 for I,J,K in the following program segment, to which statement will the program branch upon execution of statement number 50?

```
10 INPUT I,J,K
20 I=2*J-6/K
30 J=4*I-2*K
40 K=I+3*J-K
50 ON 5*I-2*J/4*K GOTO 60,80,100,120,140,160
```

Solution. Starting with I = 1, J = 2, K = 3, we see that I is changed to $2 \cdot 2 - 6/3 = 4 - 2 = 2$ by statement number 20. Similarly, J becomes $4 \cdot 2 - 2 \cdot 3 = 2$ at statement 30 and K becomes $2 + 3 \cdot 2 - 3 = 5$ at statement number 40. Thus at statement 50 the value computed is

$$5 \cdot 2 - 2 \cdot (2/4) \cdot 5 = 10 - 5 = 5$$

and the program branches to the fifth statement number (i.e., to number 140).

Example 2.5

Indicate which of the following BASIC statements are in error and state the reason why.

(a) FOR I=1 STEP 2
(b) ON X=Y+Z GOTO 30,40,70
(c) FOR I=5 TO 10
(d) L=10 : FOR I=20 TO L : PRINT I : NEXT I
(e) FOR I=10 TO 5 STEP −1

Solution. (a) In error. No upper limit for the loop is given.
(b) In error. Expression should not define a new variable.
(c) OK. (STEP=1 is understood.)
(d) Loop does not execute, since upper limit is less than lower limit.
(e) OK. (It is permissible to have the lower limit greater than the upper limit if a negative step is designated.)

2.2.7 Functions and Subroutines

Computer programming would be very tedious if it were necessary to include a procedure for the extraction of square roots, evaluation of logarithms and trigonometric functions, or calculation of a variety of other functions each time they were encountered in a program. Fortunately, many of the common functions are included in the software as *library functions,* many of which are computed using truncated series of the kind discussed in Section 1.8. The letters SQR, for example, stand for the square root function. If we have a program step that reads

```
320 A=X*SQR(3)
```

the value X times the square root of 3 is assigned to the variable A. The argument of the square root function can be a variable or an arithmetic expression:

```
780 A=X+SQR(Y*Z)
```

It is required that Y*Z be greater than or equal to zero when this step is executed. It is the positive root that is returned by the SQR function.

There are many other library functions available, and they are used in a similar way. The commonly used numeric library functions are tabulated in Appendix A.5 together with a few of the more important functions for the manipulation of string variables. General descriptions of all available library functions will be found in the manuals supplied with your computer.

It will be observed that the arguments of the trigonometric functions are all in radians, and we must include the appropriate conversion factor if required. For example, if X were in degrees and we desire the sine of X, we must write SIN(X*.01745329). It should also be noted that only the natural logarithmic function is included in the library. To get the common logarithm, we must write LOG(X)/LOG(10) (see Eq. 1.14).

It is useful for many applications to have random numbers available, and these are generated by the library functions RND. The random numbers generated by RND are in the range 0 to 1; if we wish a set of integral numbers in the range 0 to n, they may be obtained using INT(RND*(n+1)). Note that what are generated here are actually *pseudorandom numbers*. If we were to run the program a second time, we would get exactly the same set of numbers. Since the set is generated by a formula and is predictable, they are not truly random. We can, however, change them by using the statement

```
RANDOMIZE m
```

prior to the generation of numbers with RND. Here m is an integer in the range −32768 to 32768, and it reseeds the random number generator so that a different set of numbers will be obtained. The numbers are still not truly random. The same set will be obtained each time for a given value of m, so it is preferable to change the value of m each time the program is run. With some computers we can avoid manual selection of random m values each time we run the program by using the internal clock. TIMER is a read-only numeric function that returns the fractional seconds to the nearest degree possible. We reseed the random number generator using it by writing

RANDOMIZE TIMER

If a given expression will be evaluated repeatedly in the execution of a program, it may be preferable to create a *user-defined function* with the DEF FN statement. Suppose that we need to calculate expressions of the form $x^2 - x + 7$ many times. We might define the function G using

20 DEF FNG(X)=X*X−X+7

and then it would only be necessary to write FNG(6) to represent $6^2 - 6 + 7$, or 5*FNG(A)/2 to represent $(5/2) \cdot (A^2 - A + 7)$, and so forth.

Sometimes the user-defined function is too restrictive, and we need the capability of executing an entire sequence of operations repeatedly using different values of the variables. It is then convenient to use a *subroutine*. The GOSUB command is similar to the GOTO command, except that control returns to the same point in the program following the execution of the subroutine. The transfer back to the main program is made by means of a RETURN statement at the end of the subroutine.

For example, consider the program with the following structure:

```
70  :
80  :
90  :
100 GOSUB 1000
110 :
120 :
130 :
200 GOSUB 1000
210 :
220 :
230 :
999 STOP
1000 ON SGN(X)+2 GOTO 1010,1030,1050
1010 PRINT"ERROR:  ARGUMENT OF SQR FTN <0"
1020 STOP
1030 Y=0
1040 RETURN
1050 Y=.5*Y
1060 R=0
1070 S=(X/Y-Y)*.5
1080 IF S=0 OR S=R THEN RETURN
1090 Y=Y+S
1100 R=S
1110 GOTO 1070
```

Actually, we would have little occasion to use this particular subroutine since it evaluates the square root of X and returns the value as Y, and we have a library function which will do that. It does illustrate some important features of subroutines, however.

Note that the subroutine can be entered from various points in the main program, here illustrated as statements 100, 200, ... The statement number from which branching occurs to the subroutine is retained so that control can be returned to the proper location in the main program after return from the subroutine. We can branch from one subroutine to another, and so on through several levels, but care must be taken to execute a proper return from each.

Statement 999 serves only as a protective block in the illustration above. This halts execution so we cannot enter the subroutine by normal program flow in case there is an error in the main program. Statement 1000 provides a three-way branch depending on the sign of X. Only if X > 0 does control pass to statement 1050, where the actual process of square root extraction begins. The calculation is by an iterative procedure called the Newton–Raphson method, which is discussed in Chapter 3.

Example 2.6

Write a BASIC program which, for an input value of x, will calculate the value of $\ln(1 + x)$ correct to six-decimal-place accuracy based on the series expansion of Example 1.26.

Solution. First note that a general term of the series may represented by

$$(-1)^{N-1}\frac{x^N}{N}$$

The following program sums the series to the desired accuracy:

```
10 INPUT X
20 N=0 : S=1 : SN=0
30 WHILE ABS(SN-S)>.000001
40 N=N+1 : S=SN
50 SN=S+(-1)^(N-1)*X^N/N
60 WEND
70 PRINT SN : END
```

2.3 CONCEPTS OF STRUCTURED PROGRAMMING

To see how well you have mastered the BASIC language, predict the output of the following program and then run it on your computer to see if you were correct.

```
10 DATA "George","Pete","Dick","Fred","My Name Is "
20 READ N$(3),N$(1),N$(4),N$(2),M$
30 PRINT M$;
40 K=1
50 GOTO 170
```

```
60 L=L+1
70 ON L GOTO 90,120,170
80 FOR I=1 TO L
90 N$(I)=N$(I+1)
100 IF I=2 THEN GOTO 140
110 NEXT I
120 PRINT N$(I)
130 GOTO 50
140 PRINT N$(K)
150 GOTO 190
160 I=I+1
170 L=K+2
180 GOTO 60
190 END
```

Did you get it right? Congratulations. You probably were saying to yourself, "There's got to be a better way of doing this!" There certainly is, and in fact, the principal purpose of the above was to illustrate how *not* to write a program.

The recommended method of programming embodies the use of various *program structures,* which are self-contained units. Branching back and forth between them is to be avoided. If one writes a flowchart for a program such as that above, there are many interconnecting lines which result in what has come to be known as "Spaghetti Bowl" logic. The unconditional GOTO statement is used very sparingly in structured programming. In fact, purists advocate that it never be used, although going to this extreme can lead to some awkward procedures. A few of the more important ideas that should improve program design are outlined below.

2.3.1 Top-Down Design

At present the *flowchart* is used more often than any other design structure. We will not discuss flowchart construction, however, because its use is now discouraged by professional programmers. It leads to the Spaghetti Bowl logic we wish to avoid. What is recommended in its place is a *tree structure,* such as

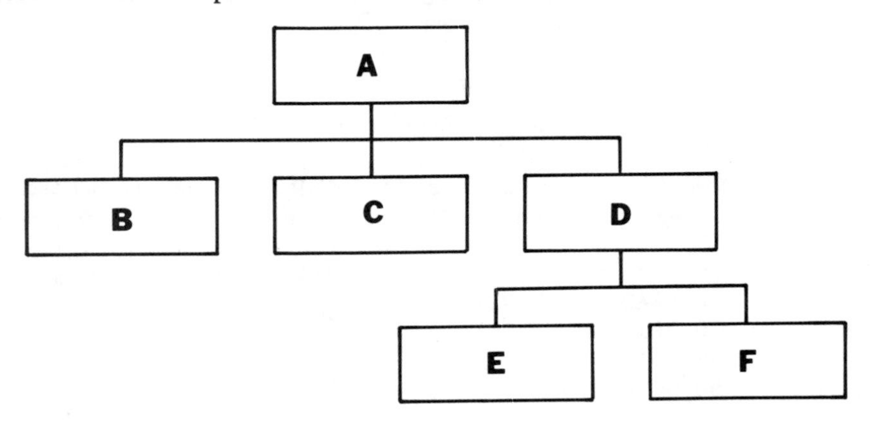

Each of these blocks represents a design element that performs a given function. Input and output to the block are stipulated, and its role is specified. It is viewed as a separate entity that can be programmed and checked with test data independent of the remainder of the program. The function of each individual block is separately verified; the blocks are then assembled to complete the overall task. It can be a very difficult problem to locate the source of an error if testing of the individual components has not been carried out in advance.

The essence of *top-down* design is to begin with a statement of the overall task that must be performed. This is then broken down into smaller and smaller segments, and the tree structure should clearly show which blocks are used by other blocks to complete the problem. They should function independently, such that changes within one block do not force changes to be made in another block. Such design features make a program easier to follow, so that subsequent changes, either by the original programmer or by others, can be made more easily.

2.3.2 Control Structures

There is a set of basic logic constructs called *control structures,* which can be used to build a program that conforms to the independent blocklike design referred to above. Each should have a single entry point and exit point. Unconditional branching is avoided except where required to advance to the next logical level, and backward branching is avoided entirely.

The following are the control structures in common use:

1. SEQUENCE—the simple case of normal forward flow through a sequence of two or more operations.
2. IFTHENELSE—allows for a branch to either one of two operations depending on a specified condition.
3. DOWHILE—an operation or sequence of operations that is performed as long as a condition is true.
4. DOUNTIL—provides a branch back to repeat an operation or sequence of operations as long as a condition is true.
5. CASE—provides a multiple branch, depending on a specified condition, with subsequent exit of the structure from a common point.

These control structures are readily associated with the BASIC language statements previously encountered. The WHILE-WEND sequence accomplishes the DOWHILE control structure, and the computed GOTO statement obviously meets the basic requirement of the CASE control structure. Numbers 3 and 4 are quite similar; they differ only in that the test is made before the sequence of operations in the case of the DOWHILE structure, but it is made after the operations are performed in the DOUNTIL structure. This ensures that the calculation will be made at least once in the latter case. The FOR-NEXT loop may be used to accomplish control structures

3 or 4 in many cases, but the use of WHILE-WEND is recommended if the number of times the sequence is to be repeated is not known at the outset, such as in the case of an iterative calculation.

2.3.3 Pseudocoding

Once the basic problem has been defined and a tentative tree structure drawn, we are prepared to begin thinking through the *algorithm,* or sequence of program steps, which will be used to carry out each program segment. This should be done before the actual process of coding begins. The rules for preparing a *pseudocode* are purposely rather vague, as a loose structure that readily allows for modification as the process develops is what is desired. The pseudocode is a step-by-step description of the computation, using a mixture of easily interpreted English-language statements and BASIC program statements.

If the exact process is known, the BASIC statements in their usual uppercase form may be included. If the definition of variables or precise sequence of operations is not yet defined, they are written using lowercase characters. The steps that are written in the pseudocode often represent several statements in the final code. For instance, if we wish to compute the average of M numbers, we might represent the algorithm in pseudocode form as follows:

A. Input M
B. Calculate the sum $S = a_1 + a_2 + \cdots + a_M$
C. Calculate A=S/M
D. Print A
E. Stop

This pseudocode may be sufficient for the purpose, but on the other hand, we may wish to have a more detailed representation of the sequence of operations required. The pseudocode may be subjected to a process of *stepwise refinement* until it contains sufficient detail for the efficient generation of the actual BASIC code. In this case, for example, we may include further elaboration of step B by adding

B.1 LET S=0
B.2 For I=1, 2, . . . , M, add A(I) to S

Only after the pseudocode is refined to the point where the programmer completely understands the logic should the actual program be written.

2.3.4 Program Optimization

We generally like to have an algorithm that performs a given calculation as rapidly as possible. However, there may be times when it is preferable to sacrifice speed to maintain clarity, particularly if the program is very short anyway.

It is vital that adequate documentation, both external and internal, be maintained. By external documentation we mean a record that shows the logic of the calculation (e.g., the pseudocode from which it was constructed), gives references to books or other sources where the methods are described, and defines the variables employed, particularly for input and output. The internal documentation consists of the REMARK statements that identify key variables and program segments. Studies have shown that in the case of system programs, they are typically used and modified through six generations of programmers. The application programs we write may not enjoy such longevity, but we still need to be aware that most programs will require later modification, either by the original programmer or others. Coming back to a program that has not been used for some time, even for the originator, can be a very frustrating experience if the documentation is inadequate.

It might be necessary to strike a compromise regarding program documentation in some cases. The liberal use of REMARK statements facilitates later use of the program, but as pointed out earlier, they take up space and slow the execution. In recent years the speed of computers has increased considerably, and there has been a concomitant reduction in hardware costs. As a consequence, the development and efficient use of good software is frequently the limiting factor. Thus it is generally preferable to err in the direction of what appears to be too much documentation rather than too little. In some cases the optimization of computational efficiency makes a program appear more obscure and esoteric; then careful documentation, even though brief, may be considered essential.

Once a program has been constructed and shown to function properly, it might then be appropriate to consider changes that will speed the execution, if they can be accomplished without sacrifice of the structuring described above. For example, if we have many calculations involving only integers, we can improve the execution time by restricting them to integral values with an appropriate type declaration. Similarly, double-precision declarations should be used only when single precision is clearly insufficient. If a numerical constant occurs repeatedly, it is preferable to define it as a variable rather than writing it as a constant each time it appears.

As a general rule, the speed of operations is in reverse order to the hierarchy given in Section 2.2.2. For example, multiplications and divisions are slower than additions and subtractions, and so forth. Thus we generally prefer to write Y=X*X or U=V+V rather than Y=X^2 or U=2*V. This may also be beneficial in minimizing the loss of accuracy due to round-off errors.

Care should also be given to the factorization of algebraic expressions in such a way as to minimize time-consuming arithmetic operations. For example, the expression $y = ax^3 + bx^2 + cx + d$ could be represented in a BASIC computer program by the statement

```
Y=A*X^3+B*X^2+C*X+D
```

but it would be better written as

```
Y=X*(X*(X*A+B)+C)+D
```

Frequently used numbers should be introduced through READ and DATA statements, rather than entering them from the keyboard. Care should also be given to make sure that only the essential variables that change within a FOR-NEXT loop are included in it. Subscript searches for elements in an array can be time consuming and they should be kept to a minimum, particularly in lengthy FOR-NEXT loops. For example, we might write

```
160 FOR I=1 TO 100
170 A(I)=B(I)+C(I+1)
180 D(I)=B(I)-C(I+1)
190 NEXT I
```

but

```
160 FOR I=1 TO 100
170 J=I+1
180 X=B(I)
190 Y=C(J)
200 A(I)=X+Y
210 D(I)=X-Y
220 NEXT I
```

is better. Keeping the number of program statements to a minimum does not always lead to the greatest efficiency in the long run.

2.4 COMPUTER GRAPHICS

The computer can be used for graphical display of the result of a given experiment or simulation. This can be more useful than the raw data. Relationships are often revealed which can be spotted only with difficulty by review of a long list of numbers that might be generated. There are almost endless possibilities; they are limited only by the imagination of the programmer.

2.4.1 Using the Line Printer

At times a hard copy of a display may be desired where no high-resolution graphics capability is available. Then the line printer can be used. Many printers have a set of graphical display characters that can improve the appearance, but we will describe a display that uses only normal print characters. Because the number of print positions is restricted, the display has limited resolution, but this is not too serious for many purposes. The efficient coding of a display using a line printer makes use of manipulations of string variables. Since this is not our major concern, such operations will be kept to a minimum. The example given is a simple code to display a spectrum (Figures 2.2 and 2.3). A few brief remarks to clarify the statements not previously encountered are included.

```
10 WIDTH LPRINT 132
20 LPRINT CHR$(27)"1"CHR$(12)
30 LPRINT CHR$(15)
40 DIM X%(36),Y(36)
50 INPUT"INPUT SPECTRUM LABEL";A$
60 FOR I-1 TO 36
70 READ X%(I),Y(I)
80 NEXT I
90 LPRINT : LPRINT : LPRINT
100 A$-"Sectrum Label : "+A$
110 LPRINT A$
120 LPRINT : LPRINT
130 B$-SPACE$(50)+"Absorbance"
140 LPRINT B$
150 S$-SPACE$(5)
160 C$-SPACE$(24)
170 D$-S$+"0"+C$+"1"+C$+"2"+C$+"3"+C$+"4"
180 LPRINT D$
190 E$-" 200 "
200 FOR I-1 TO 4
210 E$-E$+CHR$(43)+STRING$(24,CHR$(45))
220 NEXT I
230 E$-E$+CHR$(43)
240 LPRINT E$
250 I-1
260 FOR J-35 TO 1 STEP -1
270 Y%-INT(Y(J)*25)
280 IF I-5 THEN 310
290 F$-S$+CHR$(124)
300 GOTO 330
310 F$-STR$(X%(J))+" "+CHR$(43)
320 I-0
330 IF Y%-0 THEN 350
340 F$-F$+SPACE$(Y%-1)+CHR$(42)
350 LPRINT F$
360 I-I+1
370 NEXT J
380 DATA 340,.029,336,.1,332,.3,328,.6,324,1.05,320,1.4
390 DATA 316,1.63,312,1.68,308,1.71,304,1.715,300,1.69
400 DATA 296,1.64,292,1.6,288,1.54,284,1.45,280,1.38
410 DATA 276,1.31,272,1.21,268,1.1,264,1.,260,.84
420 DATA 256,.74,252,.58,248,.45,244,.47,240,.55,236,.72
430 DATA 232,.8,228,.77,224,.64,220,.5,216,.34,212,.21
440 DATA 208,.15,204,.09,200,.04
```

Figure 2.2 Program for the display of a spectrum using a line printer.

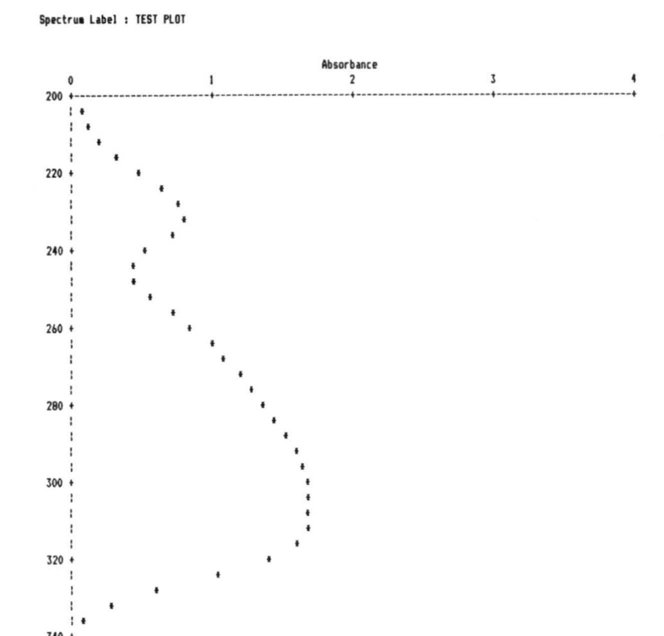

Figure 2.3 Spectral display generated using the program of Figure 2.2.

Statement 10 of the program shown in Figure 2.2 is used to reset the width of the output line. Without this, the computer automatically inserts a carriage return after 80 characters. We reset the font for compressed mode in statement 30; otherwise, we would only have 80 characters per line and our resolution would be severely limited. Statement 20 sets the left margin to 12 spaces. The format needed for statements 20 and 30 may be different for other printers (an Epson LX printer was used to print Figure 2.3).

SPACE$(m) is a statement that creates a string of m spaces. This is used to insert an appropriate spacing in the print variable, so the print characters will be properly located on the plot. The print characters are generated by supplying their ASCII codes n and converting them to the corresponding string variables through use of the function STR$(n). Those used are n = 43 for +, n = 45 for −, n = 124 for ¦, and n = 42 for *. The function STRING$(m,CHR$(n)) generates a string variable m spaces long filled with the character whose ASCII code is n. The number 25 in statement 270 is a scale factor that can be changed. You should now be able to understand how the program of Figure 2.2 operates and be able to use similar procedures to generate other plots using a line printer.

2.4.2 Using the CRT Display

To illustrate, we will generate the curve for the titration of a weak acid with a strong base. The display will be described for the IBM PC using the BASICA software. There are generally several alternative ways to accomplish a given display feature, but we will limit ourselves to those which are quite general and applicable, with but limited modification, to other models.

To be specific, let us assume that we have a volume V_a of an acetic acid solution, whose concentration is C_a. This solution is titrated with sodium hydroxide of concentration C_b; the volume of base added at any stage of the titration will be represented by V_b. The acid dissociation constant of acetic acid is $K_a = 1.8 \times 10^{-5}$.

For any addition of base up to the equivalence point, the acid is partially neutralized so that we in essence have a mixture of acetic acid and sodium acetate, which constitutes a buffer system. The Henderson–Hasselbalch equation,

$$\text{pH} = \text{p}K_a + \log \frac{[\text{salt}]}{[\text{acid}]}$$

is applicable, where [acid] is the concentration of the weak acid and [salt] is the concentration of the salt (sodium acetate in this case). The moles of acid m_a present in the initial solution is $(V_a)(C_a)$. Addition of a given volume of base V_b gives $m_b = (V_b)(C_b)$ moles of base. Neutralization results in m_b moles of salt, and $m_a - m_b$ moles of acid. Since this is in a volume of $V_a + V_b$, we write

$$\text{pH} = \text{p}K_a + \log \frac{\dfrac{m_b}{V_a + V_b}}{\dfrac{m_a - m_b}{V_a + V_b}}$$

$$= \text{p}K_a + \log \frac{m_b}{m_a - m_b}$$

At the equivalence point, where $(C_a)(V_a) = (C_b)(V_b)$, we have just a solution of sodium acetate, and its concentration is $m_a/(V_a + V_b)$. The hydrolysis reaction at the equivalence point is

$$\begin{array}{cccc} \text{OAc}^- & + \text{H}_2\text{O} \rightleftharpoons & \text{HOAc} & + \text{OH}^- \\ \dfrac{m_a}{V_a + V_b} - x & & x & x \end{array}$$

and the hydrolysis constant is given by (see Example 1.1)

$$K_b = \frac{K_w}{K_a} = \frac{x^2}{\dfrac{m_a}{V_a + V_b} - x}$$

We expect that x will be very small, so we can write

$$x \approx \sqrt{\frac{K_w}{K_a} \frac{m_a}{V_a + V_b}} = [OH^-]$$

and pH = pK_w − pOH.

Beyond the equivalence point the pH of the solution will be dominated by the presence of excess sodium hydroxide. Since this is a strong base the hydroxide ion concentration can readily be calculated assuming complete dissociation, and from this the pH is readily obtained. The amount of excess base is $m_b - m_a$, and it is in a volume of $V_a + V_b$, so $[OH^-] = (m_b - m_a)/(V_a + V_b)$ and

$$\text{pH} = \text{p}K_w + \log \frac{m_b - m_a}{V_a + V_b}$$

Now that we have sketched how the pH is calculated at any stage of the titration, we are ready to examine a simple computer program that carries out the calculations and presents a graphical display of the titration curve. Such a program is given in Figure 2.4. Before we consider the program in detail, we need to note some features of the screen layout. For the IBM PC there are two graphics modes, either a medium-resolution mode with 40 characters per line or a high-resolution mode with 80 characters per line. By the statement SCREEN 2 (see statement 30) we select the high-resolution mode. Statement 40 is the command CLS, which clears the screen prior to the beginning of the plot.

We can move the curser to any desired position on the screen using the LOCATE command. Then a PRINT command causes printing to begin at that location. Look, for example, at statement number 110 of the program. The curser is moved by the LOCATE command to row 12 and column 1 of the screen, and the characters "pH" are printed in columns 1 and 2. This is a label for the ordinate of the plot. A similar technique is used in statement 160 to label the abscissa. Note the use of the semicolon at the end of the PRINT statements. This prevents scrolling of the image on the screen, which would otherwise occur for any characters written on the last line. The last line number is 25 on the IBM PC, and it is ordinarily reserved for the designation of the *soft keys*. The soft keys are used to enter an entire sequence of frequently used characters with a single keystroke. The soft key line is suppressed in the current program, so we can use all 25 lines for the graphical display. This is done with the KEY OFF command in statement 20.

With the screen set to the high-resolution graphics mode, it has a grid of 640 horizontal points by 200 vertical points. The actual numbering of these points is from 0 to 639 across, and 0 to 199 down, with the point 0,0 occurring in the upper left-hand corner. The individual points (often called *pixels*) are addressed by their respective X and Y coordinates. The command PSET(X,Y) causes the point with coordinates X,Y to be lit. See, for example, the FOR-NEXT loop statements 50 to 70. This lights all the points along a horizontal line from 50,180 to 550,180. This constitutes the x-axis of our plot. The loop 80–100 lights all the points in a vertical column from 50,12 to 50,180; this constitutes the y-axis. The loop 120–150 places labels on the ordinate to

```
10 'TITRATION OF A WEAK ACID WITH A STRONG BASE
20 KEY OFF
30 SCREEN 2
40 CLS
50 FOR I=50 TO 550      'DRAW X-AXIS
60 PSET(I,180)
70 NEXT I
80 FOR I=12 TO 180      'DRAW Y-AXIS
90 PSET(50,I)
100 NEXT I
110 LOCATE 12,1 : PRINT"pH";                      'LABEL Y-AXIS
120 FOR I=0 TO 14 STEP 2
130 J=(180-I*12)/8
140 LOCATE J,3 : PRINT I;
150 NEXT I
160 LOCATE 25,30 : PRINT"Volume of Base";   'LABEL X-AXIS
170 FOR I=0 TO 50 STEP 10
180 J=(10*I+50)/8
190 LOCATE 24,J : PRINT I;
200 NEXT I
210 READ VA,CA,CB,KA      'READ PARAMETERS
220 MA=VA*CA
230 PKA=-LOG(KA)/LOG(10)
240 FOR VB=1 TO 50 STEP .1     'GENERATE TITRATION CURVE
250 MB=VB*CB
260 IF ABS(MA-MB) > .0001 THEN 300
270 OH=SQR(MA*1E-14/KA/(VA+VB))      'EQUIVALENCE PT
280 PH=14+LOG(OH)/LOG(10)
290 GOTO 340
300 IF MB>MA THEN 330
310 PH=PKA+LOG(MB/(MA-MB))/LOG(10)      'BEFORE EQUIVALENCE PT
320 GOTO 340
330 PH=14+LOG((MB-MA)/(VA+VB))/LOG(10)      'BEYOND EQUIVALENCE PT
340 PSET(10*VB+50,180-PH*12)
350 NEXT VB
360 DATA 50,.1,.2,1.8E-5
370 GOTO 370
```

Figure 2.4 BASIC program for graphical CRT display of the titration curve for a weak acid titrated with a strong base.

represent the pH, and the loop 170–200 places labels on the abscissa to represent the volume of base added. This completes the initial graph layout; the actual calculation of the titration curve begins with the reading of parameters in statement 210.

The logic from this point will not be described, but you should be able to follow the program flow using the principles previously given. Study the program carefully until you are sure you understand what is accomplished at every step. Note in statements 260 and 300 that THEN GOTO n has been replaced by THEN n, which is a legitimate contraction in BASIC. Note also the form of the test made in statement 260. This is necessary because of the way numbers are handled by the computer. At the equivalence point the number of millimoles of acid (MA) and of base (MB) should be identically equal, but they may actually differ in the sixth or seventh decimal place, and a simple test such as IF MA=MB THEN 270 ELSE 300 will probably never be satisfied. The program could be made more efficient [e.g., by evaluating LOG(10) once and for all outside the FOR-NEXT loop, etc.], but it has purposely been kept simple for the sake of clarity.

After you have studied the program thoroughly so that you understand it well, run it on the computer and watch the titration curve develop on the screen as the program is executed. Figure 2.5 shows what the screen should look like when the graph is completed; it is an image of the display transferred to the line printer using a program known as PCPLOT®. Note the last statement of the program listed in Figure 2.4. It is an "infinite loop" which keeps executing indefinitely so that a program interrupt will not appear on the screen to interfere with the display that has been generated. After you have viewed the graph sufficiently to satisfy your needs, you will need to return to the command mode using the BREAK key. Then you can modify the program to run with various other conditions (see Exercise 2.12).

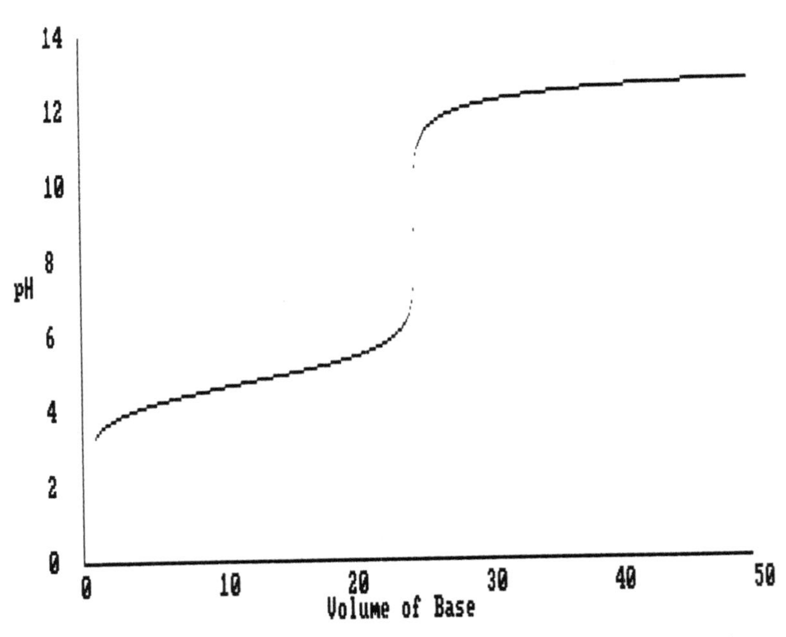

Figure 2.5 Graphical simulation of the titration of a weak acid with a strong base. The display was generated using the program of Figure 2.4, and transferred to a line printer using PCPLOT®.

SUGGESTED ADDITIONAL READING

1. Robert J. Bent and George C. Sethares, *Microsoft® BASIC-Programming the IBM PC*, Brooks/Cole Publishing Co., Monterey, Calif., 1985.
2. Peter C. Jurs, Thomas L. Isenhour, and Charles L. Wilkins, *BASIC Programming for Chemists*, John Wiley & Sons, Inc., New York, 1987.
3. Gary L. Richardson, Charles W. Butler, and John D. Tomlinson, *A Primer on Structured Program Design*, Petrocelli Books, Inc., Princeton, N.J., 1980.
4. Leonard J. Soltzberg, *Computer Strategies for Chemistry Students*, Houghton Mifflin Company, Boston, 1987.
5. Gloria Harrington Swann, *Top-Down Structured Design Techniques*, Petrocelli Books, Inc., Princeton, N.J., 1978.

EXERCISES

2.1. Indicate which of the following are not valid variable representations in BASIC and state the reason why.

(a) S2
(b) BG
(c) 3Z
(d) ON
(e) IF
(f) S2T

2.2. Indicate which of the following are not valid constant representations in BASIC and state the reason why.

(a) 5.7704
(b) 6 × 10−3
(c) 8E−04
(d) 1,000
(e) −7E+20
(f) 10000

2.3. Indicate which of the following BASIC statements are in error and state the reason why.

(a) N=N+1
(b) FOR J=1 TO 5
(c) A=B^−2
(d) A+B=C+D
(e) PRINT A,B;C
(f) ON Z=1 GOTO 40
(g) X=A*−B
(h) Y=C^D^E
(i) R=(C+3F)*2
(j) G3=1,587.2*8.47
(k) QZT=(((3+7)/(8+3+9))/(10+99)

2.4. What is the computed value of X for each of the following BASIC arithmetic replacement statements, given that A = 6, B = 4, C = 3, D = 2, E = 8, F = 5, and G = 7?

(a) X=B/C*D

(b) X=B/(C*D)
(c) X=B^D/C+D*E*G−F
(d) X=B^D/(C+D)*(E*G−F)
(e) X=C*(A+((F/G)−D))
(f) X=A^B/C+D/E*F/G
(g) X=C+D*F^2
(h) X=A*(B+C/D−E)
(i) X=(A*(B+1)−G)/(C*(D−(E+F)))
(j) X=A+E−C*G/F−C

2.5. What is the value of F determined by each of the following program segments?

(a)
```
10 DATA 2,4,5,3,6,7
20 READ A,B,C,D
30 F=B*C/A+D
```

(b)
```
10 DATA 2,4,5
20 READ A,B
30 DATA 3,6,7
40 READ C,D,E
50 F=B*C^A+D*E/A
```

(c)
```
10 DATA 1,4,5,3,6,7
20 READ A,B,C
30 READ D,E
40 F=A+B+C*D/E
```

(d)
```
10 DATA 2,4,5
20 READ A,B
30 RESTORE
40 DATA 3,6,7
50 READ C,D,E
60 F=C^A-B^A/D/E
```

(e)
```
10 DATA 2,7,5
20 READ C,B,A
30 B=4
40 F=B*A*A^B/C
```

2.6. What is the value of Y printed by the following BASIC program?

```
10 READ A,B
20 READ C,D
30 RESTORE
40 READ E,F
50 ON E-3 GOTO 60,80
60 Y=B/F/A*C^2
70 GOTO 90
80 Y=B*B/F*D
```

```
90 PRINT Y
100 DATA 5,8,20,4E10,4,70
110 END
```

2.7. What is the value of Y=FIX(X) for each of the following values of X?

(a) 2.0
(b) 5.7
(c) 0.5
(d) −3.0
(e) −3.5
(f) 1.7
(g) −0.5

2.8. What is the value of Z=INT(X) for each of the values of X given in Exercise 2.7?

2.9. The length of a vector r in terms of the Cartesian coordinates of its end points (x_1,y_1,z_1) and (x_2,y_2,z_2) is given by

$$r = [(x_2 - x_1)^2 + (y_2 - y_1)^2 + (z_2 - z_1)^2]^{1/2}$$

Write a BASIC statement that gives r in terms of the Cartesian coordinates.

2.10. Enter the program listed in Figure 2.1 and run it for several different values of the input parameters. Modify the program and run it several times with each of the branching options described in Section 2.2.6.

2.11. To which statement does control pass in each of the following program segments?

(a)
```
50 X=5
60 Y=X+1
70 Z=X^2-Y^2
80 IF Z<>0 THEN GOTO 180
90 GOTO 200
```

(b)
```
300 N=11/3
310 NN=INT(N+.5)
320 ON NN GOTO 330,380,430,480,600
```

(c)
```
160 A=INT(-2.8)-FIX(-2.1)
170 ON SGN(A)+2 GOTO 220,240,260
```

(d)
```
60 M=.9
70 MS=M*M
80 Z=5*MS-3*M
90 IF M>0 AND MS>M OR Z<0 THEN GOTO 110
100 GOTO 140
```

2.12. Enter the program of Figure 2.4 in the computer, and run it for several different values of the input parameters. Note in particular what happens to the end point as K_a decreases, with all other parameters held constant. Write a modification of the program for a solution of a weak base titrated with a strong acid.

2.13. An integral we will encounter later may be written in the form

$$\frac{1}{\sqrt{2\pi}} \int_0^{z_1} e^{-z^2/2}\, dz$$

This is known as the normal probability integral. It cannot be integrated in closed form, but the integrand may be expanded in series and the resulting series integrated term by term to obtain

$$\frac{z_1}{2}\sqrt{\frac{2}{\pi}}\left[1 - \frac{z_1^2}{2 \cdot 1! \cdot 3} + \frac{z_1^4}{2^2 \cdot 2! \cdot 5} - \frac{z_1^6}{2^3 \cdot 3! \cdot 7} + \cdots\right] \qquad (z_1^2 < \infty)$$

Write a subroutine that will return the proper values of this integral correct to four significant digits for arbitrary values of the parameter z_1.

2.14. Suppose that you wish to determine the cube root of many numbers in the range 0 to 1. This can be done using a series expansion of the function

$$f(x) = (1 - x)^{1/3} \qquad 0 \leq x \leq 1$$

(a) Derive the Maclaurin series expansion for the function $f(x)$.

(b) Write a subroutine that will return values of the cube root correct to four significant figures for arbitrary values of x in the range $0 \leq x \leq 1$.

2.15. Study the subroutine given in Section 2.2.7. Does it conform to good structured programming practice? If not, make changes that will improve its structure.

chapter three

GRAPHICAL AND NUMERICAL METHODS

One would usually prefer exact analytic solutions to problems wherever possible. They can be rather cumbersome in many cases, however, and then a good approximation might be better. There are times when an approximate analytic solution is preferred even though an exact numerical solution may be possible. Relationships between the variables that enhance our understanding might be spotted by examination of the analytic form, but they can be easily obscured in a list of numbers.

A convenient approximate solution may not be obvious, and even if it is, it may not be sufficiently accurate to represent the behavior in which we are interested. We may then have no recourse except to a graphical or numerical solution. In this chapter we discuss some of the graphical and numerical techniques commonly used by the chemist.

3.1 GRAPHING TECHNIQUES

Estimation of the roots of equations by construction of a graph is generally not the method of choice, but we may find it necessary in some cases as a last resort. If so, there are a couple points, even though trivial, which should be borne in mind. We often deal with numbers that are exceedingly large or small, and a rescaling by change of variables may be advantageous. Furthermore, if we have to find the root(s) of $f(x) = 0$, it might be useful to break up the function such that $g(x) = h(x)$. In this case we find the value of x where the two curves intersect, and this might be much quicker than finding the root of $f(x) = 0$ since the functions $g(x)$ and $h(x)$ are generally simpler.

When dealing with chemical equilibria, it is generally easier to graph logarithms of the concentrations rather than the concentrations themselves. There are two reasons for this. The first is that the concentrations usually vary over a large range (i.e., over several orders of magnitude) and a graphical display is difficult except on a log plot. The second reason is that if we take the logarithm of an equilibrium expression, say,

$$K = \frac{[X]^x[Y]^y}{[A]^a[B]^b}$$

we have

$$\log K = x \log[X] + y \log[Y] - a \log[A] - b \log[B]$$

The equation is linear in each of the log terms, with slopes that depend on the coefficients in the balanced equation.

As an illustration, consider a sulfide precipitation of Cd^{2+} and Zn^{2+}, each at 10^{-3} *M*. It is found from solubility product tables that $K_{sp} = 1.6 \times 10^{-28}$ ($pK_{sp} = 27.8$) for CdS, and $K_{sp} = 1.6 \times 10^{-24}$ ($pK_{sp} = 23.8$) for ZnS. We can quickly sketch a log plot of these concentrations as functions of $[S^{2-}]$, as shown in Figure 3.1. Note that each metal ion concentration remains at -3 on this log scale until the sulfide concentration $K_{sp}/10^{-3}$ is reached, following which the metal ion concentration falls with slope -1. The equations of the lines are given by

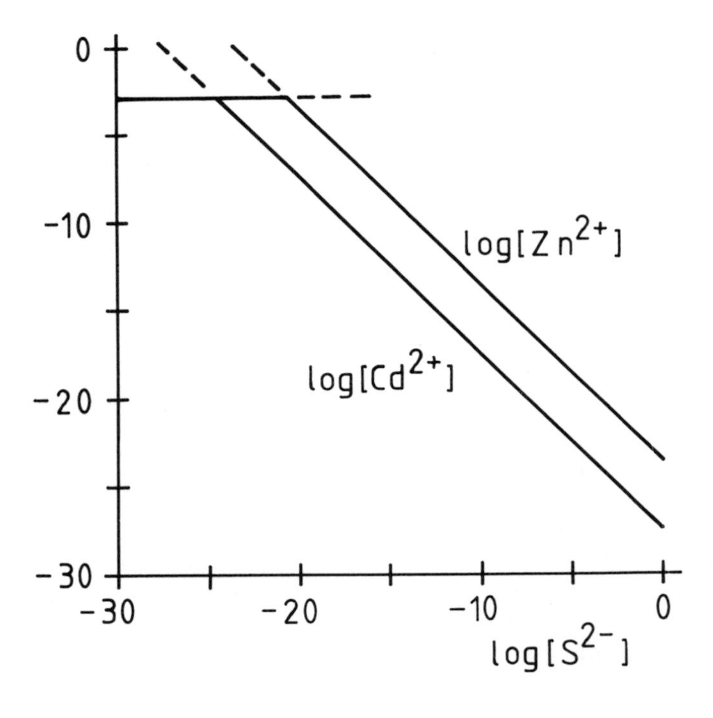

Figure 3.1 Logarithmic plot of Zn^{2+} and Cd^{2+} concentrations as functions of the sulfide concentration.

$$\log[Cd^{2+}] = -pK_{sp} - \log[S^{2-}] = -27.8 - \log[S^{2-}]$$

$$\log[Zn^{2+}] = -pK_{sp} - \log[S^{2-}] = -23.8 - \log[S^{2-}]$$

It is apparent from the graph that $\log[Cd^{2+}] = -7$ at $\log[S^{2-}] = -20.8$, where ZnS begins to precipitate, so that quantitative separation is possible. Similar graphs can readily be sketched for other precipitation reactions.

The graphing of acid-base equilibria can be advantageous at times. We will assume as an example that we have an aqueous solution of a weak acid HA at concentration C_a. If we neglect the dissociation of H_2O, the material balance condition ($C_a = [HA] + [A^-]$) and the equilibrium constant ($K_a = [H^+][A^-]/[HA]$) can be combined to give

$$[HA] = \frac{C_a[H^+]}{K_a + [H^+]} \tag{3.1}$$

and

$$[A^-] = \frac{K_a C_a}{K_a + [H^+]} \tag{3.2}$$

We will first consider [HA]. At high pH values, $[H^+] \ll K_a$, so $[HA] \approx C_a[H^+]/K_a$, or

$$\log[HA] \approx \log C_a + pK_a - pH \tag{3.3}$$

On the other hand, in acidic solutions $[H^+] \gg K_a$, so $[HA] \approx C_a$ and

$$\log[HA] \approx \log C_a \tag{3.4}$$

We see that log[HA] is constant in acidic solutions, but falls linearly with pH in basic solutions. These two lines intersect at the *system point,* that is, the point (pK_a, $\log C_a$) (see Figure 3.2).

The curve does not actually go through the system point, since neither of the approximations above is valid at this particular pH. The system point in fact corresponds with the condition $[H^+] = K_a$. Then from Eq. 3.1 we see that $[HA] = C_a/2$, or

$$\log[HA] = \log C_a - \log 2$$
$$= \log C_a - 0.3$$

Thus the curve passes through a point 0.3 log unit below the system point, as shown in Figure 3.2.

Similar considerations apply to the graph of $\log[A^-]$ using Eq. 3.2. However, here it will be observed that $\log[A^-]$ is constant above the system point but falls linearly with pH below it.

Now that the form of these lines is understood, we can quickly sketch the complete diagram for a real case. An example is shown in Figure 3.3 for a 0.1 *M* solution of acetic acid where the system point is at $pK_a = 4.757$, $\log C_a = 1$. Note the inclusion of the lines $\log[H^+] = -pH$ and $\log[OH^-] = -14 + pH$. We can read the concentra-

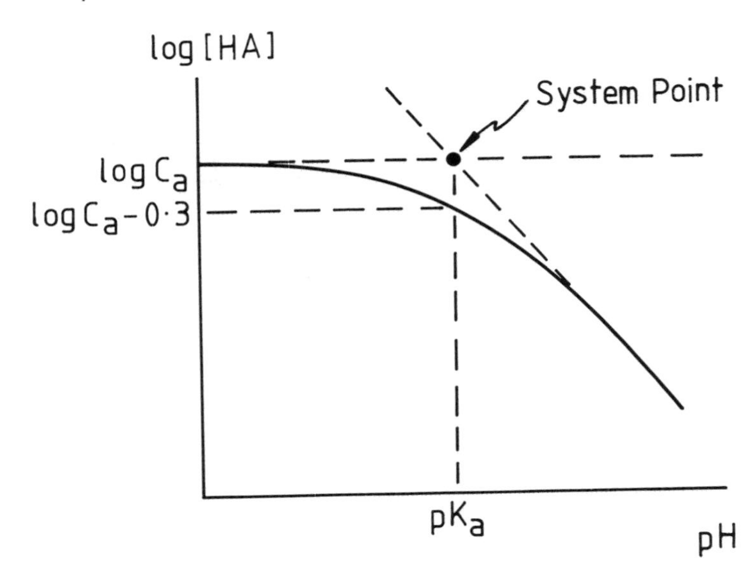

Figure 3.2 Logarithmic plot of [HA] as a function of pH for a weak acid. The two straight-line segments given by Eqs. 3.3 and 3.4 intersect at the system point.

tions (or rather the logarithms of the concentrations) of each of the solution species $[H^+]$, $[OH^-]$, $[OAc^-]$, and [HOAc] at any arbitrary pH from the figure.

A figure such as that given in Figure 3.3 can be particularly useful for a quick estimation of the pH when it is not given. Suppose that we want to know the pH of a 0.1 M solution of acetic acid. We make use of the *proton condition,* which is essentially a material balance condition for the protons. For an aqueous solution of acetic acid, the species introduced are H_2O and HOAc. We call this the *zero level*. When equilibrium is attained the sum of the concentrations of all species that have gained protons must equal the sum of the concentrations of all species that have lost protons. At equilibrium we have H_3O^+ (abbreviated H^+), which has gained a proton, and OAc^- and OH^-, which have lost protons. Therefore, the proton condition is

$$[H^+] = [OAc^-] + [OH^-]$$

and we must find a point on the figure for which

$$\log[H^+] = \log([OAc^-] + [OH^-]) \tag{3.5}$$

If we start at the right-hand side of Figure 3.3 and move along the $\log[H^+]$ line, the first intersection we find is with the $\log[OH^-]$ line. At the point of intersection (pH = 7) we see that $\log[OAc^-] \gg \log[OH^-]$, so Eq. 3.5 is not satisfied. We therefore reject this possibility and move along to the next intersection (i.e., with the $\log[OAc^-]$ line). Here we find that $[OH^-]$ is about eight orders of magnitude smaller than $[OAc^-]$ at this point, so it is a very good approximate solution of Eq. 3.5. The pH at the point of intersection is 2.8.

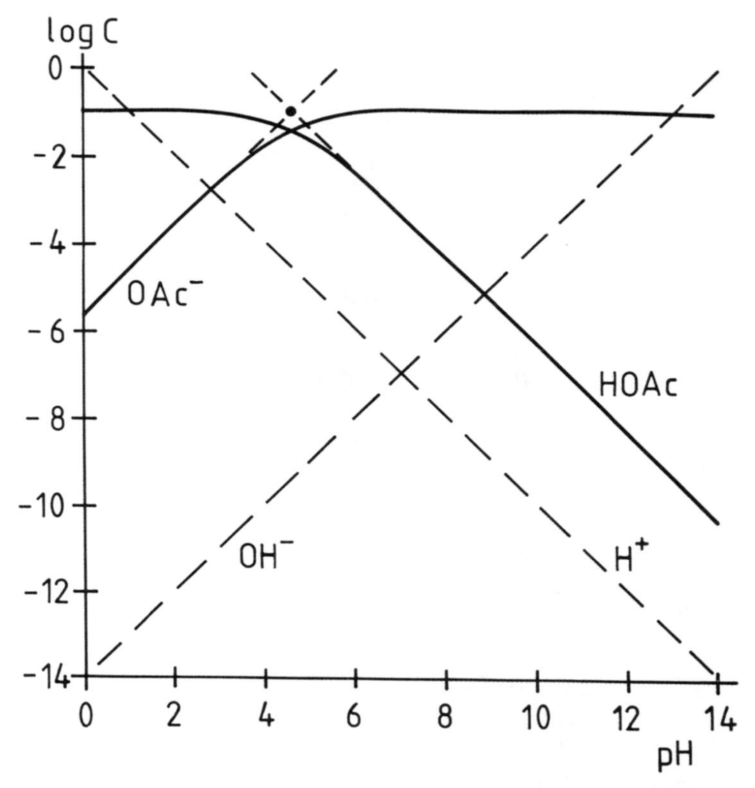

Figure 3.3 Logarithmic concentration diagram for a 0.1 *M* aqueous acetic acid solution as a function of pH.

Now assume that we have a 0.1 *M* solution of sodium acetate, and we desire to know its pH. The same equilibria are involved so Figure 3.3 still applies. However, the zero level is now H_2O and OAc^-, so the proton condition becomes

$$[H^+] + [HOAc] = [OH^-]$$

We proceed as before, but this time along the $\log[OH^-]$ line. The intersection that approximately satisfies this proton condition is with the $\log[HOAc]$ line, where the $[H^+]$ concentration is about four orders of magnitude smaller than [HOAc]. Therefore, the pH in this case is about 8.8.

Generalization of the foregoing procedure to include mixtures of acids and bases is not difficult. Let us assume that we have a 0.01 *M* solution of ammonium fluoride. There are now two competing equilibria,

$$NH_4^+ \rightleftharpoons NH_3 + H^+$$

$$F^- + H_2O \rightleftharpoons HF + OH^-$$

In this case we need to add a second system point at (9.245, −2) for the ammonium system ($K_a = 5.7 \times 10^{-10}$) to that at (3.17, −2) for the hydrofluoric acid system ($K_a = 6.7 \times 10^{-4}$), and the diagram is as shown in Figure 3.4.

We will use Figure 3.4 to determine the pH of a 0.01 *M* solution of NH_4F. The zero level in this case is H_2O, NH_4^+, and F^-, so the proton condition is

$$[H^+] + [HF] = [OH^-] + [NH_3]$$

We must find an intersection in Figure 3.4 such that

$$\log([H^+] + [HF]) = \log([OH^-] + [NH_3])$$

is approximately valid. At the intersection of the $\log[NH_3]$ and $\log[HF]$ curves we see that $[OH^-] \ll [NH_3]$ and $[H^+] \ll [HF]$, and the condition is fulfilled. The pH at the intersection is 6.25.

The diagram for a 0.1 *M* aqueous solution of carbonic acid, which is an example of a diprotic acid ($pK_1 = 6.352$, $pK_2 = 10.33$), is shown in Figure 3.5. Note that there

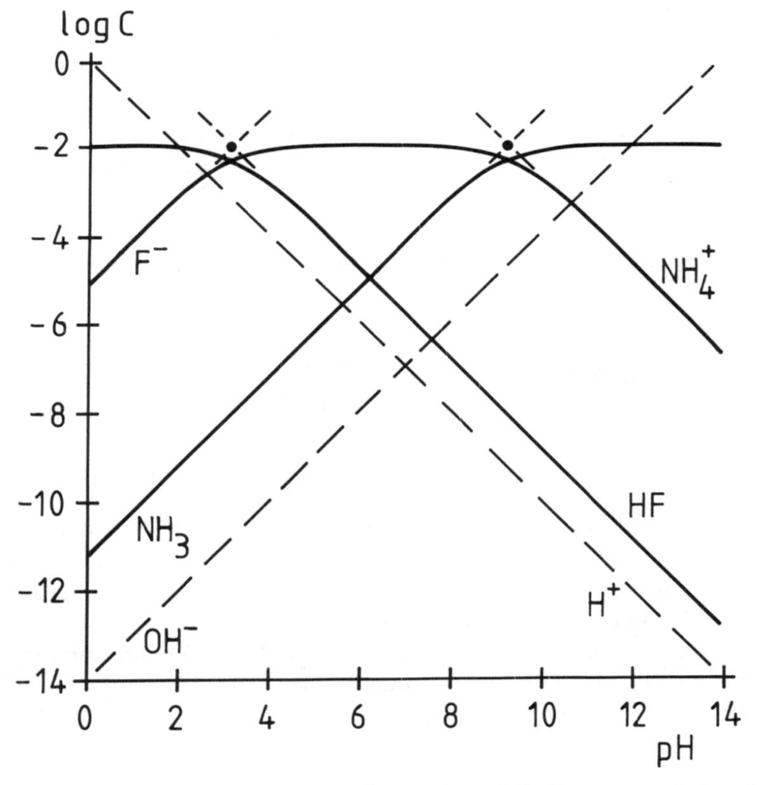

Figure 3.4 Logarithmic concentration diagram for a 0.01 *M* aqueous solution of NH_4F as a function of pH.

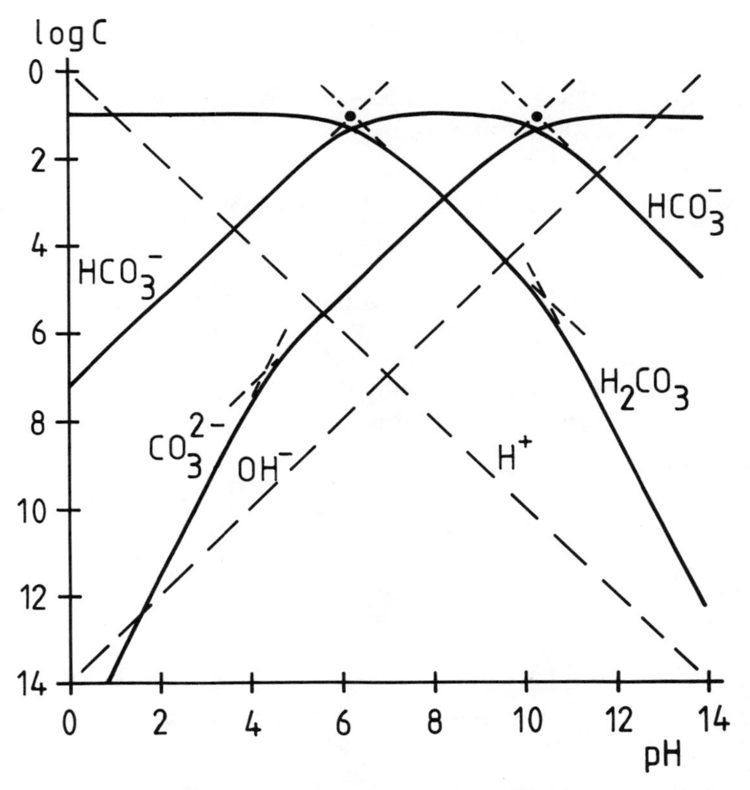

Figure 3.5 Logarithmic concentration diagram for a 0.1 M aqueous solution of H_2CO_3 as a function of pH.

are two system points (pK_1, log C) and (pK_2, log C). At very low pH values the hydrogen ion concentration is large and most of the carbonic acid is in the form H_2CO_3. Therefore, $[CO_3^{2-}]$ is negligible and the equilibrium of concern is

$$H_2CO_3 \rightleftharpoons H^+ + HCO_3^-$$

The left-hand portion of the diagram represents this equilibrium, and it is just the same as that for a monoprotic acid.

At high pH values the hydrogen ion concentration is small and the equilibrium shifts so that the principal process is

$$HCO_3^- \rightleftharpoons H^+ + CO_3^{2-}$$

and once again the diagram in the high-pH range is similar to that of a monoprotic acid, but centered about the second system point. Joining these two systems as in Figure 3.5 shows that HCO_3^- is dominant in the mid-pH range, falling off at both higher and lower pH values.

At high pH values the $[CO_3^{2-}]$ is nearly constant and equal to 0.1, and we have

$$K_1K_2 = \frac{[H^+]^2[CO_3^{2-}]}{[H_2CO_3]}$$

or

$$\log[H_2CO_3] = pK_1 + pK_2 + \log[CO_3^{2-}] - 2pH$$
$$= 15.68 - 2pH$$

Thus we see that the slope of the $\log[H_2CO_3]$ versus pH curve changes from -1 to -2 at high pH values. Similarly, at very low pH values where $[H_2CO_3] \approx 0.1$, the slope of the $\log[CO_3^{2-}]$ curve becomes $+2$.

The construction of logarithmic concentration diagrams for equilibria involving complex ions can be carried out in a similar way.

Example 3.1

Construct a logarithmic concentration diagram for a 0.0316 *M* solution of acetic acid, and use it to estimate the acetate concentration of a 0.0316 *M* solution of acetic acid.

Solution. Since log 0.0316 = -1.5, the system point is at (4.757, -1.5) and the diagram is as follows:

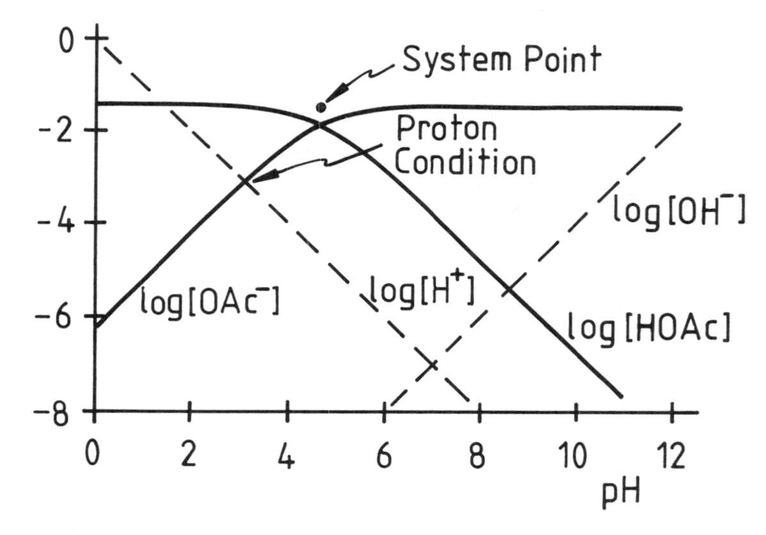

In this case the zero level is (H_2O, HOAc), so the proton condition may be written

$$[H^+] = [OH^-] + [OAc^-]$$

We see that $[OH^-]$ is several orders of magnitude smaller than $[OAc^-]$, where the $\log[OAc^-]$ line intersects the $\log[H^+]$ line, so this point satisfies the proton condition to a high degree of accuracy. At this point

$$\log[OAc^-] \approx -3.1$$

or

$$[\mathrm{OAc}^-] \approx 8 \times 10^{-4}$$

3.2 NUMERICAL APPROXIMATION OF THE ROOTS OF EQUATIONS

The roots of linear equations and quadratic equations are easily found by familiar formulas. There are even explicit formulas for the roots of cubic equations, but they are sufficiently cumbersome that it is questionable whether they are of much worth for practical applications. Crude approximations may be derived by graphical means as we have seen, but for most practical purposes, we rely on *iterative techniques*. The successful application of an iterative technique implies that we have some suitable starting estimate of the root, and this is used in a successive refinement process until the value is known to the desired degree of accuracy. The actual calculation is likely to be carried out by means of a hand-held calculator or computer, and we need to be concerned that the methods we develop will yield the result in an efficient manner.

An iterative procedure is said to be *convergent* if we can generate a series of approximate roots $x_0, x_1, x_2, \ldots$ in sequence, and there exists an integer n for which the error

$$|x_m - x_t| < \epsilon$$

for all $m \geq n$, where ϵ is a small positive quantity and x_t is the true value of the root. A *divergent* iterative process, on the other hand, is one for which $|x_{m+1} - x_m|$ increases continuously as m increases.

The definition of convergence above is of limited use in practice, because the precise value of the root is generally not known at the outset. Thus it is impossible to say when n has become large enough. In practice we use a termination criterion of the sort

$$\left|\frac{x_{m+1} - x_m}{x_{m+1}}\right| < \epsilon'$$

rather than an absolute convergence condition. Similarly, allowance is usually made for a given procedure being nonconvergent or slow to converge by placing an upper limit on the number of iterations that are allowed.

3.2.1 The Method of Successive Approximations

The method of successive approximations is an iterative technique which, because of its simplicity, is used a great deal by the chemist. It is referred to as a *fixed-point method*. According to the mathematical definition, a fixed point is any point that is not changed by the transformation under consideration. For example, $x = 3$ is a fixed point of the transformation $T(x) = 4x - 9$.

In order to set up a general problem for solution by the method of successive approximations we solve the equation in such a way that the independent variable is recalculated in terms of some function $f(x)$; that is, $x = f(x)$. To begin the procedure we make the best possible estimate of the solution, that is, of the value of x that satisfies the equation. We call this x_0. We evaluate $f(x_0)$, which defines a new value of x, which we call x_1. We now insert this value x_1 in the function $f(x)$ to generate still another approximation of x, which we call x_2, and so forth. In general, then, we have

$$x_{m+1} = f(x_m) \tag{3.6}$$

If the change in x with each cycle gets progressively smaller, we have a convergent iteration and we will ultimately find that $x_{m+1} = x_m$. On the other hand, we might find that the iteration is divergent, as signaled by the changes in x becoming progressively larger. We do indeed quite often find that the method either does not converge or is slow to converge, and several modifications have been developed which are designed to accelerate the convergence and thereby improve the applicability. It can be shown that the method gives a convergent series only if $|f'(x_t)| < 1$, that is, when the absolute value of the derivative of the function evaluated at the correct value of x is less than 1. This limitation is not as severe as might at first appear, although it is sometimes necessary to reconstruct the problem so that solution is possible. There are often quite a number of different ways in which the equation may be solved to obtain a fixed-point form $x = f(x)$.

As an example of solution by the method of successive approximations, let us calculate the pH of an aqueous solution of a weak acid. The dissociation of the water and of the acid are both considered to be significant sources of protons. We obtain

$$K_a = \frac{[H^+]\left([H^+] - \dfrac{K_w}{[H^+]}\right)}{C_a - [H^+] + \dfrac{K_w}{[H^+]}} \tag{3.7}$$

which may be rearranged to give

$$[H^+] = \sqrt{K_a\left(C_a - [H^+] + \frac{K_w}{[H^+]}\right) + K_w} \tag{3.8}$$

The H^+ coming from the dissociation of water is small, and for most aqueous solutions of weak acids $[H^+] \ll C_a$. Thus

$$[H^+]_0 = \sqrt{K_a C_a} \tag{3.9}$$

can be used as a first approximation to start the procedure.

For a specific illustration we will calculate the pH of a 0.200 M solution of formic acid, for which $K_a = 1.77 \times 10^{-4}$. The first approximation is

$$[H^+]_0 = \sqrt{(0.200)(1.77 \times 10^{-4})}$$
$$= 5.950 \times 10^{-3}$$

Inserting this value in Eq. 3.8 and continuing gives

$$[H^+]_1 = 5.861 \times 10^{-3}$$

$$[H^+]_2 = 5.862 \times 10^{-3}$$

The procedure has quickly converged, and the corresponding pH is 2.23.

As indicated above, the convergence is not always as rapid as in the illustration above. If, for example, the hydrogen ion concentration and total acid concentration were of the same order of magnitude, we might expect the method to produce some oscillations because of the subtraction in Eq. 3.8. Try using the method of successive approximations as outlined above to solve for the pH of an aqueous solution which is $2.00 \times 10^{-4}\ M$ in formic acid ($K_a = 1.77 \times 10^{-4}$).

3.2.2 The Newton-Raphson Method

Suppose that we have an approximate root x_0 of the function $f(x)$. Expansion of the function in a Taylor's series about the point x_0 gives

$$f(x) = f(x_0) + \frac{f'(x_0)}{1!}(x - x_0) + \frac{f''(x_0)}{2!}(x - x_0)^2 + \cdots$$

If x_0 is a good approximation of the root, $x - x_0$ is a small quantity for values of x near this root and the series converges rapidly. We can therefore neglect higher-order terms to obtain

$$f(x) \approx f(x_0) + (x - x_0)f'(x_0)$$

We are looking for the value of x for which $f(x) = 0$. We therefore solve

$$0 = f(x_0) + (x - x_0)f'(x_0)$$

for x to get

$$x = x_0 - \frac{f(x_0)}{f'(x_0)}$$

The value of x given by this formula should be a better approximation of the root than x_0 itself. It is still only an approximation because of the neglect of higher-order terms in the Taylor's series expansion, so we again have the basis for an iterative process. It may be represented by the formula

$$x_{m+1} = x_m - \frac{f(x_m)}{f'(x_m)} \tag{3.10}$$

It is instructive to see how this process may be interpreted graphically. In Figure 3.6 we see a function $f(x)$ plotted in the region of its root. The slope m at the point x_0 is given exactly by $f'(x_0)$, but it is also given by

$$m = \frac{f(x_0) - 0}{x_0 - x_1}$$

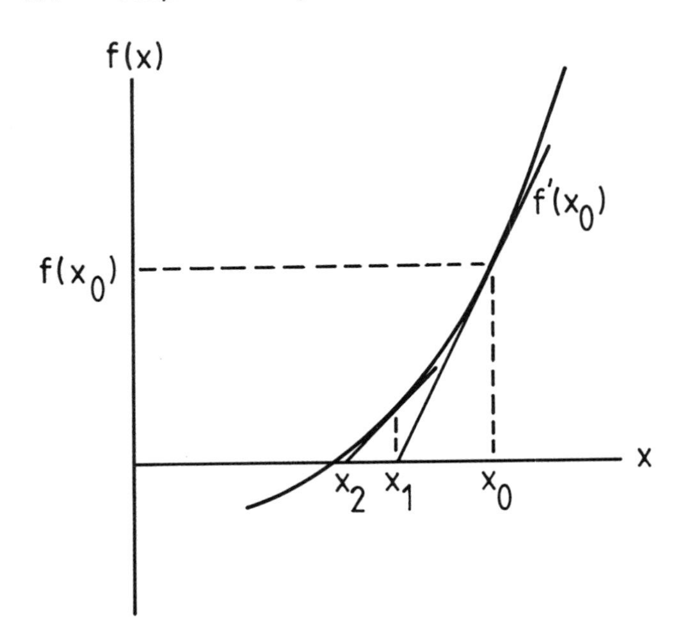

Figure 3.6 Geometric interpretation of the Newton–Raphson method for determining the root of the function $f(x)$.

where x_1 is the value of the independent variable for which $f(x)$ would be equal to zero if it were a linear function. Equating these two expressions for the slope gives

$$f'(x_0) = \frac{f(x_0)}{x_0 - x_1}$$

or

$$x_1 = x_0 - \frac{f(x_0)}{f'(x_0)}$$

Once this improved estimate of the root is obtained it is put into the expression

$$x_1 - \frac{f(x_1)}{f'(x_1)}$$

to obtain a still better estimate x_2, and so forth. We see that the method involves a linear extrapolation of the function using its slope at the point of evaluation to obtain an improved estimate of the root.

The Newton–Raphson method is much more highly convergent than simple iterative methods such as successive approximations. In fact, the number of significant figures may be expected to double at each stage as the root is approached. It is convergent for all functions except those for which $f'(x) = 0$ and $f''(x) = 0$ at the root.

We will illustrate the Newton–Raphson method using

$$[H^+]^3 + K_a[H^+]^2 - (K_w + K_aC_a)[H^+] - K_aK_w = 0$$

which is the equivalent of Eq. 3.7, to solve for the pH of a 2.00×10^{-4} M solution of formic acid. We again use Eq. 3.9 to start the procedure:

$$[H^+]_0 = \sqrt{(2.00 \times 10^{-4})(1.77 \times 10^{-4})}$$
$$= 1.881 \times 10^{-4}$$

The next estimate is

$$[H^+]_1 = [H^+]_0 - \frac{[H^+]_0^3 + K_a[H^+]_0^2 - (K_w + K_aC_a)[H^+]_0 - K_aK_w}{3[H^+]_0^2 + 2K_a[H^+] - (K_w + K_aC_a)}$$
$$= 1.425 \times 10^{-4}$$

Continuing, we find

$$[H^+]_2 = 1.235 \times 10^{-4}$$
$$[H^+]_3 = 1.196 \times 10^{-4}$$
$$[H^+]_4 = 1.194 \times 10^{-4}$$
$$[H^+]_5 = 1.194 \times 10^{-4}$$

and pH $= 3.92$. If the attempt was made to solve this problem by the method of successive approximations, as suggested above, it will undoubtedly be agreed that the Newton–Raphson method gives far superior convergence in this case.

The Newton–Raphson method has many applications. It leads to a widely used method for finding the roots of numbers. Suppose that we need to know the kth root of the number N. It is given by the solution of the equation

$$f(x) = x^k - N = 0$$

Inserting this in Eq. 3.10 gives

$$x_{m+1} = x_m - \frac{x_m^k - N}{kx_m^{k-1}}$$
$$= \frac{1}{k}\left[x_m\left(\frac{N}{x_m^k} + k - 1\right)\right] \tag{3.11}$$

As an example, assume that we are on a desert island without a calculator having square root functions, and we need to know the square root of 3 very accurately. (Don't ask why!) Equation 3.11 in this case becomes

$$x_{m+1} = \frac{1}{2}\left(\frac{3}{x_m} + x_m\right) \tag{3.12}$$

If we start the procedure with $x_0 = 1$, we find, using Eq. 3.12,

$$x_1 = 2.$$
$$x_2 = 1.75$$

$$x_3 = 1.732142857$$

$$x_4 = 1.732050810$$

$$x_5 = 1.732050808$$

Similarly, if we need the fifth root of π, Eq. 3.11 becomes

$$x_{m+1} = \frac{1}{5}\left[x_m\left(\frac{\pi}{x_m^5} + 4\right)\right]$$

Again starting the procedure with $x_0 = 1$ gives

$$x_1 = 1.428318531$$

$$x_2 = 1.293620977$$

$$x_3 = 1.259260005$$

$$x_4 = 1.257280369$$

$$x_5 = 1.257274116$$

$$x_6 = 1.257274116$$

3.2.3 The Regula Falsi and Secant Methods

Both the method of successive approximations and the Newton–Raphson method are *single-position methods,* in that they use the properties of the function at a single point as a basis for approximation of the root. We will now consider some simple *double-position methods*.

We assume that we have some function $y = f(x)$, and that we know the values of the function y_{m-1} and y_m for the two values of the independent variable x_{m-1} and x_m. If y has a well-behaved derivative at $y = 0$, it should be given approximately by

$$\left(\frac{dy}{dx}\right)_{y=0} \approx \frac{y_m - y_{m-1}}{x_m - x_{m-1}} \approx \frac{y_m - 0}{x_m - x_{m+1}}$$

where x_{m+1} is the value of the independent variable for which $y = 0$. Solving the above for x_{m+1} gives

$$x_{m+1} = x_m - y_m\left(\frac{x_m - x_{m-1}}{y_m - y_{m-1}}\right) \tag{3.13}$$

which provides the basis for another iterative method. We see in this case that we need two points to start the process, but for each new cycle, it is only necessary to evaluate the function at one new point as the old ones are relabeled.

If we now compare Eq. 3.13 with the formula for the Newton–Raphson method (Eq. 3.10), we see that they are very similar. The difference is that we have replaced the factor $1/f'(x_m)$ of the Newton–Raphson method by an approximate finite difference

$$\frac{1}{f'(x_m)} \approx \frac{x_m - x_{m-1}}{y_m - y_{m-1}} = \frac{\Delta x}{\Delta y}$$

Two distinct iterative procedures have been developed using Eq. 3.13. The first requires that x_m and x_{m-1} be chosen such that y_m and y_{m-1} are of different sign, that is, the root lies between them. We may then carry out the calculation by dropping the end point with the same sign as the function at x_{m+1} in the next cycle. This guarantees that the root remains bracketed as the calculation progresses, and it is known as the *regula falsi method* (rule of false position method). On the other hand we may always keep the two most recent values in the next step. Geometrically, we use as the next estimate the intersection between the x-axis and the chord or secant passing through two points on the curve. This is therefore known as the *secant method*. It will generally have superior convergence when compared with the regula falsi method, although there are certain functions for which long extrapolations will lead to divergence.

The regula falsi and secant methods have a couple of obvious disadvantages. They require two points to initiate the process, and the two points must be sufficiently close to the desired root to produce a convergent sequence. In spite of these shortcomings the methods are often used in place of the Newton–Raphson method for actual calculations. Even though a finite difference is used to approximate the actual slope, the convergence with the secant method is often almost as rapid. Only the function itself must be evaluated, whereas the Newton–Raphson method requires an evaluation of the derivative of the function as well. This requires considerable extra time, particularly when dealing with a fairly complex function. The methods should not be pushed too far, since round-off errors will obviously become severe near the root as the interval becomes very small.

3.2.4 The Bisection Method

There are times when the methods described above may be ill-conditioned for one reason or another, and one desires a simple method that is essentially foolproof, even though it may not be particularly elegant. The *bisection method* is a systematic trial-and-error method that can be used to develop a simple computational algorithm.

We assume as usual that our problem is structured such that we have some function $f(x)$, and we seek the value of x for which $f(x) = 0$. Let us assume that we have found two values of x, say x_1 and x_2, such that $f(x_1) \cdot f(x_2) < 0$. In other words, the function at x_1 and at x_2 is of different sign, and we know that the root must lie between these two extremes. We set $x_3 = \frac{1}{2}(x_1 + x_2)$ and calculate the function $d = f(x_2) \cdot f(x_3)$. Now if $|d| < \epsilon$, where ϵ is a small positive quantity that we take to be the maximum permissible error, we have found the root. Otherwise, we check the sign of d and if we find that $d < 0$, we know that the root is bracketed by the values x_2 and x_3, so we set $x_1 = x_3$ and repeat the cycle. If, on the other hand, $d > 0$, we can be sure that the root lies between x_1 and x_3, so we set $x_2 = x_3$ before repeating the cycle.

Because of its simplicity, the bisection method can be applied in many circumstances where some of the more sophisticated methods fail. If there is a single real root, it will generally be located with any desired degree of accuracy, since there is no limit to the number of times the interval may be subdivided. One will encounter difficulty starting the procedure if there is an even number of roots within the interval $[x_1, x_2]$, however. There are also times when a root lies right on the x-axis, but the function never actually crosses. The method will obviously fail in this case, but it can then be applied to the derivative of the function, which must cross the axis at that point.

It is impossible to say that any one of the foregoing methods is superior to the others, since the convergence in a given case is entirely dependent on the nature of the function whose root is desired. We may expect in general that the Newton–Raphson, regula falsi, and secant methods will have more rapid convergence if a good approximate root or set of roots is available to begin the iterative procedure. In some of the more elaborate schemes, a simpler method is used for the first few iterations to arrive at a good starting value, and then an algorithm with more rapid convergence is used after that. The important point to remember is that the best method for hand calculation is not always the best method for computer calculation.

Example 3.2

Use the method of successive approximations to find the value of $[Ag^+]$ that satisfies the equation

$$[Ag^+]^3 + 0.020[Ag^+]^2 = 3.0 \times 10^{-5}$$

Solution. There are several different ways in which we may set up the problem for solution, and a trial-and-error choice is probably the best way to find a quick solution. If we use

$$[Ag^+] = \sqrt{\frac{3.0 \times 10^{-5} - [Ag^+]^3}{0.020}}$$

we find that

$$[Ag^+]_0 = 0$$

$$[Ag^+]_1 = 3.87 \times 10^{-2}$$

$$[Ag^+]_2 = \text{imaginary}$$

and this approach must be rejected. If, on the other hand, we use

$$[Ag^+] = \sqrt[3]{3.0 \times 10^{-5} - 0.020[Ag^+]^2}$$

or

$$[Ag^+] = \sqrt{\frac{3.0 \times 10^{-5}}{[Ag^+] + 0.020}}$$

a uniform convergence is obtained. In the latter case, once again starting with an initial value of $[Ag^+]_0 = 0$, we have

$$[Ag^+]_1 = 3.87 \times 10^{-2}$$

$$[Ag^+]_2 = 2.26 \times 10^{-2}$$

$$[Ag^+]_3 = 2.65 \times 10^{-2}$$

$$[Ag^+]_4 = 2.54 \times 10^{-2}$$

$$[Ag^+]_5 = 2.57 \times 10^{-2}$$

The reader can readily verify that $|f'(0.0257)| < 0$ is satisfied by the second and third forms but not by the first.

Example 3.3

If the dissociation of water is neglected, the hydrogen ion concentration of a diprotic acid solution is given by the equation

$$[H^+]^3 + K_1[H^+]^2 + (K_1K_2 - K_1C_a)[H^+] - 2K_1K_2C_a = 0$$

Calculate the hydrogen ion concentration of a 0.100 M aqueous succinic acid ($K_1 = 6.21 \times 10^{-5}$, $K_2 = 2.32 \times 10^{-6}$) solution correct to three significant digits.

Solution. We will solve this problem using the Newton–Raphson method, in which case

$$f([H^+]) = [H^+]^3 + 6.21 \times 10^{-5}[H^+]^2 - 6.21 \times 10^{-6}[H^+] - 2.88 \times 10^{-11}$$

$$f'([H^+]) = 3[H^+]^2 + 1.242 \times 10^{-4}[H^+] - 6.21 \times 10^{-6}$$

As the initial approximation we will use $[H^+]_0 = \sqrt{K_1C_a}$ and we have

$$[H^+]_0 = 2.49 \times 10^{-3}$$

$$f = 3.57 \times 10^{-10}$$

$$f' = 1.27 \times 10^{-5}$$

$$[H^+]_1 = 2.46 \times 10^{-3}$$

$$f = 5.90 \times 10^{-12}$$

$$f' = 1.23 \times 10^{-5}$$

$$[H^+]_2 = 2.46 \times 10^{-3}$$

Example 3.4

Write a BASIC computer program that will locate the root of the equation

$$10x^3 + 6x^2 - 70x - 600 = 0$$

correct to at least four decimal places using the bisection method.

Solution. By insertion of several trial values of x in the left-hand side we readily find that the root lies between $x = 4$ and $x = 5$. Thus we take $x_1 = 4$ and $x_2 = 5$ as starting points. Because of the complicated x-dependence represented by this function, it is not convenient to use ϵ itself as a convergence criterion, so we simply terminate the calculation when the difference of the x values between successive calculations is less than 0.0001. A program that solves this problem and the result of executing it are listed below.

```
10 REM     PROGRAM BISECT
20 DEF FNF(X)=X*(X*(X*10+6)-70)-600
30 I=0 : X1=4 : X2=5
40 WHILE (X2-X1)>=.0001
50 X3=(X1+X2)/2
60 D=FNF(X2)*FNF(X3)
70 I=I+1
80 LPRINT I,X3,D
90 IF I<=20 THEN 120
100 LPRINT"FAILED TO CONVERGE IN ";I;" ITERATIONS
110 STOP
120 IF D<0 THEN X1=X3 ELSE X2=X3
130 WEND
140 LPRINT"CONVERGED IN ";I;" ITERATIONS TO X = ";X3
```

```
 1             4.5            52987.5
 2             4.25          -2527.945
 3             4.375          5416.04
 4             4.3125         539.7579
 5             4.28125       -58.65355
 6             4.296875       39.14032
 7             4.289063      -2.800594
 8             4.292969       4.155189
 9             4.291016       .2522904
 10            4.290039      -6.453424E-02
 11            4.290528      -1.176721E-02
 12            4.290772       1.462249E-02
 13            4.29065        5.112142E-04
 14            4.290589      -1.806319E-04
CONVERGED IN  14  ITERATIONS TO X =  4.290589
```

Example 3.5

In a very important historical development, Planck was able to show that his equation for the distribution of blackbody radiation

$$u_\lambda = \frac{8\pi hc}{\lambda^5} \frac{1}{e^{hc/\lambda kT} - 1}$$

leads to the experimentally observed Wien displacement law, which gives the relationship between the absolute temperature T and the wavelength λ_{max}, for which the radiation density is a maximum. Derive this relationship by finding the value of λ for which u_λ is a maximum according to the equation above.

Solution. The calculation can be simplified by change of variable to

$$x = \frac{hc}{\lambda kT}$$

or

$$\lambda = \frac{hc}{xkT}$$

Insertion in Planck's equation gives

$$u_\lambda = \frac{8\pi hc}{\left(\frac{hc}{xkT}\right)^5} \frac{1}{e^x - 1} = \frac{8\pi k^5 T^5}{h^4 c^4} \frac{x^5}{e^x - 1}$$

This is differentiated and equated to zero to find the optimum value of x:

$$\frac{du_\lambda}{dx} = \frac{8\pi k^5 T^5}{h^4 c^4} [5x^4(e^x - 1)^{-1} + x^5(-1)(e^x - 1)^{-2}e^x] = 0$$

or

$$\frac{5x^4}{e^x - 1} = \frac{x^5 e^x}{(e^x - 1)^2}$$

$$xe^x = 5(e^x - 1)$$

By comparison of the left-hand side with the right-hand side of this equation for various values of x, it is quickly found that the root lies between $x = 4$ and $x = 5$. We solve for x as follows:

$$x = \frac{5(e^x - 1)}{e^x}$$

and use the method of successive approximations beginning with $x_0 = 5$ to obtain

$$x_0 = 5$$

$$x_1 = 4.966$$

$$x_2 = 4.965$$

Putting this value of x in the equation above gives

$$\frac{hc}{\lambda_{\text{max}} kT} = 4.965$$

or

$$\lambda_{\text{max}} T = \frac{hc}{4.965k} = \text{constant}$$

Wien's displacement law states that the absolute temperature times the wavelength where a maximum occurs in the spectral distribution function is a constant. The Planck formula is consistent with this only for a certain small discrete value of h, which in turn implies that energy can be exchanged only in fixed units called *quanta*. This led to the development of quantum mechanics, and it is an excellent example of how the use of a mathematical model to describe a physical system can lead to new insights.

3.3 INTERPOLATION

Interpolation is a process by which we make use of $f(x)$ values, which are tabulated for several different values of the independent variable x, in order to determine $f(x)$ for some nontabular value of x that lies within the range of the table. An interpolation formula, once developed, can be put to a variety of uses. We might wish to supply a missing tabular value for the function $f(x)$. For example, we may know $f(x)$ at $x = 2, 3, 5$, and 6, but need to know it at $x = 4$. A common use of the interpolation formula is the estimation of values for the function $f(x)$ at values of x intermediate between those for which the data are tabulated. The interpolation of logarithmic or trigonometric tables is a common example. It should not be implied that the interpolation provides any improvement in accuracy over that which is determined by the tabulated data itself, but with a suitable formula we can expect to know the data for intermediate values approximately as well as for the tabulated points.

Other useful applications of interpolation formulas include error location and correction, the smoothing of data (to which reference will be made later), and analytic substitution (the use of a smooth analytic function to represent numerical data). Interpolation formulas can also be useful for short-range extrapolations, but this should be done very cautiously since one generally has no assurance that the function has any validity outside the range of points used in constructing the interpolation formula.

A large number of interpolation formulas have been developed and each has its own distinct advantages, depending on the circumstances involved. Our treatment is not meant to be exhaustive, only to describe some of the more useful formulas that illustrate the principles involved.

3.3.1 The Lagrange Interpolation Formula

The first interpolation formula we discuss is that due to Lagrange,

$$
\begin{aligned}
f(x) = {} & \frac{(x - x_1)(x - x_2) \cdots (x - x_n)}{(x_0 - x_1)(x_0 - x_2) \cdots (x_0 - x_n)} f(x_0) \\
& + \frac{(x - x_0)(x - x_2) \cdots (x - x_n)}{(x_1 - x_0)(x_1 - x_2) \cdots (x_1 - x_n)} f(x_1) \\
& + \cdots + \frac{(x - x_0)(x - x_1) \cdots (x - x_{n-1})}{(x_n - x_0)(x_n - x_1) \cdots (x_n - x_{n-1})} f(x_n)
\end{aligned}
\tag{3.14}
$$

where $x_0, x_1, x_2, \ldots$ represent certain discrete values of the independent variable x for which the function $f(x)$ is known. This formula is computationally less efficient than some others, but it is nonetheless of considerable interest because of its theoretical implications and its easily remembered form. Note that the independent variable x occurs only in the numerator of the coefficients of $f(x_j)$, and that the various factors are of the form $x - x_i$, where i takes on all values from 0 to n except j. The denominators of the coefficients are the same as the numerators, except that x is replaced by x_j.

It will be observed that each of the coefficients in Eq. 3.14 is a polynomial in x of order n. Naturally, we can collect the coefficients of like powers of x, and write the interpolation formula in standard polynomial form if we choose. However, by writing the formula in the form given by Eq. 3.14, we see that if we insert the value $x = x_0$, each term except the first contains the factor $x_0 - x_0$ and vanishes. Furthermore, at $x = x_0$ the numerator and denominator of the first term are identical, so the function defined by Eq. 3.14 is equal to $f(x_0)$ at x_0. In the same way, only the second term is nonzero at x_1 and it is equal to $f(x_1)$, and so forth. Thus we see that this polynomial of degree n passes through the $n + 1$ points $f(x_0), f(x_1), \ldots, f(x_n)$ which are used to define it.

The Lagrange interpolation formula can be used to fit any arbitrary set of points, but it becomes especially useful in those cases where the x values occur at equal intervals. This is usually referred to as the case of *equally spaced abscissas*. We then define the variable x as

$$x = x_0 + ph \tag{3.15}$$

where h is the size of the interval between the various values of x.

We can set $n = 1$, in which case we are dealing only with the two points $f(x_0)$ and $f(x_1)$, where $x = x_0$ and $x = x_0 + h$, respectively. Putting these values in Eq. 3.14, we have

$$\begin{aligned} f(x_0 + ph) &= \frac{(x_0 + ph - x_0 - h)}{(x_0 - x_0 - h)} f(x_0) + \frac{(x_0 + ph - x_0)}{(x_0 + h - x_0)} f(x_1) \\ &= (1 - p)f(x_0) + pf(x_1) \end{aligned} \tag{3.16}$$

This is the *Lagrange two-point interpolation formula for equally spaced abscissas*. Obviously, it equals $f(x_0)$ at $p = 0$ and $f(x_1)$ at $p = 1$, as indicated above. This is, of course, just a line connecting the two points $f(x_0)$ and $f(x_1)$ and is nothing more than the linear interpolation formula often used for the estimation of intermediate values from mathematical tables.

Using $n = 2$, we obtain a quadratic formula passing through the three points $f(x_0), f(x_1)$, and $f(x_2)$. By setting n to progressively larger values, we extend the number of points through which the interpolation formula is fit and correspondingly reduce the interpolation errors, unless of course the data itself can be represented by a very simple formula. Appendix A.6 gives Lagrange interpolation formulas for up to five points.

3.3.2 Interpolation by Finite Differences

Let us assume that we have a function $f(x)$ and that we know values of (x_i, f_i) at certain regular intervals (i.e., for equally spaced abscissas as indicated by Eq. 3.15). These may be placed in a table as indicated by the first two columns of Figure 3.7. Differences between the values of the dependent variable may also be formed as indicated; we will see how they may be used to develop interpolation formulas.

x	$f(x)$	Δ	Δ^2	Δ^3	Δ^4
x_0	f_0				
		Δf_0			
x_1	f_1		$\Delta^2 f_0$		
		Δf_1		$\Delta^3 f_0$	
x_2	f_2		$\Delta^2 f_1$		$\Delta^4 f_0$
		Δf_2		$\Delta^3 f_1$	
x_3	f_3		$\Delta^2 f_2$		
		Δf_3			
x_4	f_4				

Figure 3.7 Finite difference table. Δ is the forward difference operator defined in the text.

We first consider the calculation of some arbitrary point between 0 and h using a general linear relationship

$$f = a + bx$$

We determine the constants by fitting the equation to the two points $(0, f_0)$ and (h, f_1). This gives $f_0 = a$ and $f_1 = a + bh$, from which we get

$$a = f_0$$

$$b = \frac{f_1 - f_0}{h}$$

We are taking x_0 to be the origin, so we may substitute $x = ph$ to get

$$f(x_0 + ph) = f_0 + (f_1 - f_0)p$$

We now introduce the difference notation

$$\Delta f_n = f_{n+1} - f_n \tag{3.17}$$

and the result may be written in the form

$$f(x_0 + ph) = f_0 + p\,\Delta f_0$$

We now repeat this process, but use a quadratic equation of the form

$$f = a + bx + cx^2$$

and determine the constants by fitting to the three points $(0, f_0)$, (h, f_1), and $(2h, f_2)$. This gives the three equations

$$f_0 = a$$

$$f_1 = a + bh + ch^2$$

$$f_2 = a + 2bh + 4ch^2$$

which may be solved for a, b, and c:

$$a = f_0$$

$$b = \frac{4f_1 - f_2 - 3f_0}{2h} = \frac{f_1 - f_0}{h} - \frac{f_2 - 2f_1 + f_0}{2h}$$

$$c = \frac{f_2 - 2f_1 + f_0}{2h^2}$$

Our equation may therefore be written

$$f = f_0 + \left(\frac{f_1 - f_0}{h} - \frac{f_2 - 2f_1 + f_0}{2h}\right)x + \frac{f_2 - 2f_1 + f_0}{2h^2}x^2$$

$$= f_0 + (f_1 - f_0)\frac{x}{h} + (f_2 - 2f_1 + f_0)\frac{x^2 - hx}{2h^2}$$

If we again substitute $x = ph$, we get

$$f(x_0 + ph) = f_0 + (f_1 - f_0)p + (f_2 - 2f_1 + f_0)\frac{p(p-1)}{2}$$

We note that

$$f_2 - 2f_1 + f_0 = (f_2 - f_1) - (f_1 - f_0) = \Delta f_1 - \Delta f_0$$

but this is the difference of two differences, so we might write

$$\Delta f_1 - \Delta f_0 = \Delta(\Delta f_0) = \Delta^2 f_0$$

Using this notation, we have

$$f(x_0 + ph) = f_0 + p\,\Delta f_0 + \frac{p(p-1)}{2}\Delta^2 f_0$$

We can continue this process indefinitely, which leads to the general result

$$f(x_0 + ph) = f_0 + p\,\Delta f_0 + \frac{p(p-1)}{2!}\Delta^2 f_0 + \frac{p(p-1)(p-2)}{3!}\Delta^3 f_0 + \cdots \tag{3.18}$$

This is known as the *forward Gregory–Newton interpolation formula*. The name derives from the fact that we start at the top of the difference table and move along the forward diagonal in order to find the interpolation coefficients.

The formula above may be written more succinctly through definition of a *forward difference operator* Δ, such that

$$\Delta(f_n + g_n) = \Delta f_n + \Delta g_n$$

$$\Delta(cf_n) = c\,\Delta f_n$$

$$\Delta^a(\Delta^b f_n) = \Delta^{a+b} f_n$$

In terms of the forward difference operator Eq. 3.18 may be written

$$f(x_0 + ph) = (1 + \Delta)^p f_0 \tag{3.19}$$

The forward Gregory–Newton interpolation formula is most useful if our principal interest lies near the top part of the table. Then the formula will be rapidly convergent because the higher-order terms are small. If, on the other hand, our interest is near the bottom of the table, the higher-order terms in the forward formula will be large and the formula will be prone to instabilities and large round-off errors. In this case it is more appropriate to interpolate along a backward diagonal. We define a new operator, ∇, called the *backward difference operator:*

$$\nabla f_n = f_n - f_{n-1}$$

and write

$$f(x_0 + ph) = (1 - \nabla)^{-p} f_0 \tag{3.20}$$

where x_0 is now interpreted to be the abscissa corresponding to the last entry in the table. This is the *backward Gregory–Newton interpolation formula,* which when expanded is of the form

$$f(x_0 + ph) = f_0 + p\,\nabla f_0 + \frac{p(p+1)}{2!}\nabla^2 f_0 + \frac{p(p+1)(p+2)}{3!}\nabla^3 f_0 + \cdots \tag{3.21}$$

Of course in actual practice a forward or backward interpolation may be carried out from intermediate points in the table simply by relabeling the data.

There are certain cases where the data with which we are presented are not at equally spaced intervals, in which case we must work with a formula based on *divided differences*. The first divided difference is defined as

$$f(x_0, x_1) = \frac{f(x_0) - f(x_1)}{x_0 - x_1}$$

and the second divided difference is the difference between two successive first divided differences,

$$f(x_0, x_1, x_2) = \frac{f(x_0, x_1) - f(x_1, x_2)}{x_0 - x_2}$$

and so forth.

When we construct a difference table in a general case, we should find that the differences become constant at some point. Then all higher-order terms in the interpolation formula vanish and the data may be faithfully represented by a polynomial of the corresponding order. Let us assume, for example, that we have constant second differences in a divided difference table. Then for any general value of the argument x we may write

$$f(x, x_0, x_1) = f(x_0, x_1, x_2) \tag{3.22}$$

By definition we have

$$f(x, x_0, x_1) = \frac{f(x, x_0) - f(x_0, x_1)}{x - x_1}$$

This may be substituted in Eq. 3.22 to give

$$f(x, x_0) = f(x_0, x_1) + (x - x_1)f(x_0, x_1, x_2) \tag{3.23}$$

Once again by definition

$$f(x, x_0) = \frac{f(x) - f(x_0)}{x - x_0}$$

and this may be inserted in Eq. 3.23 to give

$$f(x) = f(x_0) + (x - x_0)f(x_0, x_1) + (x - x_0)(x - x_1)f(x_0, x_1, x_2)$$

This procedure may be readily generalized for higher-order constant differences, which leads to *Newton's divided difference formula*,

$$\begin{aligned} f(x) = f(x_0) &+ (x - x_0)f(x_0, x_1) \\ &+ (x - x_0)(x - x_1)f(x_0, x_1, x_2) \\ &+ (x - x_0)(x - x_1)(x - x_2)f(x_0, x_1, x_2, x_3) \\ &+ \cdots \end{aligned} \tag{3.24}$$

Numerous other interpolation formulas may be found by following different paths through the difference table. Obviously, the forward and backward formulas are useful near the ends, but for intermediate values one would prefer a formula based on a horizontal path. For this case we shift the origin of the difference table as indicated in Figure 3.8. *Stirling's interpolation formula* is developed by starting at f_0 and moving to the right,

$$f(x_0 + ph) = f_0 + p\left(\frac{\Delta f_0 + \Delta f_{-1}}{2}\right) + \frac{p^2}{2}\Delta^2 f_{-1} + \cdots \tag{3.25}$$

while *Bessel's interpolation formula* starts midway between f_0 and f_1 and moves to the right:

$$\begin{aligned} f(x_0 + ph) = \frac{f_0 + f_1}{2} &+ \left(p - \frac{1}{2}\right)\Delta f_0 \\ &+ \frac{p(p - 1)}{2}\left(\frac{\Delta^2 f_{-1} + \Delta^2 f_0}{2}\right) + \cdots \end{aligned} \tag{3.26}$$

x	$f(x)$	Δ	Δ^2	Δ^3	Δ^4
x_{-2}	f_{-2}				
		Δf_{-2}			
x_{-1}	f_{-1}		$\Delta^2 f_{-2}$		
		Δf_{-1}		$\Delta^3 f_{-2}$	
x_0	f_0		$\Delta^2 f_{-1}$		$\Delta^4 f_{-2}$
		Δf_0		$\Delta^3 f_{-1}$	
x_1	f_1		$\Delta^2 f_0$		
		Δf_1			
x_2	f_2				

Figure 3.8 Finite difference table. This is the equivalent of that given in Figure 3.7, only the origin has been shifted.

Bessel's formula is one of the most widely used in modern computation. On the other hand, Stirling's formula is seldom used, but like Lagrange's formula it is of interest because it can be used to develop useful formulas for differentiation and integration.

3.3.3 Cubic Splines Interpolation

The formulas considered above are useful for hand calculation, but they are cumbersome and rarely used for computer work. There are some other difficulties that should be mentioned as well. If one is to use an interpolation formula to calculate some particular property, such as an integral, for example, it is preferable to keep the function as simple as possible. If it is too limited, it may not represent the variable in question faithfully throughout the entire range, and there will then be a large *truncation error*. In principle one can reduce the truncation error to any arbitrarily small value by taking more and more terms in the fitting function. There are limits to this, however. Round-off errors can become unacceptably large, and although a polynomial of very high degree can give a good fit through certain ranges of data, it often gives wild oscillations in others and must be rejected.

For these reasons it has become common practice in computer applications of interpolation to rely on *spline functions*. These are short piecewise functions that fit a limited portion of the curve only, but when joined together they give a faithful representation throughout. Although the application in this context is a fairly recent development, the mechanical analog has been used by drafters for many years for the analysis of beams, and the terminology has been carried over. The several splines, or flexible segments, were secured by means of weights at their points of intersection, called *knots*.

What is desired in our particular case is a spline that will span the various points to be interpolated, but which will join smoothly to adjacent splines such that there will be no anomalous discontinuities in the derivatives, integrals, or other properties that may be calculated from them. This in essence requires not only that the splines pass through the points to be interpolated, but that they have identical values and slopes at all the knots. We can be assured that the splines will have equal values at the knots by having each spline interpolate between pairs of adjacent points (i.e., by taking each data point to be a knot). Now a linear interpolation between points would involve two parameters and force each spline to join the next without any discontinuity, but the interpolate would be a composite of straight-line segments that would have a discontinuous first derivative. We can add a quadratic term and fix its value by stipulating that the derivative of each spline join to the one below it. This relates all the splines back to the value of the slope that might be arbitrarily chosen at the origin, and this leads to a very unstable situation. Thus we add still another term and then have a *cubic spline*, which does in fact give a very stable and useful interpolation function.

For a cubic spline that spans the interval $[x_i, x_{i+1}]$, we assume the following definitions:

$$h_i = x_{i+1} - x_i \qquad w = \frac{x - x_i}{h_i} \qquad \bar{w} = 1 - w$$

It will be noted that $w = 0$ and $\bar{w} = 1$ at $x = x_i$, and that $w = 1$ and $\bar{w} = 0$ at $x = x_{i+1}$ according to these definitions. The spline itself is written in terms of these variables as follows:

$$s(x) = wf_{i+1} + \bar{w}f_i + h_i^2[(w^3 - w)B_{i+1} + (\bar{w}^3 - \bar{w})B_i] \tag{3.27}$$

It will be observed that this cubic equation does pass through the points f_i at $x = x_i$ and f_{i+1} at $x = x_{i+1}$, as required.

Now we can readily differentiate Eq. 3.27 with respect to x using the chain rule; the result is

$$s'(x) = \frac{f_{i+1} - f_i}{h_i} + h_i[(3w^2 - 1)B_{i+1} - (3\bar{w}^2 - 1)B_i] \tag{3.28}$$

Continuing, we find that

$$s''(x) = 6wB_{i+1} + 6\bar{w}B_i \tag{3.29}$$

$$s'''(x) = \frac{6(B_{i+1} - B_i)}{h_i} \tag{3.30}$$

We see that $s'''(x)$ is a constant, as it of course must be, and that all higher-order derivatives are zero.

Evaluation of Eq. 3.28 at the two knots that it spans gives

$$s'_+(x_i) = f(x_i, x_{i+1}) - h_i(B_{i+1} + 2B_i) \tag{3.31}$$

$$s'_-(x_{i+1}) = f(x_i, x_{i+1}) + h_i(2B_{i+1} + B_i) \tag{3.32}$$

where $f(x_i, x_{i+1})$ is the first divided difference as defined in the preceding section. We use the notation $s'_+(x_i)$ and $s'_-(x_{i+1})$ to emphasize that these are one-sided derivatives of the cubic spline at the respective knots, and they must be equated to those of the adjacent splines at these points. We require, then, that the condition

$$s'_-(x_i) = s'_+(x_i) \qquad i = 2, 3, \ldots, n - 1 \tag{3.33}$$

be satisfied for all interior knots. Now $s'_+(x_i)$ is determined by the spline that spans the interval $[x_{i-1}, x_i]$, but its value may be found by replacing i by $i - 1$ in Eq. 3.32. This leads to

$$f(x_{i-1}, x_i) + h_{i-1}(2B_i + B_{i-1}) = f(x_i, x_{i+1}) - h_i(B_{i+1} + 2B_i)$$

and hence our condition of continuity is

$$h_{i-1}B_{i-1} + 2(h_{i-1} + h_i)B_i + h_iB_{i+1} = f(x_i, x_{i+1}) - f(x_{i-1}, x_i)$$

$$i = 2, 3, \ldots, n - 1 \tag{3.34}$$

Equation 3.34 provides $n - 2$ simultaneous linear equations, but there are n unknowns to be determined, B_i, with $i = 1, 2, \ldots, n$. Frequently, the additional two conditions needed to complete a solution are obtained by specifying that the second derivatives be equal to zero at the ends, that is, that $s''(x_1) = 0$ and $s''(x_n) = 0$. This leads to a stable solution known as the *natural cubic spline*. This is not the only procedure in common use, however, and we chose an alternative method here.

We can find a unique linear equation that passes through two data points, a unique quadratic equation that passes through three data points, a unique cubic equation that passes through four data points, and so forth. For our cubic splines the third derivative is constant, as we have seen, and it makes sense to determine the constants by matching these third derivatives with those of the cubic equations which pass through the sets of four data points at the two ends of the curve. These unique cubic equations may be determined in a variety of ways, one of which makes use of the Newton divided difference formula (Eq. 3.24). If we assume that the interval is small, so that convergence is rapid, it should be clear by examination of Eq. 3.24 that $f(x_0,x_1)$ is a good approximation of the first derivative $f'(x)$, $2f(x_0, x_1, x_2)$ is a good approximation of the second derivative $f''(x)$, $6f(x_0, x_1, x_2, x_3)$ is a good approximation of the third derivative $f'''(x)$, and so forth. Applying Eq. 3.30 at the top of the table, we have

$$6f(x_0, x_1, x_2, x_3) = \frac{6(B_2 - B_1)}{h_1} \tag{3.35}$$

and if we apply Eq. 3.30 again at the bottom of the table, we find, similarly, that

$$6f(x_{n-3}, x_{n-2}, x_{n-1}, x_n) = \frac{6(B_n - B_{n-1})}{h_{n-1}} \tag{3.36}$$

These two conditions, together with those specified by Eq. 3.34, allow us to get a complete solution to the problem. To complete the symmetry and combine the set, we multiply Eq. 3.35 through by h_1^2 and Eq. 3.36 by h_{n-1}^2. The combined result is n linear equations that must be solved simultaneously; they may be written in the form

$$\begin{pmatrix} -h_1 & h_1 & 0 & 0 & & \\ h_1 & 2(h_1+h_2) & h_2 & 0 & & \\ 0 & h_2 & 2(h_2+h_3) & h_3 & & \\ & & & \ddots & & \\ & & h_{n-2} & 2(h_{n-2}+h_{n-1}) & h_{n-1} \\ & & 0 & h_{n-1} & -h_{n-1} \end{pmatrix} \begin{pmatrix} B_1 \\ B_2 \\ B_3 \\ \vdots \\ B_{n-1} \\ B_n \end{pmatrix}$$

$$= \begin{pmatrix} h_1^2 f(x_0, x_1, x_2, x_3) \\ f(x_1, x_2) - f(x_1, x_0) \\ f(x_2, x_3) - f(x_1, x_2) \\ \vdots \\ f(x_{n-2}, x_{n-1}) - f(x_{n-3}, x_{n-2}) \\ -h_{n-1}^2 f(x_{n-3}, x_{n-2}, x_{n-1}, x_n) \end{pmatrix} \tag{3.37}$$

There are various techniques that can be used to solve this set of simultaneous linear equations, some of which are discussed in Section 3.5. Once the B_i's are known they may be put into Eq. 3.27, which then specifies the cubic spline within each interval as indicated. For practical purposes it is often more convenient to transform the cubic spline into the form

$$s(x) = f_i + b_i(x - x_i) + c_i(x - x_i)^2 + d_i(x - x_i)^3 \qquad x_i \le x \le x_{i+1} \quad (3.38)$$

where

$$b_i = \frac{f_{i+1} - f_i}{h_i} - h_i(B_{i+1} + 2B_i)$$

$$c_i = 3B_i \qquad (3.39)$$

$$d_i = \frac{B_{i+1} - B_i}{h_i}$$

Various manipulations may then be performed, such as evaluation of derivatives and integrals, using the stored sets of coefficients, b_i, c_i, and d_i.

A BASIC program that will carry out a cubic splines interpolation is shown in Figure 3.9. It is based on a FORTRAN program described by G. E. Forsythe, M. A. Malcolm, and C. B. Moler (*Computer Methods for Mathematical Computations*, Prentice-Hall, Inc., Englewood Cliffs, N.J., 1977). The test data consist of wavelengths and absorbance values transmitted directly in digital form from a spectrophotometer to a computer in this particular case. The output b_i, c_i, and d_i values were input to a graphics plotter which constructed the smooth curve through the data points as shown in Figure 3.10. The smooth cubic splines interpolate can be used to locate the precise positions of the various peaks and shoulders through differentiation and/or spectral resolution into individual absorption bands (see Section 5.6.2).

Example 3.6

Find a polynomial that represents the following data:

x	0.0	0.5	1.0	1.5	2.0	2.5	3.0
$f(x)$	−7.00	−4.75	−2.00	1.25	5.00	9.25	14.00

Solution. We first construct a difference table using the data above as follows:

x	f	Δ	Δ^2	Δ^3
0.0	−7.00			
		2.25		
0.5	−4.75		0.50	
		2.75		0.00
1.0	−2.00		0.50	
		3.25		0.00
1.5	1.25		0.50	
		3.75		0.00
2.0	5.00		0.50	
		4.25		0.00
2.5	9.25		0.50	
		4.75		
3.0	14.00			

```
10 DIM X(50),Y(50),B(50),C(50),D(50)
20 READ N  'NUMBER OF DATA POINTS
30 NM1=N-1
40 FOR I=1 TO N
50 READ X(I),Y(I)
60 NEXT I
70 D(1)=X(2)-X(1)
80 C(2)=(Y(2)-Y(1))/D(1)
90 FOR I=2 TO NM1
100 D(I)=X(I+1)-X(I)
110 B(I)=2*(D(I-1)+D(I))
120 C(I+1)=(Y(I+1)-Y(I))/D(I)
130 C(I)=C(I+1)-C(I)
140 NEXT I
150 B(1)=-D(1)
160 B(N)=-D(N-1)
170 C(1)=0
180 C(N)=0
190 C(1)=C(3)/(X(4)-X(2))-C(2)/(X(3)-X(1))
200 C(N)=C(N-1)/(X(N)-X(N-2))-C(N-2)/(X(N-1)-X(N-3))
210 C(1)=C(1)*D(1)^2/(X(4)-X(1))
220 C(N)=-C(N)*D(N-1)^2/(X(N)-X(N-3))
230 FOR I=2 TO N
240 T=D(I-1)/B(I-1)
250 B(I)=B(I)-T*D(I-1)
260 C(I)=C(I)-T*C(I-1)
270 NEXT I
280 C(N)=C(N)/B(N)
290 FOR IB=1 TO NM1
300 I=N-IB
310 C(I)=(C(I)-D(I)*C(I+1))/B(I)
320 NEXT IB
330 B(N)=(Y(N)-Y(NM1))/D(NM1)+D(NM1)*(C(NM1)+2*C(N))
340 FOR I=1 TO NM1
350 B(I)=(Y(I+1)-Y(I))/D(I)-D(I)*(C(I+1)+2*C(I))
360 D(I)=(C(I+1)-C(I))/D(I)
370 C(I)=3*C(I)
380 NEXT I
390 C(N)=3*C(N)
400 D(N)=D(N-1)
410 FOR I=1 TO N
420 LPRINT I,X(I),Y(I)
430 LPRINT B(I),C(I),D(I)
440 NEXT I
450 STOP
460 DATA 36,200,.1547,204,.1536,208,.1262,212,.1150,216,.1895,220,.2899
470 DATA 224,.4369,228,.6525,232,.8875,236,1.0569,240,1.1348,244,1.2565
480 DATA 248,1.2369,252,1.0619,256,.6675,260,.4297,264,.2916,268,.2135
490 DATA 272,.1461,276,.1051,280,.0956,284,.0988,288,.1153,292,.1487
500 DATA 296,.1946,300,.2181,304,.2474,308,.2921,312,.2889,316,.2556
510 DATA 320,.2034,324,.1455,328,.1157,332,.0915,336,.0782,340,.0710
520 END
```

Figure 3.9 BASIC program for cubic splines interpolation, including spectrophotometric data from which the curve of Figure 3.10 was generated.

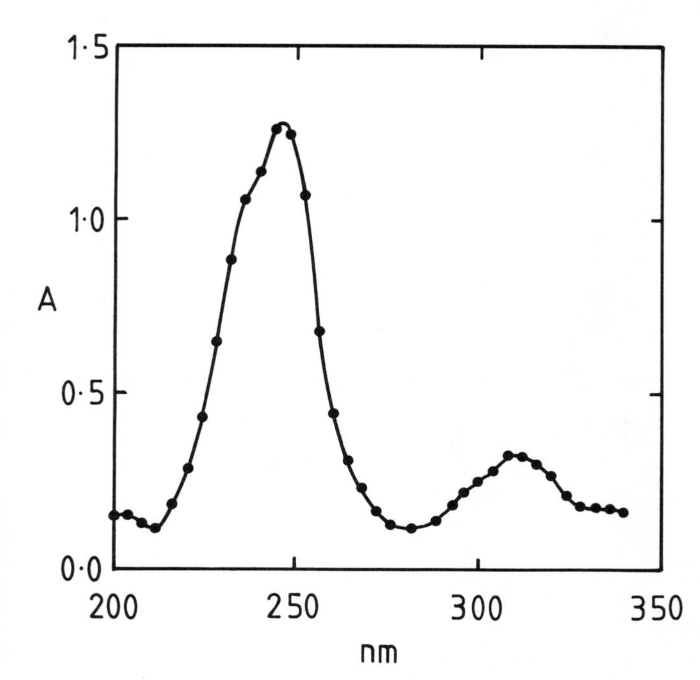

Figure 3.10 Smooth absorption curve for benzo[*c*]cinnoline generated by cubic splines interpolation of the digital data.

It will be observed that the column of third differences is zero, so according to Eq. 3.18, we can represent the data by a quadratic equation. We find that

$$f = -7 + p(2.25) + \frac{p(p-1)}{2}(0.50)$$

$$= -7 + \frac{9}{4}p + \frac{1}{4}p^2 - \frac{1}{4}p$$

We now substitute

$$p = \frac{x}{h} = \frac{x}{0.5} = 2x$$

and obtain

$$f = -7 + \frac{9}{2}x + x^2 - \frac{1}{2}x$$

$$= x^2 + 4x - 7$$

It may be readily verified that the original $(x, f(x))$ values conform to this equation.

Example 3.7

Fit a three-point Lagrange interpolation formula to the following data, and use it to determine $f(1.25)$:

x	$f(x)$
0.0	6.3
1.0	8.2
2.0	12.4

Solution. Putting $x_0 = 1.0$ and $h = 1.0$ in Eq. 3.15, we find that $p = 0.25$ when $x = 1.25$. The three-point formula according to Appendix A.6 is

$$f(x) = \frac{p(p-1)}{2} f(x_{-1}) + (1 - p^2) f(x_0) + \frac{p(p+1)}{2} f(x_1)$$

Inserting the values of $f(x)$ from the table above and the appropriate value of p, we have

$$\begin{aligned} f(1.25) &= \frac{(0.25)(-0.75)}{2}(6.3) + (1 - (0.25)^2)(8.2) + \frac{(0.25)(1.25)}{2}(12.4) \\ &= -0.59 + 7.69 + 1.94 \\ &= 9.0 \end{aligned}$$

Example 3.8

The probability density for an electron in a spherically symmetric orbital is $4\pi r^2 R^2$, where R is the radial wave function. Numerical calculations by the self-consistent field (SCF) method yield the results shown in the following table for the 1s orbital of a helium atom, where $P = rR$ and $x = r/\mu$ with $\mu = \frac{1}{2}(3\pi/4)^{2/3}(Z)^{-1/3}$. The r used here is the distance from the nucleus in atomic units (1 a.u. $= a_0 = 0.529$ Å).

x	P
0.0	0.0
0.1	0.2963
0.2	0.5153
0.3	0.6729
0.5	0.8545
0.7	0.9191
1.1	0.8759
1.5	0.7519
2.3	0.4881
3.1	0.2902
4.7	0.0902

Find where the 1s orbital of He has its maximum probability density and compare this with that of an H atom, whose 1s radial distribution function is

$$R_H = \left(\frac{1}{\pi a_0^3}\right)^{1/2} \exp\left[-\frac{r}{a_0}\right]$$

Solution. The function P is a maximum where the probability density is a maximum. We see that the maximum occurs at about $x = 0.7$, where there is a change of scale. Thus we need to use the divided difference formula (Eq. 3.24). We construct the following divided difference table:

x_0 0.5 f_0 0.8545

$f(x_0, x_1)$ 0.3230

x_1 0.7 f_1 0.9191

$f(x_0, x_1, x_2)$ −0.7183

$f(x_1, x_2)$ −0.1080

$f(x_0, x_1, x_2, x_3)$ 0.4658

x_2 1.1 f_2 0.8759

$f(x_1, x_2, x_3)$ −0.2525

$f(x_2, x_3)$ −0.3100

x_3 1.5 f_3 0.7519

and insertion of the appropriate values in Eq. 3.24 gives us

$$f(x) = 0.8545 + (x - 0.5)(0.3230) + (x - 0.5)(x - 0.7)(-0.7183) + (x - 0.5)(x - 0.7)(x - 1.1)(0.4658)$$

Differentiation of this equation gives

$$\begin{aligned} f'(x) = {} & (0.3230) + (x - 0.7)(-0.7183) + (x - 0.5)(-0.7183) \\ & + (x - 0.7)(x - 1.1)(0.4658) + (x - 0.5)(x - 1.1)(0.4658) \\ & + (x - 0.5)(x - 0.7)(0.4658) \end{aligned}$$

or by collecting terms,

$$f'(x) = 1.3974x^2 - 3.5793x + 1.9629$$

Setting this equal to zero and solving for the two values of x gives

$$x = 1.7660 \qquad x = 0.7954$$

Obviously, it is the second root that has physical significance. By insertion of $Z = 2$ in the formula for μ we find that it is equal to 0.7027. Thus the probability density is a maximum at

$$0.7954 \times 0.7027 = 0.5589 \text{ a.u.} = 0.296 \text{ Å}$$

Now for the H atom the corresponding function $P = rR$ is

$$P_{\mathrm{H}} = r\left(\frac{1}{\pi a_0^3}\right)^{1/2} \exp\left[-\frac{r}{a_0}\right]$$

Differentiating this and equating it to zero gives

$$\frac{dP_{\mathrm{H}}}{dr} = \left(\frac{1}{\pi a_0^3}\right)^{1/2}\left[\exp\left(-\frac{r}{a_0}\right) - \frac{r}{a_0}\exp\left(-\frac{r}{a_0}\right)\right] = 0$$

or

$$1 - \frac{r}{a_0} = 0$$

$$r = a_0 = 0.529 \text{ Å}$$

3.4 NUMERICAL DIFFERENTIATION AND INTEGRATION

Any of the interpolation formulas discussed in the preceding section can be differentiated or integrated term by term to give an approximation of the true derivative or integral of the function under consideration. Such methods are in common use, particularly in the case where the derivative is of concern. Some caution needs to be exercised in such procedures, however. If there are residual errors in the fitting function, from round-off or any other cause, these can become greatly amplified.

Suppose, for example, that some erratic fluctuations are imposed on the interpolating function as indicated schematically in Figure 3.11. The derivative of the inter-

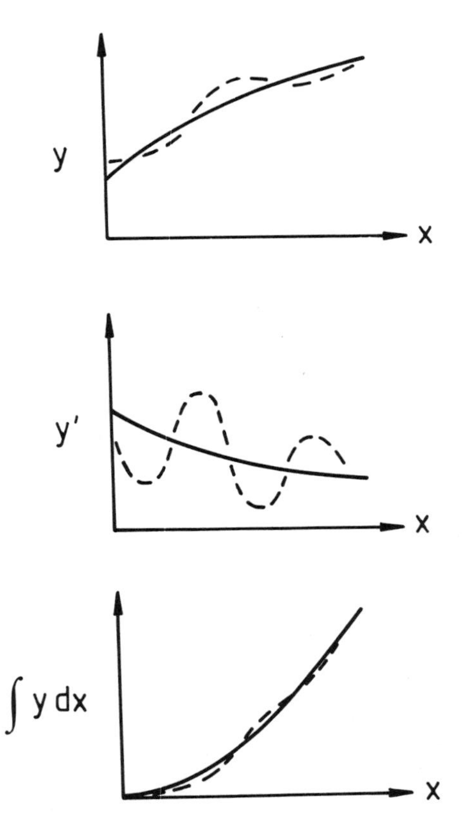

Figure 3.11 The solid curve represents the true function of x, and the dashed curve is the interpolate which is subject to spurious fluctuations. These fluctuations become amplified in the derivative but smoothed in the integral as indicated.

polate will be subject to the same erratic behavior, in fact showing changes of slope which are even more pronounced than is apparent in the original curve. Such fluctuations can be very misleading and quite unrepresentative of the true value that should characterize the function at a given value of the independent variable.

The situation is quite different, however, if one uses this same interpolate to calculate an integral. In this case the derived function is the accumulated area under the curve, and this is less susceptible to error from random fluctuations since the positive and negative contributions tend to cancel.

Thus it may be maintained that integration is a smoothing process, but just the opposite is true of differentiation. As a result, care must be exercised in the use of numerical formulas. One must be particularly cautious in dealing with derivatives of higher order, where these effects can be still further amplified. In spite of these limitations, we must often rely on such numerical techniques when no alternatives exist, but these sources of error should always be borne in mind.

3.4.1 Differentiation Formulas

The formulas that are most useful are those for equally spaced abscissas, and they are the only ones we will consider here. The independent variable x may then be expressed as given by Eq. 3.15, so the various formulas that are written in terms of p may all be differentiated using the chain rule to give

$$f' = \frac{df(x)}{dx} = \frac{df_p}{dp}\frac{dp}{dx} = \frac{1}{h}\frac{df_p}{dp} \tag{3.40}$$

The Lagrange formula (Eq. 3.14) is often used for the numerical differentiation of a function. The result for the two-point form given by Eq. 3.16 is

$$\frac{df_p}{dp} = -f(x_0) + f(x_1)$$

or

$$f' = \frac{1}{h}(f(x_1) - f(x_0)) \tag{3.41}$$

The result in this case is trivial. Since Eq. 3.16 is just the linear interpolation formula, its derivative Eq. 3.41 is an obvious expression for the slope of the line. Differentiation of higher-order Lagrange formulas for equally spaced abscissas is readily carried out and leads to more useful results, some of which may be found in Appendix A.6.

Formulas for numerical differentiation based on finite difference formulas are likewise readily obtained. For example, that based on the forward Gregory–Newton interpolation formula (Eq. 3.18) is

$$f' = \frac{1}{h}\left(\Delta f_0 + \frac{2p-1}{2!}\Delta^2 f_0 + \frac{3p^2 - 6p + 2}{3!}\Delta^3 f_0 + \cdots\right) \tag{3.42}$$

while that derived from the backward Gregory–Newton interpolation formula is

$$f' = \frac{1}{h}\left(\nabla f_0 + \frac{2p+1}{2!}\nabla^2 f_0 + \frac{3p^2+6p+2}{3!}\nabla^3 f_0 + \cdots\right) \tag{3.43}$$

3.4.2 Integration Procedures

Several formulas that can be used for numerical integration can be derived through term-by-term integration of the interpolation formulas discussed above. The more useful ones can be readily derived through alternative means which give an enhanced understanding of the nature of the process, however, and that is the principal approach used here.

The process of numerical integration is often called *quadrature*. Many numerical quadrature procedures are described in reference works. We describe only a few of those in most common use: the trapezoidal rule, Simpson's rule, and Gaussian quadrature.

The Trapezoidal Rule

A very simple technique that can be used if high accuracy is not required is based on the calculation of areas under the curve to be integrated by the construction of cords at equally spaced intervals. This is essentially the procedure followed in the development of integration formulas in Section 1.6.1. If we divide the range of the integral

$$\int_a^b f(x)\,dx$$

into n equal parts of length h, the arc of $f(x)$ for each subinterval $x_{i+1} - x_i$ is replaced by its chord and each of the areas is approximated by that of the trapezoid as shown in Figure 3.12. The area of the trapezoid is equal to the average length of the parallel sides times the perpendicular distance between them, so we have

$$A_1 = \left(\frac{f(x_0) + f(x_1)}{2}\right)h$$

$$A_2 = \left(\frac{f(x_1) + f(x_2)}{2}\right)h$$

$$\vdots$$

$$A_{n-1} = \left(\frac{f(x_{n-2}) + f(x_{n-1})}{2}\right)h$$

$$A_n = \left(\frac{f(x_{n-1}) + f(x_n)}{2}\right)h$$

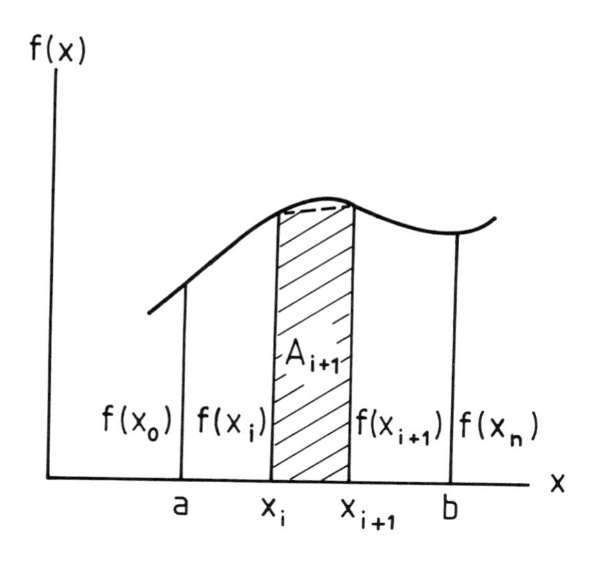

Figure 3.12 Replacement of subintervals under the curve $f(x)$ by trapezoids for integration according to the trapezoidal rule.

Summing these contributions to the total area under the curve gives the *trapezoidal rule*,

$$\int_a^b f(x)\,dx = h\left[\frac{1}{2}f(x_0) + f(x_1) + f(x_2) + \cdots + f(x_{n-1}) + \frac{1}{2}f(x_n)\right] \quad (3.44)$$

We would expect the integral calculated in this way to become more accurate as the number of quadrature points is increased.

Simpson's Rule

There are limits to how far one might want to push the trapezoidal rule. The calculation becomes more tedious as n increases, and beyond a certain point the accuracy will no longer increase because of the increase in accumulated computational error. Simpson's rule gives improved accuracy for the same number of equally spaced quadrature points.

We will evaluate the integral between $x = x_0$ and $x = x_2$ for the function $f(x)$ as shown in Figure 3.13, assuming h to be very small:

$$\int_{x_0}^{x_2} f(x)\,dx$$

A shift of axes will allow us to evaluate the integral about the midpoint of the range, x_1. We do this by means of the substitution

$$x = x_1 + r$$

with $x_0 = x_1 - h$ and $x_2 = x_1 + h$. The integral in terms of the variable r becomes

$$\int_{x_0}^{x_2} f(x)\,dx = \int_{-h}^{+h} f(x_1 + r)\,dr \quad (3.45)$$

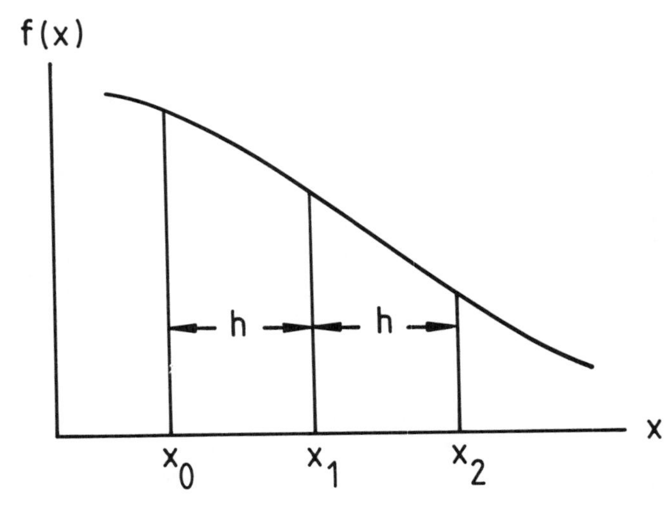

Figure 3.13 Integration of the function $f(x)$ within the limits $x = x_0$ and $x = x_2$ according to Simpson's rule. It is assumed that the interval is small.

We now expand $f(x)$ about the point x_1. Since we have shifted axes this is equivalent to the Maclaurin series (see Eq. 1.123),

$$f(x_1 + r) = f(x_1) + \frac{r}{1!} f'(x_1) + \frac{r^2}{2!} f''(x_1) + \frac{r^3}{3!} f'''(x_1) + \cdots \tag{3.46}$$

Putting Eq. 3.46 into Eq. 3.45 and integrating term by term gives

$$\left[rf(x_1) + \frac{r^2}{2} f'(x_1) + \frac{r^3}{6} f''(x_1) + \cdots \right]_{-h}^{+h}$$

Since the interval h is small, higher-order terms may be neglected and we have

$$\int_{x_0}^{x_2} f(x)\,dx \approx 2hf(x_1) + \frac{h^3}{3} f''(x_1) \tag{3.47}$$

Now from Eq. 3.46 we see that

$$f(x_0) = f(x_1 - h) = f(x_1) - hf'(x_1) + \frac{h^2}{2!} f''(x_1) + \cdots$$

and

$$f(x_2) = f(x_1 + h) = f(x_1) + hf'(x_1) + \frac{h^2}{2!} f''(x_1) + \cdots$$

so we may write

$$f(x_0) + f(x_2) \approx 2f(x_1) + h^2 f''(x_1) \tag{3.48}$$

Equation 3.48 may be solved for $f''(x_1)$ and the result inserted in Eq. 3.47 to give

$$\int_{x_0}^{x_2} f(x)\,dx \approx \frac{h}{3}\,[f(x_0) + 4f(x_1) + f(x_2)] \tag{3.49}$$

which is known as *Simpson's rule*. It is easily shown that Simpson's rule is just the integrated form of the Lagrange equation (Eq. 3.14) fit to three equally spaced points (see Exercise 3.30).

Simpson's rule may be readily generalized to accommodate a larger number of quadrature points. Suppose that the interval x_0 to x_2 is just one element in the total range of integration x_0 to x_{2n}, where $x_0 = a$ and $x_{2n} = b$. Then we must add a number of area elements, each of which is of the form given by Eq. 3.49. We have

$$\begin{aligned}\int_a^b f(x)\,dx &= \int_{x_0}^{x_2} f(x)\,dx + \int_{x_2}^{x_4} f(x)\,dx + \cdots + \int_{x_{2n-2}}^{x_{2n}} f(x)\,dx \\ &= \frac{h}{3}\,[f(x_0) + 4f(x_1) + f(x_2)] + \frac{h}{3}\,[f(x_2) + 4f(x_3) + f(x_4)] \\ &\qquad + \cdots + \frac{h}{3}\,[f(x_{2n-2}) + 4f(x_{2n-1}) + f(x_{2n})] \\ &= \frac{h}{3}\,\{f(x_0) + 4[f(x_1) + f(x_3) + \cdots + f(x_{2n-1})] \\ &\qquad + 2[f(x_2) + f(x_4) + \cdots + f(x_{2n-2})] + f(x_{2n})\}\end{aligned} \tag{3.50}$$

This is generally called the *extended Simpson's rule*.

Gaussian Quadrature

The procedure for numerical integration may be represented in a general way as a replacement of the integral by a summation:

$$\int_a^b f(x)\,dx \to \sum_{j=1}^{n} a_j f(x_j) \tag{3.51}$$

We have just seen that in the simplest case (the trapezoidal rule) the integrand is evaluated for equally spaced abscissas x_j using equal weighting factors a_j. A considerable improvement in accuracy for a given number of quadrature points is achieved with Simpson's rule, which varies the weighting factors. We now consider how we might make a further improvement in the accuracy of integration for a fixed number of quadrature points.

Let us use the Lagrange interpolation formula (Eq. 3.14) to construct a polynomial of degree $\leq n - 1$ which passes through n data points. We now represent this by a more compact notation,

$$\phi(x) = \sum_{j=1}^{n} f(x_j)\,\frac{F(x)}{(x - x_j)F'(x_j)} \tag{3.52}$$

where $F(x)$ is a polynomial of degree n whose zeros are the n points x_j; it may be written

$$F(x) = \prod_{j=1}^{n} (x - x_j) \tag{3.53}$$

The symbol Π represents the continued product over all the factors in parentheses, where j takes all values between 1 and n. Using this same notation, the derivative of $F(x)$ evaluated at the point x_j is written

$$F'(x_j) = \prod_{i \neq j} (x_j - x_i) \tag{3.54}$$

Assuming that the Lagrange formula gives a faithful representation of the function, we may write the integral as

$$\int_a^b f(x)\, dx \approx \int_a^b \phi(x)\, dx = \sum_{j=1}^{n} a_j f(x_j) \tag{3.55}$$

where

$$a_j = \frac{1}{F'(x_j)} \int_a^b \frac{F(x)}{x - x_j}\, dx \qquad j = 1, 2, \ldots, n \tag{3.56}$$

The weighting factors a_j defined by Eq. 3.56 are called the *Christoffel numbers*. For equally spaced abscissas they are given by

$$a_j = \frac{(-1)^{n-j} h}{j!(n-j)!} \int_a^b \frac{x(x-1)(x-2)\cdots(x-n)}{x - x_j}\, dx \tag{3.57}$$

We note that these weighting factors are independent of $f(x)$, so they can be calculated once and for all when the interval and the number of quadrature points are specified. The division into $n - 1$ equally spaced intervals forms the basis of the *Newton-Cotes method*, and tables of the weighting factors a_j are available in various sources. The trapezoidal rule and Simpson's rule are special cases of the Newton-Cotes formulas.

The formulation above is obviously adequate to give a truncation error

$$\int_a^b f(x)\, dx - \sum_{j=1}^{n} a_j f(x_j) \tag{3.58}$$

equal to zero for any set of data points that can be represented exactly by a polynomial of degree $\leq n - 1$. In 1814, Gauss showed that it is possible to reduce the truncation error still further for a given number of quadrature points. Stated another way, the truncation error can be made to vanish for polynomials $f(x)$ of higher degree than $n - 1$ if the a_j are suitably chosen *and* the x_j are not confined to equally spaced intervals.

Since both the a_j and x_j may be varied, there are actually $2n$ variables represented in the sum of Eq. 3.58, and this implies that the truncation error for a polynomial of degree $2n - 1$ can be reduced to zero by suitable choice of the n values of a_j and the n values of x_j. Now $F(x)$ need not be confined to the precise format of Eq. 3.53, but it must be a polynomial of degree n which vanishes at x_j, with $j = 1, 2, \ldots, n$. Since $f(x) - \phi(x) = 0$ at $x = x_j$, we can construct a polynomial of degree $2n - 1$ as follows:

$$f(x) = \phi(x) + F(x) \sum_{l=0}^{n-1} c_l x^l \tag{3.59}$$

In order that the truncation error be zero, we require that

$$\int_a^b F(x)x^l \, dx = 0 \qquad l = 0, 1, \ldots, n - 1 \tag{3.60}$$

There are several sets of polynomials that may be found in reference works which satisfy Eq. 3.60, and any one of them may be used to derive the abscissas and weighting factors for Gaussian quadrature. The most widely used set is the *Legendre polynomials,* which may be written

$$P_n(z) = \frac{1}{2^n n!} \frac{d^n}{dz^n} (z^2 - 1)^n \tag{3.61}$$

The Legendre polynomials are defined for $-1 \leq z \leq +1$, so it is necessary to standardize the range of integration from -1 to $+1$. This is done for an arbitrary integral over the range a to b by the substitution

$$x = \frac{a + b}{2} + \frac{b - a}{2} z \tag{3.62}$$

which leads to

$$\int_a^b f(x) \, dx = \frac{b - a}{2} \int_{-1}^{+1} f(z) \, dz \approx \frac{b - a}{2} \sum_{j=1}^{n} w_j f(z_j) \tag{3.63}$$

It can be shown through integration by parts l times that the Legendre polynomials defined by Eq. 3.61 are *orthogonal* over the range -1 to $+1$ to every power of z less than n. That is,

$$\int_{-1}^{+1} z^l P_n(z) \, dz = 0 \qquad l = 0, 1, \ldots, n - 1$$

Thus the condition Eq. 3.60 is satisfied by these polynomials and we can find the points at which the integrand should be evaluated by finding the values of z_j for which the Legendre polynomial of a given order is zero:

$$P_n(z) = 0 \tag{3.64}$$

Once these abscissas have been determined, we may use them in

$$w_j = \frac{1}{P_n'(z_j)} \int_{-1}^{+1} \frac{P_n(z)}{z - z_j} \, dz \tag{3.65}$$

to solve for the values of the weighting factors.

The abscissas and their corresponding weighting factors for Gaussian quadrature based on Legendre polynomials are given in Appendix A.7. The list is complete only up to eight quadrature points; more extensive tables are available in many reference works. Abscissas and weights based on other sets of polynomials may also be found. The Gaussian procedure represents the best accuracy that can be attained for a given number of quadrature points.

Example 3.9

Real gases do not behave ideally at high pressures but are described in terms of an effective pressure called the fugacity f. The fugacity is related to the pressure p through

$$\ln f = \ln p - \frac{1}{RT}\int_0^p \alpha \, dp$$

where $\alpha = RT/p - V$ is the difference between the ideal and the true molar volume V. We take $\alpha = 0.00$ at $p = 0$, but at higher pressures it is determined through gas density measurements. For H_2 gas at $T = 273$ K, the measured molar volumes are as follows:

p (atm)	V (mL mol^{-1})
200	129.1
400	71.5

Calculate the value of the fugacity f at $T = 273$ K when the pressure p is 400 atm.

Solution. The molar gas constant R is 82.05 mL atm K^{-1} mol^{-1}, and with the data given above, we can construct a table of $\alpha = RT/p - V$ values as follows:

p	α
0	0.00
200	$112.0 - 129.1 = -17.1$
400	$56.0 - 71.5 = -15.5$

We use Simpson's rule to carry out the indicated integration:

$$\ln \frac{f}{p} = -\frac{1}{(82.05)(273)}\left(\frac{200}{3}\right)[0.00 + 4(-17.1) + (-15.5)]$$
$$= 0.250$$

The antilog gives

$$\frac{f}{p} = 1.28$$

so the fugacity or effective pressure at 400 atm is

$$f = (1.28)(400 \text{ atm})$$
$$= 513 \text{ atm}$$

Example 3.10

Compare the values of the integral

$$\int_0^1 \sinh^3 x \, dx$$

obtained for five-point integration by the trapezoidal rule, Simpson's rule, and Gaussian quadrature with the true value.

Solution. This integral may be found in standard integral tables; the result is

$$I_{\text{exact}} = \left[\frac{\cosh^3 x}{3} - \cosh x\right]_0^1$$

$$= -0.318338643 + 0.666666667$$

$$= 0.348328024$$

To carry out the integration by the trapezoidal rule, we divide the range of integration into five equally spaced points with $h = 0.25$ and evaluate the integrand as follows:

x_j	$\sinh^3(x_j)$
0.00	0.00
0.25	0.0161199
0.50	0.1414984
0.75	0.5560545
1.00	1.6230678

Putting these values in Eq. 3.44 gives

$$I_{\text{trap}} = 0.25[\tfrac{1}{2}(0.00) + 0.0161199 + 0.1414984 + 0.5560545 + \tfrac{1}{2}(1.6230678)]$$

$$= 0.381302$$

and the error is

$$\frac{0.381302 - 0.348328}{0.348328} \times 100 = 9.47\%$$

by the trapezoidal rule.

Using the same equally spaced points but putting them in Eq. 3.50 gives

$$I_{\text{Simp}} = \frac{0.25}{3}[0.00 + 4(0.0161199 + 0.5560545) + 2(0.1414984) + 1.6230678]$$

$$= 0.349564$$

and the error for integration according to Simpson's rule is

$$\frac{0.349564 - 0.348328}{0.348328} \times 100 = 0.35\%$$

For the Gaussian quadrature we make the transformation

$$x = \frac{1+0}{2} + \frac{1-0}{2} z$$
$$= 0.5 + 0.5z$$

and with quadrature points listed in Appendix A.7 we construct the following table of values of the integrand together with the corresponding weighting factors:

z_j	x_j	$\sinh^3(x_j)$	w_j
−0.9061798	0.04691010	0.000103342	0.2369269
−0.5384693	0.23076535	0.01261987	0.4786287
0.0	0.5	0.1414984	0.5688889
0.5384693	0.76923465	0.6084489	0.4786287
0.9061798	0.95308990	1.3458722	0.2369269

These values are inserted in Eq. 3.63 and the result is

$$\begin{aligned} I_{\text{Gauss}} &= \tfrac{1}{2}[(0.2369269)(0.000103342) \\ &\quad + (0.4786287)(0.01261987) + (0.5688889)(0.1414984) \\ &\quad + (0.4786287)(0.6084489) + (0.2369269)(1.3458722)] \\ &= 0.3483280 \end{aligned}$$

The integral is accurate to seven significant digits, which is as good as we could expect since the tabulation of Appendix A.7 contains only seven digits.

3.5 SOLUTION OF SYSTEMS OF LINEAR EQUATIONS

Systems of linear equations appear in many contexts, and in this section we discuss a few of the more common techniques used to solve them. We assume that we are dealing with n equations involving n unknowns, and that all the equations are *linearly independent;* that is, that no one of the equations is a simple multiple of another. The equations to be solved can be represented in a general way by

$$\begin{aligned} a_{11}x_1 + a_{12}x_2 + \cdots &= c_1 \\ a_{21}x_1 + a_{22}x_2 + \cdots &= c_2 \\ &\vdots \\ a_{n1}x_1 + a_{n2}x_2 + \cdots &= c_n \end{aligned} \tag{3.66}$$

These equations may be written in compact matrix notation as

$$\mathbf{Ax} = \mathbf{c} \tag{3.67}$$

If we multiply Eq. 3.67 on the left by the inverse of $\mathbf{A}$, we have

$$\mathbf{A}^{-1}\mathbf{A}\mathbf{x} = \mathbf{I}\mathbf{x} = \mathbf{x} = \mathbf{A}^{-1}\mathbf{c} \tag{3.68}$$

and we see that in a formal way the desired solution $\mathbf{x}$ is obtained simply by multiplication of $\mathbf{c}$ by $\mathbf{A}^{-1}$. We will see later that there is a close relationship between the solution of linear equations and matrix inversion, although direct solution by matrix inversion is not particularly efficient and is seldom used unless $\mathbf{A}^{-1}$ itself is needed for another purpose as well. Note that if the system is *singular* (see Section 1.4), no stable solution will be found by this or by any other method. We assume in the following that we are dealing with nonsingular systems of equations only.

Three distinct approaches to the problem will be considered here. The Cramer's and Gauss elimination methods involve explicit solution and can be used for either hand or computer solution of smaller systems of linear equations; the Gauss–Seidel method is an iterative technique that is often used where large systems are involved.

3.5.1 Cramer's Rule

It can be shown that values of x_k which constitute a solution of the set of linear equations given as Eq. 3.66 may be written

$$x_k = \frac{|\mathbf{A}_k|}{|\mathbf{A}|} \tag{3.69}$$

where $|\mathbf{A}|$ is the determinant of the coefficients and $|\mathbf{A}_k|$ is the determinant of the coefficients with the kth column replaced by the column of constants. For example, for x_2 we write

$$x_2 = \frac{\begin{vmatrix} a_{11} & c_1 & a_{13} & \cdots \\ a_{21} & c_2 & a_{23} & \cdots \\ \vdots & & & \end{vmatrix}}{\begin{vmatrix} a_{11} & a_{12} & a_{13} & \cdots \\ a_{21} & a_{22} & a_{23} & \cdots \\ \vdots & & & \end{vmatrix}}$$

This solution is called *Cramer's rule*. Although it illustrates some important principles, the method is inefficient and not much used in practice except possibly for the very simplest of systems. The reason will become apparent if we consider a particular case. Suppose that we have a system of 20 equations which we wish to solve for 20 unknowns. This involves solution of 21 determinants of order 20. The expansion of each of these determinants by minors generates some 20! terms, each of which requires 19 multiplications. Thus there are $21 \times 19 \times 20!$ multiplications and a similar number of additions. Working with a computer that performs 10^6 multiplications per second would require approximately 3×10^7 years for the multiplications alone! Clearly, alternative methods must be used for practical solution for all but the simplest of problems.

3.5.2 The Gauss Elimination Technique

The general approach of the Gauss elimination technique is to reduce the system of equations represented by Eq. 3.66 to a triangular form

$$\begin{aligned}
a'_{11}x_1 + a'_{12}x_2 + a'_{13}x_3 + \cdots + a'_{1n}x_n &= c'_1 \\
a'_{22}x_2 + a'_{23}x_3 + \cdots + a'_{2n}x_n &= c'_2 \\
a'_{33}x_3 + \cdots + a'_{3n}x_n &= c'_3 \\
&\vdots \\
a'_{nn}x_n &= c'_n
\end{aligned}$$

Then the last equation is used to solve for x_n; this is put into the second to last to solve for x_{n-1}, and so forth. The procedure involves *forward elimination* and *back substitution,* and since the same operations are performed on the column of constants c_i to accomplish this, it is useful to first construct the *augmented matrix:*

$$\left(\begin{array}{ccccc|c}
a_{11} & a_{12} & a_{13} & \cdots & a_{1n} & c_1 \\
a_{21} & a_{22} & a_{23} & \cdots & a_{2n} & c_2 \\
\vdots & & & & & \\
a_{n1} & a_{n2} & a_{n3} & \cdots & a_{nn} & c_n
\end{array}\right) \qquad (3.70)$$

Gauss elimination consists of replacement of the second equation by a new equation derived by multiplying the first by the factor a_{21}/a_{11} and subtracting it from the second. Then the third is replaced by the new equation derived by multiplying the first by the factor a_{31}/a_{11} and subtracting it from the third, and so forth. Once the elements under a_{11} have all been set to zero in this way, a similar procedure is used for those under a_{22}, a_{33}, . . . , until at last only the upper triangle remains.

Let us illustrate the procedure with an example. Suppose that we need to solve the set of equations

$$\begin{aligned}
20x_1 + 2x_2 + 3x_3 &= 24 \\
x_1 + 8x_2 + x_3 &= 12 \\
2x_1 - 3x_2 + 15x_3 &= 30
\end{aligned} \qquad (3.71)$$

for x_1, x_2, and x_3. The first step in the solution is to construct the augmented matrix as follows:

$$\left(\begin{array}{ccc|c} 20 & 2 & 3 & 24 \\ 1 & 8 & 1 & 12 \\ 2 & -3 & 15 & 30 \end{array}\right)$$

We then multiply line 1 by 1/20 and subtract it from line 2 and this becomes the new line 2. Similarly, line 3 is replaced by 2/20 times line 1 subtracted from line 3. The result at this stage of the calculation is

$$\left(\begin{array}{ccc|c} 20 & 2 & 3 & 24 \\ 0 & 7.9000 & 0.8500 & 10.8000 \\ 0 & -3.2000 & 14.7000 & 27.6000 \end{array}\right)$$

We next set the (3,2) matrix element to zero by multiplying the new line two by $-3.2000/7.9000$ and subtracting the result from line 3 to get a new line 3, and the triangularized system of equations is now

$$\left(\begin{array}{ccc|c} 20 & 2 & 3 & 24 \\ 0 & 7.9000 & 0.8500 & 10.8000 \\ 0 & 0 & 15.0443 & 31.9747 \end{array}\right)$$

From the last equation we have

$$x_3 = \frac{31.9747}{15.0443} = 2.1254$$

This result is used in the second equation to give

$$x_2 = \frac{10.8000 - (0.8500)(2.1254)}{7.9000} = 1.1384$$

Finally, from the first equation we obtain

$$x_1 = \frac{24 - (3)(2.1254) - (2)(1.1384)}{20} = 0.7674$$

which completes the solution. These values may be inserted in Eq. 3.71 to verify that we have indeed found the correct solution.

This technique can be generalized for larger sets of equations. The procedure is simple and it is readily reduced to a computer algorithm.

The accuracy of the result obtained by the Gauss elimination procedure will depend on the relative sizes of the matrix elements involved. It works well as long as the diagonal elements, called the *pivots,* are large in magnitude compared to the others so that the absolute value of the multiplicative factor is always less than or equal to 1. However, if the magnitude of the multiplicative factor is very large, round-off errors are severe because the subtraction of large numbers to obtain small differences is involved. This can be avoided by interchanging rows if necessary to place the largest element of a given column on the diagonal so that the magnitude of the multiplicative factor will always be less than or equal to 1. This is called *partial pivoting*, and it is usual to include it in any routine that is intended for general use.

The *Gauss–Jordan method* is similar to that described above, except that the elimination process is extended to diagonalize the **A** matrix completely so that no back substitution is required:

$$\begin{pmatrix} a_{11}'' & 0 & 0 & \cdots \\ 0 & a_{22}'' & 0 & \cdots \\ 0 & 0 & a_{33}'' & \cdots \\ \vdots & \vdots & \vdots & \\ & & & a_{nn}'' \end{pmatrix} \begin{pmatrix} x_1 \\ x_2 \\ x_3 \\ \vdots \\ x_n \end{pmatrix} = \begin{pmatrix} c_1'' \\ c_2'' \\ c_3'' \\ \vdots \\ c_n'' \end{pmatrix} \tag{3.72}$$

It is interesting to note that each equation then defines a single unknown, and each equation is simply divided through by a_{ii}' to find the value of that particular element x_i. When this is done the $\mathbf{A}''$ matrix is just the unit matrix **I**. Now since $\mathbf{A}^{-1}\mathbf{A} = \mathbf{I}$, it follows that if these same operations used to generate **I** are performed on the unit matrix, the inverse $\mathbf{A}^{-1}$ will result.

To illustrate, let us further augment the example matrix used above to include the unit matrix,

$$\left(\begin{array}{ccc|c|ccc} 20 & 2 & 3 & 24 & 1 & 0 & 0 \\ 1 & 8 & 1 & 12 & 0 & 1 & 0 \\ 2 & -3 & 15 & 30 & 0 & 0 & 1 \end{array}\right)$$

and carry through the full Gauss–Jordan transformation. The result is

$$\left(\begin{array}{ccc|c|ccc} 20 & 0 & 0 & 15.3741 & 1.0349 & -0.3281 & -0.1851 \\ 0 & 7.9000 & 0 & 8.9934 & -0.04321 & 0.9771 & -0.05650 \\ 0 & 0 & 15.0443 & 31.9747 & -0.1203 & 0.4051 & 1 \end{array}\right)$$

We divide these equations through by 20, 7.9000, and 15.0443, respectively, to obtain

$$\left(\begin{array}{ccc|c|ccc} 1 & 0 & 0 & 0.7674 & 0.05175 & -0.01641 & -0.009256 \\ 0 & 1 & 0 & 1.1384 & -0.005470 & 0.1237 & -0.007152 \\ 0 & 0 & 1 & 2.1254 & -0.007993 & 0.02692 & 0.06647 \end{array}\right)$$

The fourth column is just the set of x_i, which checks with the result obtained above by

Gauss elimination with back substitution, and the last three columns contain the inverse matrix $\mathbf{A}^{-1}$. This may be checked by multiplying by the original $\mathbf{A}$ matrix, and we do indeed obtain the unit matrix as required:

$$\begin{pmatrix} 20 & 2 & 3 \\ 1 & 8 & 1 \\ 2 & -3 & 15 \end{pmatrix} \begin{pmatrix} 0.05175 & -0.01641 & -0.009256 \\ -0.005470 & 0.1237 & -0.007152 \\ -0.007993 & 0.02692 & 0.06647 \end{pmatrix} = \begin{pmatrix} 1.000 & 0.000 & 0.000 \\ 0.000 & 1.000 & 0.000 \\ 0.000 & 0.000 & 1.000 \end{pmatrix}$$

This discussion of the Gauss–Jordan method was included here because of its conceptual simplicity. It shows that the processes of solving linear equations and matrix inversion are indeed closely related as asserted above. As a practical matter, however, Gauss elimination with back substitution is generally preferred because fewer arithmetic operations are involved, and there are more efficient methods available for matrix inversion as well, although we will not go into them here.

3.5.3 The Gauss–Seidel Method

For some problems, particularly where very large sets of linear equations are involved, it is preferable to use an iterative procedure rather than one based upon an explicit solution. The Gauss–Seidel method makes use of an approximate solution with successive refinement by direct substitution. Let us again solve the example set defined by Eq. 3.71 to illustrate the method. We will use

$$\mathbf{x}^{(0)} = \begin{pmatrix} 0 \\ 0 \\ 0 \end{pmatrix}$$

as an initial guess. Putting these into Eq. 3.71 leads to

$$x_1^{(1)} = \frac{24 - 0 - 0}{20} = 1.2000$$

$$x_2^{(1)} = \frac{12 - (1.2000)(1) - 0}{8} = 1.3500$$

$$x_3^{(1)} = \frac{30 - (1.2000)(2) - (1.3500)(-3)}{15} = 2.1100$$

as the first iteration. These results are again inserted into Eq. 3.71 and the result is

$$x_1^{(2)} = \frac{24 - (1.3500)(2) - (2.1100)(3)}{20} = 0.7485$$

$$x_2^{(2)} = \frac{12 - (0.7485)(1) - (2.1100)(1)}{8} = 1.1427$$

$$x_3^{(2)} = \frac{30 - (0.7485)(2) - (1.1427)(-3)}{15} = 2.1287$$

and so forth. Note that the approximate values of x_i are updated within each cycle as soon as they are generated, rather than keeping them constant for the cycle and then changing them all at once for the next iteration. This is an important feature of the Gauss–Seidel method and it leads to much more rapid convergence than would otherwise be obtained.

We should also note that the system we are dealing with as an example has larger terms on the diagonal than elsewhere, and this is also an important feature. It can be shown that the Gauss–Seidel procedure is always convergent for a system of equations structured such that the magnitude of the diagonal terms is greater than the sum of the magnitudes of the off-diagonal terms (a *diagonally dominant* matrix). Others may converge as well, but the convergence will not be as rapid.

Example 3.11

Assume that we have a solution containing three materials X_1, X_2, and X_3, whose molar absorptivities at three different wavelengths λ_1, λ_2, and λ_3 are

	Molar Absorptivities (ϵ) at:		
	λ_1	λ_2	λ_3
X_1	600	50	80
X_2	70	900	100
X_3	40	75	750

If a given solution containing the three components has measured absorbances at λ_1, λ_2, and λ_3 of 0.63, 0.47, and 0.82, respectively, what is the concentration of each component?

Solution. To the extent that Beer's law is obeyed, the total absorbance is the sum of that for each of the individual components. For a particular wavelength i we can write

$$A_i = \epsilon_{i1}[X_1] + \epsilon_{i2}[X_2] + \epsilon_{i3}[X_3]$$

For our case this leads to the following set of equations for the three wavelengths:

$$600[X_1] + 70[X_2] + 40[X_3] = 0.63$$

$$50[X_1] + 900[X_2] + 75[X_3] = 0.47$$

$$80[X_1] + 100[X_2] + 750[X_3] = 0.82$$

We can write these in augmented matrix form as

$$\left(\begin{array}{ccc|c} 600 & 70 & 40 & 0.63 \\ 50 & 900 & 75 & 0.47 \\ 80 & 100 & 750 & 0.82 \end{array}\right)$$

Naturally, in this case the wavelengths are chosen at or near the (hopefully unique) maxima for the various components so that the matrix is diagonally dominant and readily solved. With the Gauss elimination procedure we have

$$\left(\begin{array}{ccc|c} 600 & 70 & 40 & 0.63 \\ 0 & 894.17 & 71.67 & 0.4175 \\ 0 & 0 & 737.40 & 0.6937 \end{array}\right)$$

and back substitution gives the results

$$[X_3] = \frac{0.6937}{737.40} = 9.4 \times 10^{-4}$$

$$[X_2] = \frac{0.4175 - (9.4 \times 10^{-4})(71.67)}{894.17} = 3.9 \times 10^{-4}$$

$$[X_1] = \frac{0.63 - (9.4 \times 10^{-4})(40) - (3.9 \times 10^{-4})(70)}{600} = 9.4 \times 10^{-4}$$

The solution may be checked by showing that these values satisfy the original equations.

3.6 SOLUTION OF EIGENVALUE PROBLEMS

The boundary conditions imposed on many of the equations of importance in the physical sciences limit the possible solutions to a set of discrete values. Wave equations are a familiar example, which admit only certain solutions corresponding to standing waves. The German adjective *eigen* is generally used to denote the characteristic solutions of these problems. In general the equations can be cast in the form

$$\hat{\mathcal{O}} f = ef \tag{3.73}$$

where $\hat{\mathcal{O}}$ is known as a mathematical *operator* and e is the *eigenvalue*. Solutions of Eq. 3.73 do not exist for arbitrary values of e, only for the discrete eigenvalues. For each eigenvalue e there is an associated function f, which is known as an *eigenfunction* of the operator $\hat{\mathcal{O}}$. There may be several eigenfunctions f corresponding to the same eigenvalue e. If there are n such functions, we say that there is an n-fold *degeneracy*.

A case of special interest to the chemist is where the operator $\hat{\mathcal{O}}$ represents the energy of a system written in terms of position coordinates and the momenta associated with them. The classical equations of motion written in this way are known as Hamilton's equations, and when the classical Hamiltonian function is converted into a quantum mechanical operator, it is known as the *Hamiltonian operator* $\hat{\mathcal{H}}$. The exact form of $\hat{\mathcal{H}}$ need not concern us here; what is important to note is that the eigenvalue E of the Hamiltonian operator $\hat{\mathcal{H}}$ represents the energy of the system. We generally write

$$\hat{\mathcal{H}}\psi = E\psi \tag{3.74}$$

to represent this relationship. This is the well-known (time-independent) Schrödinger wave equation (see Section 1.7). ψ is called the *wavefunction*, and it is defined such that ψ^2 is a probability function, corresponding classically to the intensity of the wave. The Hamiltonian operator belongs to a special class called *Hermitian operators*. The eigenvalues of Hermitian operators are real, although the eigenfunctions are not always so. Combinations of wavefunctions can be taken which are real, and these can then be used without loss of generality as a *basis set* for the calculation. In the subsequent discussion we will assume that we are dealing with real wavefunctions.

In many cases our principal interest is in obtaining the lowest energy E that is a solution of Eq. 3.74, that is, the *ground state*. One common method of solution makes use of the *variation method*. We multiply Eq. 3.74 on the left by ψ and divide through by ψ^2 to obtain

$$E = \frac{\psi \hat{\mathcal{H}} \psi}{\psi^2} \tag{3.75}$$

We can insert a wavefunction ψ in Eq. 3.75 and calculate E. If the wavefunction ψ is the true ground state eigenfunction, then the value of E calculated in this way is constant. For an arbitrary function, however, the value of E is not constant but varies as the coordinates change. Equation 3.75 may be integrated over the entire configuration space of the system to obtain

$$\langle E \rangle = \frac{\int \psi \hat{\mathcal{H}} \psi \, d\tau}{\int \psi^2 \, d\tau} \tag{3.76}$$

The numerator is basically just the local energy for a given set of coordinates weighted by the probability ψ^2, and the denominator is the total probability, so we recognize Eq. 3.76 as an expression for the average energy of the system. It can easily be demonstrated that a value of $\langle E \rangle$ calculated in this way can never be lower than the true ground state energy of the system. This result should be intuitively obvious, since we would not expect to be able to construct a ground state wavefunction which is better than any that nature could devise!

The variation method consists of writing a trial wavefunction ψ with certain parameters in it. We minimize the energy with respect to each parameter. According to the variation theorem, this is the best approximation to the ground state energy that can be calculated with a wavefunction of this particular form. If the trial wavefunction ψ is well chosen, we may in fact calculate the true ground state energy by this procedure. This is possible for very simple systems such as the hydrogen atom, but it cannot be done for complex atoms or molecules.

Although we cannot expect to obtain exact solutions for molecules, we can obtain good approximate solutions in many cases by expansion of the wavefunction ψ in terms of a simple set of functions. A common technique is to represent a molecular orbital (MO) ψ as a linear combination of atomic orbitals (LCAO) ϕ_i:

$$\psi = \sum_i c_i \phi_i \tag{3.77}$$

Here the c_i are the expansion coefficients that are to be determined by minimization of the calculated energy $\langle E \rangle$. Insertion of Eq. 3.77 in Eq. 3.76 gives

$$\langle E \rangle = \frac{\int \sum_i c_i \phi_i \hat{\mathcal{H}} \sum_j c_j \phi_j \, d\tau}{\int \sum_i c_i \phi_i \sum_j c_j \phi_j \, d\tau} = \frac{\sum_i \sum_j c_i c_j \int \phi_i \hat{\mathcal{H}} \phi_j \, d\tau}{\sum_i \sum_j c_i c_j \int \phi_i \phi_j \, d\tau}$$

We generally represent the *resonance integral* by

$$H_{ij} \equiv \int \phi_i \hat{\mathcal{H}} \phi_j \, d\tau$$

and the *overlap integral* by

$$S_{ij} \equiv \int \phi_i \phi_j \, d\tau$$

so that the equation above may be written

$$\sum_i \sum_j c_i c_j H_{ij} = \langle E \rangle \sum_i \sum_j c_i c_j S_{ij}$$

Implicit differentiation with respect to c_i gives

$$\sum_j c_j H_{ij} = \frac{\partial \langle E \rangle}{\partial c_i} \left(\sum_i \sum_j c_i c_j S_{ij}\right) + \langle E \rangle \sum_j c_j S_{ij}$$

where $i = 1, 2, 3, \ldots$. Since we wish to minimize the energy with respect to each of the coefficients, we set $\partial \langle E \rangle / \partial c_i = 0$, in which case we obtain a set of equations, called the *secular equations*, one for each value of i. The secular equations are of the form

$$\begin{aligned}
(H_{11} - \langle E \rangle S_{11})c_1 + (H_{12} - \langle E \rangle S_{12})c_2 + \cdots + (H_{1n} - \langle E \rangle S_{1n})c_n &= 0 \\
(H_{21} - \langle E \rangle S_{21})c_1 + (H_{22} - \langle E \rangle S_{22})c_2 + \cdots + (H_{2n} - \langle E \rangle S_{2n})c_n &= 0 \\
&\vdots \\
(H_{n1} - \langle E \rangle S_{n1})c_1 + (H_{n2} - \langle E \rangle S_{n2})c_2 + \cdots + (H_{nn} - \langle E \rangle S_{nn})c_n &= 0
\end{aligned} \tag{3.78}$$

This is a set of n linear equations, but there are in fact $n + 1$ unknowns (c_1, c_2, $\ldots$ c_n, $\langle E \rangle$), so a straightforward solution using the techniques of Section 3.5 is not possible. This becomes obvious, for example, by application of Cramer's rule. Since the constants on the right-hand side are all zeros, the only solution that can be obtained in this way is the trivial solution for which $c_1 = c_2 = \cdots = c_n = 0$. However, a nontrivial solution is possible if the determinant of the coefficients itself is equal to zero, which, in terms of the Cramer's solution, is an indeterminate form. The *secular determinant* is

$$\begin{vmatrix} (H_{11} - \langle E\rangle S_{11}) & (H_{12} - \langle E\rangle S_{12}) & \cdots & (H_{1n} - \langle E\rangle S_{1n}) \\ (H_{21} - \langle E\rangle S_{21}) & (H_{22} - \langle E\rangle S_{22}) & \cdots & (H_{2n} - \langle E\rangle S_{2n}) \\ \vdots & & & \\ (H_{n1} - \langle E\rangle S_{n1}) & (H_{n2} - \langle E\rangle S_{n2}) & \cdots & (H_{nn} - \langle E\rangle S_{nn}) \end{vmatrix} = 0 \qquad (3.79)$$

Expansion of the secular determinant gives an nth-order polynomial in $\langle E\rangle$, known as the *characteristic equation*. The characteristic equation has n roots, or values of $\langle E\rangle$ for which it is satisfied, and these n value of $\langle E\rangle$ are the eigenvalues. Associated with each is a set of coefficients $c_1, c_2, \ldots, c_n$ which define the eigenfunction. It will be noted that there are only $n - 1$ linearly independent secular equations once a value of $\langle E\rangle$ is fixed, so that a complete solution is impossible. The magnitude of the c's cannot be determined, only their $n - 1$ relative values $c_2/c_1, c_3/c_1, \ldots, c_n/c_1$. We generally require that their magnitude be fixed by the condition $c_1^2 + c_2^2 + \cdots + c_n^2 = 1$. This condition is called *normalization*, and it is a convenient choice for the magnitude. Since the square of the wavefunction is interpreted as a probability, each of the terms c_i^2 is then a measure of the relative contribution of that particular basis function ϕ_i to the total.

For Hermitian operators with real basis functions it is found that $H_{ij} = H_{ji}$, and it is apparent that $S_{ij} = S_{ji}$. Furthermore, we can chose the basis functions in such a way that $S_{ij} = 0$ for $i \neq j$ in which case they are *orthogonal*. They are generally constructed such that $S_{ij} = 1$ for $i = j$; that is, they are also normalized. This is called an *orthonormal basis set*, and its use greatly simplifies the process of solving the secular equations. These two conditions can be summarized by $S_{ij} = \delta_{ij}$, where δ_{ij}, known as the *Kronecker delta*, is defined to be 1 for $i = j$ and 0 for $i \neq j$.

In the above we have used the example of a molecular orbital calculation to introduce the features of eigenvalue problems, but they occur in many other contexts as well. Such equations are often represented in a compact matrix notation

$$(\mathbf{A} - \lambda\mathbf{I})\mathbf{c} = 0 \qquad (3.80)$$

which we will now adopt for our subsequent discussion. Here the eigenvalue of the matrix **A** is represented by λ. **I** is the unit matrix, and **c** is a column matrix representing the eigenfunction. It can be regarded as a vector in n-dimensional space defining the contribution of each of the n "coordinates" for a particular eigenvalue λ_i. Thus it is usual to call it the *eigenvector*. The eigenvectors corresponding to two different discrete eigenvalues are orthogonal; that is, $\sum_k c_{ik}c_{jk} = 0$. Those for degenerate eigenvalues may not be orthogonal by direct solution, but they can be made so by a process called *orthogonalization*.

A simple example will be used to illustrate the above. Assume **A** to be a symmetric matrix of order 3,

$$\mathbf{A} = \begin{pmatrix} 2 & 1 & 0 \\ 1 & 3 & 2 \\ 0 & 2 & -5 \end{pmatrix} \tag{3.81}$$

Then the corresponding secular determinant is

$$\begin{vmatrix} 2-\lambda & 1 & 0 \\ 1 & 3-\lambda & 2 \\ 0 & 2 & -5-\lambda \end{vmatrix}$$

Expansion of the secular determinant gives the characteristic equation

$$\lambda^3 - 24\lambda + 33 = 0$$

The eigenvalues and their corresponding eigenvectors are

$$\lambda_1 = -5.4793 \qquad \lambda_2 = 1.5219 \qquad \lambda_3 = 3.9574$$

$$\mathbf{c}_1 = \begin{pmatrix} 0.03114 \\ -0.23294 \\ 0.97199 \end{pmatrix} \qquad \mathbf{c}_2 = \begin{pmatrix} 0.89439 \\ -0.42764 \\ -0.13114 \end{pmatrix} \qquad \mathbf{c}_3 = \begin{pmatrix} 0.44621 \\ 0.87342 \\ 0.19502 \end{pmatrix}$$

These solutions were obtained by solving the characteristic equation numerically for the eigenvalues, and these were inserted in Eq. 3.80 to find c_2/c_1 and c_3/c_1. The normalization condition $c_1^2 + c_2^2 + c_3^2 = 1$ was then used to complete the solution. It is noted that the sum of the diagonal elements (called the *trace*) is not changed in this process (the *diagonal sum rule*), which provides a useful check on the accuracy of the eigenvalues: $-5.4793 + 1.5219 + 3.9574 = 0.0000 = 2 + 3 - 5$.

Solution of the secular determinant by expansion is always possible, but it becomes impractical for large eigenvalue problems, and alternative techniques must be used. We will now outline two other methods of solution.

3.6.1 The Power Iteration Method

There are cases where only the eigenvalue of largest absolute value is needed. For example, in many quantum mechanical calculations our only interest is in the ground state. The *power iteration method*, which we will now describe, is then a convenient approach.

If we multiply any arbitrary vector $\mathbf{v}$ by the eigenvalue matrix $\mathbf{A}$ as defined above, we generate some new vector $\mathbf{u}$:

$$\mathbf{u} = \mathbf{A}\mathbf{v}$$

In general, **u** differs from **v** in both its magnitude and in its direction in the n-dimensional vector space that it spans. The magnitude as we have seen is of little concern since the vectors can be of arbitrary length and they are usually subjected to a normalization step in any case. The specification of the direction, however, is of vital importance. Now if **v** happens to be an exact eigenvector of **A**, then

$$\mathbf{u} = \mathbf{A}\mathbf{v} = \lambda\mathbf{v}$$

In this case the direction is not changed through multiplication by **A**. Only the magnitude is changed by the factor λ.

Rather than using the original basis functions to define the vector **v**, we can just as well use the eigenvectors of **A**. This follows from the fact that there are n of them, and they are mutually orthogonal. They are constructed as linear combinations of the original basis functions, which is tantamount to a rotation of coordinates in the n-dimensional vector space, so they may simply be regarded as an alternative set of orthogonal coordinates. Thus we may write

$$\mathbf{v} = x_1\mathbf{c}_1 + x_2\mathbf{c}_2 + x_3\mathbf{c}_3 + \cdots \tag{3.82}$$

where $\mathbf{c}_1, \mathbf{c}_2, \mathbf{c}_3, \ldots$ are the eigenvectors of the matrix **A**, and $x_1, x_2, x_3, \ldots$ are numerical constants.

If we multiply Eq. 3.82 from the left by the matrix **A**, the result may be written

$$\begin{aligned}\mathbf{u} = \mathbf{A}\mathbf{v} &= x_1\lambda_1\mathbf{c}_1 + x_2\lambda_2\mathbf{c}_2 + x_3\lambda_3\mathbf{c}_3 + \cdots \\ &= \lambda_1\left[x_1\mathbf{c}_1 + \frac{\lambda_2}{\lambda_1}x_2\mathbf{c}_2 + \frac{\lambda_3}{\lambda_1}x_3\mathbf{c}_3 + \cdots\right]\end{aligned}$$

We repeat this process n times and the result is

$$\mathbf{u}_n = \lambda_1^n\left[x_1\mathbf{c}_1 + \left(\frac{\lambda_2}{\lambda_1}\right)^n x_2\mathbf{c}_2 + \left(\frac{\lambda_3}{\lambda_1}\right)^n x_3\mathbf{c}_3 + \cdots\right]$$

Now if λ_1 is the eigenvalue of largest absolute value, the ratios (λ_i/λ_1) are less than 1 in magnitude, and with n sufficiently large, all terms except the first will eventually become negligible. In the limit, $\mathbf{u}_n \to (\lambda_1^n x_1)\mathbf{c}_1$ and we see that the direction of the vector **u** no longer changes with subsequent multiplication by **A**, but only its magnitude. Thus the procedure involves an iterative calculation in which each step corresponds to multiplication by a higher power of the matrix **A**—hence the name "power iteration method." If the vector is normalized at each step, convergence is indicated when a constant vector is obtained, and the constant multiplicative factor is the desired eigenvalue λ_1.

Example 3.12

Write a BASIC computer program to solve by the power iteration method for the eigenvalue of largest absolute value and its associated eigenvector correct to five significant digits, and apply it to the eigenvalue matrix **A** given by Eq. 3.81.

Solution. It is first noted that whenever an eigenvalue matrix is solved, at least one of the eigenvalues will be larger than the largest diagonal element, and at least one of the eigenvalues will be smaller than the smallest diagonal element (provided that the off-diagonal elements are not all equal to zero). Thus, we expect that the eigenvalue of greatest absolute value for the matrix given by Eq. 3.81 should be somewhat less than -5. The A_{33} diagonal element is the one that is nearest in value to λ_1, and we therefore expect the third basis function to represent the dominant contribution. Thus (0,0,1) is a logical choice for an initial trial vector $\mathbf{v}$. The following program is designed to find λ_1 and $\mathbf{c}_1$ by the power iteration method, and when it is executed, it converges to the values of λ_1 and $\mathbf{c}_1$ given in the text above.

```
10 REM--PROGRAM PIM--SOLN OF EIGENVALUE MATRIX BY POWER ITERATION METHOD
20 READ N                                          'ORDER OF MATRIX TO BE SOLVED
30 FOR I=1 TO N : FOR J=1 TO N : READ A(I,J) : NEXT J : NEXT I
40 FOR I=1 TO N : READ V(I) : NEXT I              'READ INITIAL TRIAL VECTOR
50 PRINT"ITER. NO.","ERROR","EIGENVALUE"
60 M=0 : ER=1 : WHILE ER>.00001  'M = ITERATION NUMBER, ER = ERROR LIMIT
70 NF=0 : FOR I=1 TO N : U(I)=0                   'NF = NORMALIZATION FACTOR
80 FOR J=1 TO N : U(I)=U(I)+A(I,J)*V(J) : NEXT J
90 NF=NF+U(I)*U(I) : NEXT I : E=U(3)/V(3)                  ' E = EIGENVALUE
100 NF=SQR(NF) : FOR I=1 TO N : U(I)=U(I)/NF : NEXT I      'NORMALIZATION
110 ER=ABS((ABS(U(3))-ABS(V(3)))/V(3)) : M=M+1 : PRINT M,ER,E
130 FOR I=1 TO N : V(I)=U(I) : NEXT I     'REDEFINE V FOR NEXT ITERATION
140 WEND
150 PRINT : PRINT"CONVERGED IN";M;"ITERATIONS TO EIGENVALUE =";E
160 PRINT"AND EIGENVECTOR ="; : FOR I=1 TO N : PRINT V(I); : NEXT I
170 DATA 3,2,1,0,1,3,2,0,2,-5,0,0,1
```

3.6.2 The Jacobi Rotation Method

Assume that we have a real symmetric matrix $\mathbf{A}$ whose eigenvalues λ_i are desired:

$$\begin{vmatrix} A_{11}-\lambda & A_{12} & A_{13} & \cdots & A_{1n} \\ A_{12} & A_{22}-\lambda & A_{23} & \cdots & A_{2n} \\ A_{13} & A_{23} & A_{33}-\lambda & \cdots & A_{3n} \\ \vdots & & & & \\ A_{1n} & A_{2n} & A_{3n} & \cdots & A_{nn}-\lambda \end{vmatrix} = 0$$

Presumably we could perform some operation on $\mathbf{A}$ that would reduce it to the diagonal form

$$\begin{vmatrix} A'_{11} - \lambda & 0 & 0 & \cdots & 0 \\ 0 & A'_{22} - \lambda & 0 & \cdots & 0 \\ 0 & 0 & A'_{33} - \lambda & \cdots & 0 \\ \vdots & & & & \\ 0 & 0 & 0 & & A'_{nn} - \lambda \end{vmatrix} = 0$$

in which case we have $\lambda_1 = A'_{11}$, $\lambda_2 = A'_{22}$, . . . , $\lambda_n = A'_{nn}$ as the roots. This *matrix diagonalization* can always be carried out in principle by a *similarity transformation* of the form

$$\mathbf{S}^{-1}\mathbf{A}\mathbf{S} = \mathbf{D} \tag{3.83}$$

where $\mathbf{D}$ is a diagonal matrix. We do not know the matrix $\mathbf{S}$ at the outset, but the process may be viewed as a generalized rotation of coordinates in an n-dimensional space. The diagonalization is usually carried out in practice by an iterative procedure known as the *Jacobi rotation method,* which we will now describe.

First let us consider a rotation of coordinates in a plane, as depicted in Figure 3.14. By simple trigonometry we can see that the new coordinates x', y' are related to the old x, y by

$$x = \cos\theta\, x' + \sin\theta\, y'$$
$$y = -\sin\theta\, x' + \cos\theta\, y'$$

or

$$\begin{pmatrix} x \\ y \end{pmatrix} = \begin{pmatrix} \cos\theta & \sin\theta \\ -\sin\theta & \cos\theta \end{pmatrix} \begin{pmatrix} x' \\ y' \end{pmatrix}$$

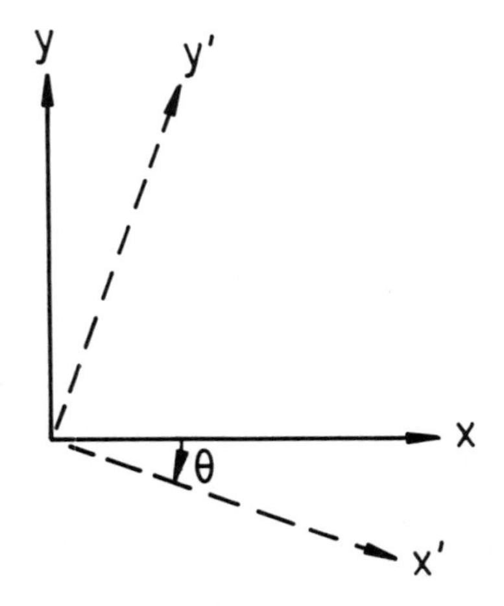

Figure 3.14 Rotation of planar Cartesian coordinates through the angle θ.

Note that in this special case

$$\begin{pmatrix} \cos\theta & -\sin\theta \\ \sin\theta & \cos\theta \end{pmatrix}\begin{pmatrix} \cos\theta & \sin\theta \\ -\sin\theta & \cos\theta \end{pmatrix} = \begin{pmatrix} 1 & 0 \\ 0 & 1 \end{pmatrix}$$

That is, the inverse is the same as the transpose, which, by definition, makes this an *orthogonal transformation*. Now performing this rotation of coordinates by a similarity transformation on the arbitrary matrix

$$\begin{pmatrix} A_{11} & A_{12} \\ A_{12} & A_{22} \end{pmatrix}$$

gives

$$\begin{pmatrix} \cos\theta & -\sin\theta \\ \sin\theta & \cos\theta \end{pmatrix}\begin{pmatrix} A_{11} & A_{12} \\ A_{12} & A_{22} \end{pmatrix}\begin{pmatrix} \cos\theta & \sin\theta \\ -\sin\theta & \cos\theta \end{pmatrix}$$

$$= \begin{pmatrix} A_{11}\cos^2\theta + A_{22}\sin^2\theta - 2A_{12}\sin\theta\cos\theta & (A_{11} - A_{22})\sin\theta\cos\theta + A_{12}(\cos^2\theta - \sin^2\theta) \\ (A_{11} - A_{22})\sin\theta\cos\theta - A_{12}(\sin^2\theta - \cos^2\theta) & A_{11}\sin^2\theta + A_{22}\cos^2\theta + 2A_{12}\sin\theta\cos\theta \end{pmatrix}$$

We see that the off-diagonal elements vanish if

$$(A_{11} - A_{22})\sin\theta\cos\theta + A_{12}(\cos^2\theta - \sin^2\theta) = 0$$

Using the trigonometric identities

$$2\sin\theta\cos\theta = \sin 2\theta$$

$$\cos^2\theta - \sin^2\theta = \cos 2\theta$$

we find that the diagonalization is accomplished with θ chosen such that

$$\frac{A_{11} - A_{22}}{2}\sin 2\theta + A_{12}\cos 2\theta = 0$$

$$\tan 2\theta = \frac{2A_{12}}{A_{22} - A_{11}} \tag{3.84}$$

When dealing with larger matrices, the Jacobi rotation is performed one step at a time starting with the largest off-diagonal matrix element and reducing it to zero, then on to the next largest, and so forth. The rotation matrix itself is constructed from a unit matrix by replacing the A_{pp}, A_{qq}, A_{pq}, and A_{qp} elements by $\cos\theta$, $\cos\theta$, $\sin\theta$, and $-\sin\theta$, where p and q define the row and column in which the largest off-diagonal element is located and θ is the angle for the rotation as defined by

$$\tan 2\theta = \frac{2A_{pq}}{A_{qq} - A_{pp}} \tag{3.85}$$

Note that as this procedure is carried out, all the matrix elements along the row p and the column q are altered, so an element that is set to zero in one step of the diagonalization process will no longer be zero after subsequent operations are performed. What we do find, however, is that as each operation is performed the magnitude of the diagonal elements increases at the expense of the off-diagonal ones, so that if this iterative process is carried far enough, the matrix may be diagonalized to any desired degree of accuracy. If we accumulate the product of each of the individual Jacobi rotation matrices,

$$\mathbf{S} = \prod_i \mathbf{S}_i \tag{3.86}$$

it is found that the matrix **S** is just the matrix required to diagonalize the initial matrix **A** in one step by a similarity transformation as indicated by Eq. 3.83, and it contains all the eigenvectors $\mathbf{c}_i$ as *columns*. Because we are dealing with a generalized rotation of coordinates, we would expect to maintain the total magnitude of the roots involved. Thus the trace is invariant under the transformation, and this provides an important check on the accuracy of the calculation, as noted previously. Furthermore, the sums of the squares of the matrix elements in a given row or column of the transformation matrix **S** are equal to unity; that is, they are normalized.

We will illustrate the Jacobi matrix diagonalization process through the first few stages for the matrix

$$\begin{pmatrix} 8 & 2 & 6 \\ 2 & 3 & 5 \\ 6 & 5 & 12 \end{pmatrix}$$

In this case we see that the 1,3 matrix element (the 6) is the largest of the off-diagonal elements, so we have

$$\tan 2\theta = \frac{2(6)}{12 - 8} = 3$$

or

$$\begin{aligned} 2\theta &= 71.565 \\ \theta &= 35.783 \\ \cos\theta &= 0.81124 \\ \sin\theta &= 0.58471 \end{aligned}$$

Therefore, the $\mathbf{S}_1$ matrix is

$$\begin{pmatrix} 0.81124 & 0 & 0.58471 \\ 0 & 1 & 0 \\ -0.58471 & 0 & 0.81124 \end{pmatrix}$$

and the desired similarity transformation is

$$\begin{pmatrix} 0.81124 & 0 & -0.58471 \\ 0 & 1 & 0 \\ 0.58471 & 0 & 0.81124 \end{pmatrix} \begin{pmatrix} 8 & 2 & 6 \\ 2 & 3 & 5 \\ 6 & 5 & 12 \end{pmatrix} \begin{pmatrix} 0.81124 & 0 & 0.58471 \\ 0 & 1 & 0 \\ -0.58471 & 0 & 0.81124 \end{pmatrix}$$

with the result

$$\begin{pmatrix} 3.6754 & -1.3011 & 0 \\ -1.3011 & 3 & 5.2256 \\ 0 & 5.2256 & 16.3246 \end{pmatrix}$$

Now the largest off-diagonal element is the 2,3 element, so

$$\tan 2\theta = \frac{2(5.2256)}{16.3246 - 3} = 0.78436$$

or

$$2\theta = 38.1092$$
$$\theta = 19.0546$$
$$\cos\theta = 0.94521$$
$$\sin\theta = 0.32647$$

This gives

$$\begin{pmatrix} 1 & 0 & 0 \\ 0 & 0.94521 & 0.32647 \\ 0 & -0.32647 & 0.94521 \end{pmatrix}$$

as the $\mathbf{S}_2$ matrix and the similarity transformation leads to

$$\begin{pmatrix} 3.6754 & -1.2298 & -0.42476 \\ -1.2298 & 1.1951 & 0 \\ -0.42476 & 0 & 18.1295 \end{pmatrix}$$

Performing one more rotation on the 1,2 element, for which

$$\tan 2\theta = \frac{2(-1.2298)}{1.1951 - 3.6754} = 0.99162$$

gives the result

$$\begin{pmatrix} 4.1818 & 0 & -0.39277 \\ 0 & 0.68873 & -0.16172 \\ -0.39277 & -0.16172 & 18.1295 \end{pmatrix}$$

This process is repeated until the off-diagonal elements are as small as desired. Obviously, the hand calculation is rather tedious, but the procedure is readily reduced to a computer algorithm.

In adapting the above for computer calculation, one small modification is usually made to prevent division by zero when the two elements A_{pp} and A_{qq} are equal (see Eq. 3.85). We define

$$\mu = -A_{pq} \qquad \nu = \frac{A_{pp} - A_{qq}}{2}$$

and

$$\omega = \mathrm{sn}(\nu)\frac{\mu}{\sqrt{\mu^2 + \nu^2}}$$

where

$$\begin{aligned} \mathrm{sn}(\nu) &= +1 \qquad \text{if} \quad \nu \geq 0 \\ &= -1 \qquad \text{if} \quad \nu < 0 \end{aligned}$$

We then calculate $\sin\theta$ and $\cos\theta$ by

$$\sin\theta = \frac{\omega}{[2 + 2(1 - \omega^2)^{1/2}]^{1/2}} \tag{3.87}$$

and

$$\cos\theta = (1 - \sin^2\theta)^{1/2} \tag{3.88}$$

and division by zero is thereby avoided.

Example 3.13

Using the BASIC Jacobi matrix diagonalization program given in Figure 3.15, the diagonalization begun above was completed and the result is

$$\lambda_1 = 4.1708 \qquad \lambda_2 = 0.68723 \qquad \lambda_3 = 18.1420$$

$$\mathbf{c}_1 = \begin{pmatrix} 0.83786 \\ -0.35170 \\ -0.41749 \end{pmatrix} \qquad \mathbf{c}_2 = \begin{pmatrix} 0.13835 \\ 0.87664 \\ -0.46083 \end{pmatrix} \qquad \mathbf{c}_3 = \begin{pmatrix} 0.52806 \\ 0.32835 \\ 0.78316 \end{pmatrix}$$

Verify the solution, the diagonal sum, and normalization conditions described in the text above, and show that the similarity transformation indicated by Eq. 3.83 diagonalizes the original matrix.

Solution. The solution may be verified by putting the eigenvalues together with their associated eigenvectors, one by one, into Eq. 3.80. The result is

$$(8 - 4.1708)(0.83786) + (2)(-0.35170) + (6)(-0.41749) = 0.0000$$

$$(2)(0.13835) + (3 - 0.68723)(0.87664) + (5)(-0.46083) = 0.0000$$

```
10 REM--PROGRAM JAC--DIAGONALIZATION OF A REAL SYMMETRIC MATRIX BY THE
20 REM--JACOBI METHOD.  ONLY THE UPPER TRIANGULAR PORTION OF THE MATRIX
30 REM--IS GIVEN.  N IS THE ORDER OF THE MATRIX.  ON RETURN FROM THE
40 REM--SUBROUTINE THE EIGENVALUES ARE CONTAINED IN LAMDA, AND THE
50 REM--EIGENVECTORS ARE STORED COLUMNWISE IN S.
60 DATA 3,8,2,6,3,5,12
70 READ N
80 FOR I=1 TO N : FOR J=I TO N
90 READ A(I,J) : A(J,I)=A(I,J) : NEXT J : NEXT I
100 GOSUB 200
110 FOR I=1 TO N : PRINT LAMDA(I)
120 FOR J=1 TO N : PRINT S(J,I), : NEXT J
130 PRINT : NEXT I
140 STOP
200 IN=0 : V=0
210 FOR I=1 TO N : V=V-A(I,I)*A(I,I)
220 FOR J=1 TO N : S(I,J)=0 : V=V+A(I,J)*A(I,J) : NEXT J
230 S(I,I)=1 : NEXT I
240 V1=SQR(V) : VF=V1*1E-08
260 V1=V1/N
270 FOR Q=2 TO N : QM1=Q-1
280 FOR P=1 TO QM1 : APQ=A(P,Q) : MU=-APQ
290 IF V1>=ABS(MU) THEN 460
300 IN=1 : APP=A(P,P) : AQQ=A(Q,Q)
310 NU=(APP-AQQ)/2
320 IF NU<0 THEN MU=-MU
330 OMEGA=MU/SQR(MU*MU+NU*NU)
340 SI=OMEGA/SQR(2*(1+SQR(1-OMEGA*OMEGA)))
350 CO=SQR(1-SI*SI)
360 FOR I=1 TO N : AIP=A(I,P) : AIQ=A(I,Q)
370 A(I,P)=CO*AIP-SI*AIQ : A(I,Q)=SI*AIP+CO*AIQ
380 AIP=S(I,P) : AIQ=S(I,Q)
390 S(I,P)=CO*AIP-SI*AIQ : S(I,Q)=SI*AIP+CO*AIQ
400 NEXT I
410 AIP=2*APQ*SI*CO
420 A(P,P)=CO*CO*APP+SI*SI*AQQ-AIP
430 A(Q,Q)=SI*SI*APP+CO*CO*AQQ+AIP
440 A(P,Q)=0 : A(Q,P)=0
450 FOR I=1 TO N : A(P,I)=A(I,P) : A(Q,I)=A(I,Q) : NEXT I
460 NEXT P : NEXT Q
470 IF IN=0 THEN 490
480 IN=0 : GOTO 270
490 IF V1>VF THEN 260
500 FOR I=1 TO N : APP=A(I,I) : LAMDA(I)=APP : AIP=SQR(ABS(APP))
510 FOR P=1 TO N : A(P,I)=S(P,I)*AIP
520 NEXT P : NEXT I
530 RETURN
```

Figure 3.15 BASIC program JAC, for diagonalization of an eigenvalue matrix by the Jacobi rotation method.

$$(6)(0.52806) + (5)(0.32835) + (12 - 18.1420)(0.78316) = -0.0001$$

The diagonal sum (trace invariance) condition may be written

$$4.1708 + 0.68723 + 18.1420 = 8 + 3 + 12$$
$$23.0000 = 23$$

and the normalization conditions are

$$(0.83786)^2 + (-0.35170)^2 + (-0.41749)^2 = 1.0000$$
$$(0.13835)^2 + (0.87664)^2 + (-0.46083)^2 = 1.0000$$
$$(0.52806)^2 + (0.32835)^2 + (0.78316)^2 = 1.0000$$

Finally, we can easily construct $\mathbf{S}^{-1}$ since it is just the transpose of $\mathbf{S}$ (i.e., it is obtained by simply interchanging rows and columns). The similarity transformation Eq. 3.83 is then

$$\begin{pmatrix} 0.83786 & -0.35170 & -0.41749 \\ 0.13835 & 0.87664 & -0.46083 \\ 0.52806 & 0.32835 & 0.78316 \end{pmatrix} \begin{pmatrix} 8 & 2 & 6 \\ 2 & 3 & 5 \\ 6 & 5 & 12 \end{pmatrix} \begin{pmatrix} 0.83786 & 0.13835 & 0.52806 \\ -0.35170 & 0.87664 & 0.32835 \\ -0.41749 & -0.46083 & 0.78316 \end{pmatrix}$$
$$= \begin{pmatrix} 4.1708 & 0.0000 & 0.0000 \\ 0.0000 & 0.68723 & 0.0000 \\ 0.0000 & 0.0000 & 18.1420 \end{pmatrix}$$

Example 3.14

The π-electron levels in unsaturated hydrocarbons can be crudely calculated using the simple or *Hückel molecular orbital* (HMO) method, according to which the *Coulomb integral*

$$H_{ii} = \int \phi_i \hat{\mathcal{H}} \phi_i \, d\tau$$

is assumed to be identical for each carbon atom, represented by the parameter α, and the *resonance integral*

$$H_{ij} = \int \phi_i \hat{\mathcal{H}} \phi_j \, d\tau$$

is taken to be equal to a parameter β for nearest-neighbor atoms and zero otherwise. Here we treat only the π electrons explicitly; they are assumed to move in the fixed core made up of all the nuclei and the σ electrons. Thus for butadiene $CH_2{=}CH{-}CH{=}CH_2$, for example, the eigenvalue matrix that must be solved is

$$\mathbf{A} = \begin{pmatrix} \alpha & \beta & 0 & 0 \\ \beta & \alpha & \beta & 0 \\ 0 & \beta & \alpha & \beta \\ 0 & 0 & \beta & \alpha \end{pmatrix}$$

Typical values of α and β as derived by fitting to spectroscopic and thermodynamic data are $\alpha \approx -146$ kJ mol^{-1} and $\beta \approx -84$ kJ mol^{-1}. Solve the eigenvalue matrix for butadiene using the Jacobi diagonalization program of Figure 3.15.

(a) At what wavelength is the lowest transition among the π levels expected to occur?

(b) How much delocalization energy is predicted for this molecule?

(c) How does the number of nodes in the wavefunction correlate with the energy of the molecular orbital?

Solution. Inserting the values of α and β in the eigenvalue matrix given above and solving with the Jacobi matrix diagonalization program given in Figure 3.15 gives the following results:

$$\lambda_1 = -10.085 \qquad \lambda_2 = -281.91 \qquad \lambda_3 = -94.085 \qquad \lambda_4 = -197.91$$

$$\mathbf{c}_1 = \begin{pmatrix} 0.37175 \\ -0.60150 \\ 0.60150 \\ -0.37175 \end{pmatrix} \quad \mathbf{c}_2 = \begin{pmatrix} 0.37175 \\ 0.60150 \\ 0.60150 \\ 0.37175 \end{pmatrix} \quad \mathbf{c}_3 = \begin{pmatrix} -0.60150 \\ 0.37175 \\ 0.37175 \\ -0.60150 \end{pmatrix} \quad \mathbf{c}_4 = \begin{pmatrix} -0.60150 \\ -0.37175 \\ 0.37175 \\ 0.60150 \end{pmatrix}$$

It is noted that α occurs in each diagonal term, and the levels are symmetrically disposed above and below its value. Thus if we subtract α from each of the eigenvalues, we have the following schematic energy-level diagram with the four π electrons assigned as indicated according to molecular *aufbau* principles:

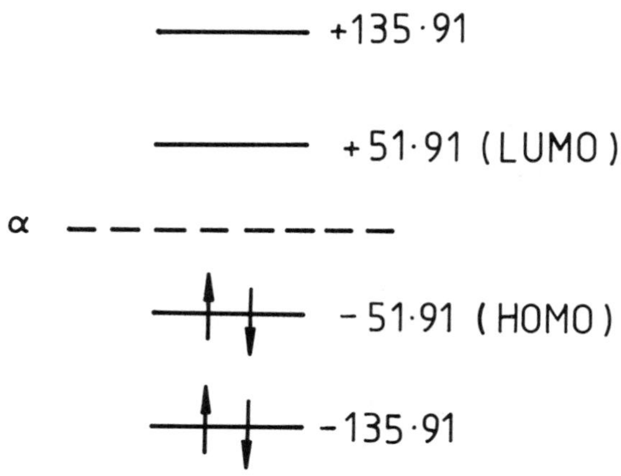

+135·91

+51·91 (LUMO)

α - - - - - - - -

↑↓ −51·91 (HOMO)

↑↓ −135·91

(a) The lowest π-electron transition is from the highest occupied molecular orbital (HOMO) to the lowest unoccupied molecular orbital (LUMO), which are indicated above. The transition is predicted to occur at $51.91 - (-51.91) = 103.83$ kJ mol^{-1}, which corresponds to a wavelength of 1152 nm.

(b) The secular determinant that must be solved for a completely localized electron, as for example in the ethylene molecule, is

$$\begin{vmatrix} \alpha - \lambda & \beta \\ \beta & \alpha - \lambda \end{vmatrix}$$

This has the roots $\lambda = \alpha \pm \beta$ and since both α and β are negative, the root with the + sign is lowest. Thus the energy of a localized π electron is $-84\ \text{kJ mol}^{-1}$ below the value α, and the delocalization energy (described as resonance stabilization energy in valence bond terms) is

$$2(-135.91) + 2(-51.91) - 4(-84) = -40\ \text{kJ mol}^{-1}$$

(c) The electron density is determined by the square of the wavefunction. If the wavefunction changes sign in moving along the molecular chain, there is a point where the density is equal to zero and this constitutes a *node*. By inspection of the wavefunction of the lowest-energy state, we see that the components are all positive, so this is a nodeless wavefunction. The next-higher state shows one sign change, and it has one node; the next has two nodes, and the highest level has three nodes. This is a general result: The higher the energy level, the greater the number of nodes in the wavefunction.

3.7 NUMERICAL SOLUTION OF DIFFERENTIAL EQUATIONS

The chemist is often confronted with the solution of differential equations, and the techniques discussed in Chapter 1 are not always applicable. In particular, we sometimes need the solution of sets of differential equations. For example, a complete description of the kinetics of many complex reactions requires the simultaneous solution of systems of coupled first-order differential equations. In most cases the writing of such equations is almost trivial. For example, the reaction scheme

$$\mathrm{A} \underset{k_2}{\overset{k_1}{\rightleftharpoons}} \mathrm{B} \xrightarrow{k_3} \mathrm{C} \xrightarrow{k_4} \mathrm{D}$$

requires solution of the simultaneous equations

$$\frac{d[\mathrm{A}]}{dt} = -k_1[\mathrm{A}] + k_2[\mathrm{B}]$$

$$\frac{d[\mathrm{B}]}{dt} = k_1[\mathrm{A}] - (k_2 + k_3)[\mathrm{B}]$$

$$\frac{d[\mathrm{C}]}{dt} = k_3[\mathrm{B}] - k_4[\mathrm{C}]$$

$$\frac{d[\mathrm{D}]}{dt} = k_4[\mathrm{C}]$$

While the writing of these equations for a given case is quite simple, the solution certainly is not. We will now describe some techniques that can be used to obtain a solution in numerical form. The solutions generally depend on the specification of a set of *initial conditions*, which for the example above implies known values of the various concentrations at time $t = 0$.

The discussion that follows will be confined to first-order differential equations, but the methods are more general. This is a result of the fact that differential equations of higher order can always be expressed as systems of first-order equations. For example, suppose that we have a second-order equation of the form

$$\frac{d^2y}{dx^2} = f\left(\frac{dy}{dt}, y, x\right) \tag{3.89}$$

If we define the variable $z = dy/dx$, we may replace Eq. 3.89 by the equivalent set of first-order equations

$$\begin{aligned} \frac{dz}{dx} &= f(z, y, x) \\ \frac{dy}{dx} &= z \end{aligned} \tag{3.90}$$

This technique can be readily generalized to include equations of even higher order.

For simplicity we will assume that we are dealing with a single first-order differential equation that may be written in the general form

$$\frac{dy}{dx} = F(x, y) \tag{3.91}$$

The techniques used to obtain a numerical solution can be readily expanded to include any number of such simultaneous equations. Since Eq. 3.91 as written defines the slope of y as a function of x, we may write

$$F(x, y) \approx \frac{\Delta y}{\Delta x} = \frac{y_{m+1} - y_m}{x_{m+1} - x_m}$$

If we take the integration interval to be $h = x_{m+1} - x_m$, then

$$y_{m+1} = y_m + hF(x_m, y_m) \tag{3.92}$$

We see from this equation that from a known value of y_m at x_m we can solve for y_{m+1}; this value of y_{m+1} together with $F(x_{m+1}, y_{m+1})$ can then be used in Eq. 3.92 to solve for y_{m+2}, and so forth. Continuing in this way we can generate the entire solution for an arbitrary range of x values based on an initial value y_0 at x_0. This technique is known as *Euler's method*.

Euler's method is not much used in practice because it is too simplistic and prone to error. Because each value of y depends directly on the preceding one, even rather small errors in the initial stages of the integration can propagate to an unacceptable level very quickly. It will be observed that the Euler formula is in fact just the first two terms of a Taylor series expansion,

$$y_{m+1} = y_m + h\left(\frac{dy}{dx}\right)_{x=x_m} + \frac{h^2}{2!}\left(\frac{d^2y}{dx^2}\right)_{x=x_m} + \frac{h^3}{3!}\left(\frac{d^3y}{dx^3}\right)_{x=x_m} + \cdots$$

$$= y_m + hF(x_m, y_m) + \frac{h^2}{2}F'(x_m, y_m) + \frac{h^3}{6}F''(x_m, y_m) + \cdots \tag{3.93}$$

and it is obvious that the integration by this same general approach could be carried to any desired degree of accuracy by retaining a sufficient number of terms in the expansion. The Taylor series method can be used in a few cases where the higher-order derivatives are particularly easy to evaluate, but it, like Euler's method, is of very limited practical value despite its conceptual simplicity.

3.7.1 Runge-Kutta Methods

It is possible to accomplish the same purpose as the Taylor's series expansion while avoiding the evaluation of the higher-order derivatives, and this forms the basis of the Runge-Kutta methods. The problem is analogous to that considered in Section 3.4.2, where Simpson's rule of integration was developed. In this case we write

$$y_{m+1} = y_m + h(ak_1 + bk_2 + \cdots)$$

Now there are various ways of choosing the constants $a, b, \ldots$ and the factors $k_1, k_2, \ldots$ such that the Taylor's series result is matched to an arbitrary order. This gives rise to a family of solutions called the Runge-Kutta formulas. One of the most widely used includes four terms

$$y_{m+1} = y_m + \frac{h}{6}(k_1 + 2k_2 + 2k_3 + k_4) \tag{3.94}$$

with

$$\begin{aligned}
k_1 &= F(x_m, y_m) \\
k_2 &= F\left(x_m + \frac{1}{2}h, y_m + \frac{1}{2}hk_1\right) \\
k_3 &= F\left(x_m + \frac{1}{2}h, y_m + \frac{1}{2}hk_2\right) \\
k_4 &= F(x_m + h, y_m + hk_3)
\end{aligned} \tag{3.95}$$

It can be shown that this fourth-order Runge-Kutta formula is the equivalent of the Taylor's series expansion to fourth order, and as such it represents a very accurate solution. In essence, rather than using a single value of the slope at x_m, y_m to extrapolate to the new point as is done by the Euler method, corrections to the slope (as represented by the terms k_2, k_3, and k_4) are made using intermediate points as indicated by Eqs. 3.94 and 3.95.

It is noted that in the special case where $F(x,y)$ is actually a function of x only, the Runge–Kutta formula represented by Eq. 3.94 reduces to Simpson's rule. This should not be surprising since the derivation of the two is very similar.

As an illustration of the use of the fourth-order Runge–Kutta technique to solve a set of coupled differential equations, we solve the complex kinetics problem given in the text above using the BASIC program RKC shown in Figure 3.16. The numerical solution is generated at 25 time intervals using a step size $h = 0.05$ for the rate constants $k_1 = 1$, $k_2 = 0.5$, $k_3 = 0.3$, and $k_4 = 0.1$, with initial concentrations $[A]_0 = 0.75$, $[B]_0 = [C]_0 = [D]_0 = 0$. The output for these conditions is shown; changes of the rate constants and/or initial concentrations can readily be made in the DATA statement (300), and the differential equations (statements 250–280) can easily be modified to treat other reaction mechanisms.

```
10 REM--PROGRAM RKC--FOURTH ORDER RUNGE-KUTTA INTEGRATION OF
20 REM---------------COUPLED DIFFERENTIAL EQUATIONS
30 READ NR,NC     'NR=NUMBER OF RATE CONSTS, NC=NUMBER OF COMPONENTS
40 FOR K=1 TO NR : READ K(K) : NEXT K              ' READ RATE CONSTS
50 FOR K=1 TO NC : READ C(K) : NEXT K             'READ INITIAL CONCNS
60 READ H,N                    'H=STEP SIZE, N=NUMBER OF TIME INTERVALS
70 M=1/H : T=0      'M=NUMBER OF INTEGRATION STEPS PER TIME INTERVAL
80 LPRINT"CALCULATION OF KINETICS FOR RATE CONSTANTS"
90 FOR K=1 TO NR : LPRINT K(K), : NEXT K : LPRINT
100 LPRINT : LPRINT" T","    [A]","    [B]","    [C]","    [D]"
110 LPRINT T, : FOR K=1 TO NC : LPRINT C(K), : NEXT K
120 FOR I=1 TO N : FOR J=1 TO M                            'MAIN LOOP
130 FOR K=1 TO NC : X(K)=C(K) : NEXT K
140 GOSUB 250 : FOR K=1 TO NC : D(K)=Y(K) : NEXT K
150 FOR K=1 TO NC : X(K)=C(K)+Y(K)/2 : NEXT K
160 GOSUB 250 : FOR K=1 TO NC : D(K)=D(K)+2*Y(K) : NEXT K
170 FOR K=1 TO NC : X(K)=C(K)+Y(K)/2 : NEXT K
180 GOSUB 250 : FOR K=1 TO NC : D(K)=D(K)+2*Y(K) : NEXT K
190 FOR K=1 TO NC : X(K)=C(K)+Y(K) : NEXT K
200 GOSUB 250 : FOR K=1 TO NC : D(K)=D(K)+Y(K) : NEXT K
210 FOR K=1 TO NC : C(K)=C(K)+D(K)*H/6 : NEXT K
220 T=T+H : NEXT J
230 LPRINT INT(T+.05), : FOR K=1 TO NC : LPRINT C(K), : NEXT K
240 NEXT I : STOP
250 Y(1)=-K(1)*X(1)+K(2)*X(2)     'DIFFERENTIAL EQUATIONS SUBROUTINE
260 Y(2)=K(1)*X(1)-(K(2)+K(3))*X(2)
270 Y(3)=K(3)*X(2)-K(4)*X(3)
280 Y(4)=K(4)*X(3)
290 RETURN
300 DATA 4,4,1,.5,.3,.1,.75,0,0,0,.05,25
```

(a)

Figure 3.16 (a) BASIC program RKC for the solution of a set of coupled first-order differential equations by the fourth-order Runge-Kutta method.

```
CALCULATION OF KINETICS FOR RATE CONSTANTS
 1              .5             .3             .1

T               [A]            [B]            [C]            [D]
0               .75            0              0              0
1               .475317        .193115        7.533093E-02   6.236951E-03
2               .3260973       .2547205       .148593        2.058896E-02
3               .2394597       .2585137       .2105296       4.149664E-02
4               .1850409       .2388051       .258765        6.738855E-02
5               .1480154       .2112876       .293842        9.685459E-02
6               .1210075       .1828975       .3173956       .1286989
7               .100236        .1564889       .3313342       .1619403
8               8.367208E-02   .133044        .337491        .1957922
9               7.015619E-02   .1127133       .3374937       .2296361
10              .058973        9.530144E-02   .3327302       .2629945
11              4.964392E-02   8.049038E-02   .3243578       .2955069
12              4.182472E-02   6.793895E-02   .3133272       .3269081
13              3.525336E-02   .0573247       .3004107       .3570102
14              .0297222       4.835921E-02   .2862298       .3856875
15              2.506255E-02   4.079138E-02   .2712807       .4128641
16              2.113516E-02   .0344057       .2559548       .4385029
17              1.782403E-02   2.901864E-02   .2405584       .4625975
18              1.503204E-02   2.447457E-02   .2253271       .4851648
19              1.267759E-02   2.064183E-02   .2104393       .5062398
20              .010692        1.740919E-02   .1960271       .5258701
21              9.017434E-03   1.468275E-02   .1821847       .5441136
22              7.60516E-03    1.238328E-02   .1689764       .5610335
23              6.41408E-03    1.044391E-02   .1564426       .5766978
24              5.409545E-03   8.808262E-03   .1446043       .5911763
25              4.562337E-03   7.428777E-03   .133468        .6045393
```

(b)

Figure 3.16 (*Continued*) (b) Sample output for BASIC program RKC.

Example 3.15

Write a BASIC computer program that will generate a numerical solution by the Runge-Kutta method for the first-order differential equation

$$\frac{dy}{dx} = x^2 - 3y$$

at intervals of $h = 0.1$ in the range $x = 0$ to $x = 2$, using the initial conditions $x_0 = 0$ and $y_0 = 2$.

Solution. A simple program making use of a fixed value of h is as follows:

```
10 DEF FND(X,Y)=X*X-3*Y
20 INPUT X,Y,H,N
30 LPRINT "XO =";X,"YO =";Y,"H =";H,"N =";N
40 LPRINT : LPRINT "  X","Y"
50 LPRINT USING "#.####      ";X,Y
60 FOR I=1 TO N
70 K1=H*FND(X,Y)
80 K2=H*FND(X+H/2,Y+K1/2)
90 K3=H*FND(X+H/2,Y+K2/2)
100 K4=H*FND(X+H,Y+K3)
110 X=X+H
120 Y=Y+(K1+2*(K2+K3)+K4)/6
130 LPRINT USING "#.####      ";X,Y
140 NEXT I
```

Execution of the program above with $h = 0.1$ and $n = 20$ gives the following results:

```
XO = 0          YO = 2          H = .1          N = 20

  X               Y
0.0000          2.0000
0.1000          1.4820
0.2000          1.1000
0.3000          0.8205
0.4000          0.6187
0.5000          0.4761
0.6000          0.3791
0.7000          0.3177
0.8000          0.2844
0.9000          0.2735
1.0000          0.2811
1.1000          0.3040
1.2000          0.3401
1.3000          0.3875
1.4000          0.4452
1.5000          0.5122
1.6000          0.5877
1.7000          0.6714
1.8000          0.7628
1.9000          0.8616
2.0000          0.9677
```

A second run with $h = 0.05$ and $n = 40$ shows that the solution above is accurate to ± 0.0001.

3.7.2 Predictor-Corrector Methods

Euler, Taylor's series, and Runge–Kutta integrations are all examples of *single-step methods*. By this we mean that each point is generated from a single preceding point. This is not the most efficient approach for computer evaluation, however. Since the solution exists for all prior points, they can be stored and used to make the extrapolation. In other words, a calculation based upon several prior points is preferable to one involving evaluation of several functions at a given point. The reason for the saving in computer time is clear; it is only necessary to make a single evaluation of each new point since the previous ones are already available.

To illustrate, let us apply the Euler formula Eq. 3.92, but double the interval of integration to obtain

$$y_{m+1} = y_{m-1} + 2hF(x_m, y_m) \tag{3.96}$$

We see that this makes use of the solution at $m - 1$ and m to predict the solution at $m + 1$. Solution by the Euler method is crude, but once we have an approximate solution this may be used to improve the result. We might, for example, retain the Euler form but use a mean value of the slope calculated using the value of y_{m+1} obtained from Eq. 3.96 to get

$$y_{m+1} = y_m + h\left(\frac{F(x_m, y_m) + F(x_{m+1}, y_{m+1})}{2}\right) \tag{3.97}$$

This would presumably yield an improved value of y_{m+1}, and once this is obtained it can be reinserted in Eq. 3.97 to get a still better estimate, and so forth. This iterative process can be continued in principle to self-consistency, that is, until there is no further change in y_{m+1}—in practice, this means until

$$|y_{m+1}^{(i+1)} - y_{m+1}^{(i)}| < \epsilon$$

where ϵ is a small positive quantity representing the convergence criterion.

Equation 3.96 is an explicit equation for y_{m+1} called the *predictor equation*, and Eq. 3.97 is an implicit equation in y_{m+1} called the *corrector equation*. A single application of Eq. 3.96 predicts the value of y_{m+1}, and then several applications of the corrector equation 3.97 are made to improve the value to the desired degree of accuracy.

The equations above were introduced because they illustrate the principles of the predictor-corrector method in a simple way, but in practice one generally uses higher-order formulas for improved accuracy. One of the most popular is the *Adams–Moulton scheme*:

$$\begin{aligned} &\text{predictor: } y_{m+1} = y_m + \frac{h}{24}(55y'_m - 59y'_{m-1} + 37y'_{m-2} - 9y'_{m-3}) \\ &\text{corrector: } y_{m+1} = y_m + \frac{h}{24}(9y'_{m+1} + 19y'_m - 5y'_{m-1} + y'_{m-2}) \end{aligned} \tag{3.98}$$

The predictor-corrector method has two very important advantages over the Runge–Kutta method. Even with the iteration using the corrector formula the time required is apt to be considerably less. Furthermore, because of the nature of the iterative calculation, one has an immediate indication of the level of accuracy achieved at each step. With the Runge–Kutta method, the only recourse is to cut down the size of the interval h and repeat the calculation to see what effect this has on the result.

An advantage of the Runge–Kutta method is that it is self-starting. The predictor-corrector method, on the other hand, requires that some other technique be used to obtain the first few points before the predictor formula can be applied. Sophisticated programs often use the Runge–Kutta technique for the initial points and then switch to the predictor-corrector method to complete the integration.

3.8 MONTE CARLO METHODS

Much of what we deal with in the study of chemistry may properly be described as *stochastic processes*, which depend upon events occurring in a strictly random fashion governed by the laws of chance. Some examples are the collision of molecules in the gas phase, the diffusion of solute molecules to form a uniform solution, the drift of molecules through a chromatographic column, or the unimolecular reaction of excited molecules. The events are entirely random on a molecular level, and we can never predict at the outset just which molecule may react or collide. Even so, when dealing with fantastically large numbers of molecules, such as Avogadro's number, statistical laws become so precise that we can indeed regard chemistry as an exact science, that is, as long as we are dealing with the behavior of matter in bulk.

Numerical techniques have been developed for the solution of many diverse problems which are based solely on the laws of chance. They have come to be known as *Monte Carlo techniques* by analogy to processes of gambling in a casino, which we would like to believe are also governed strictly by the laws of chance. We will briefly sketch a few chemical applications of Monte Carlo techniques in the following. In addition to problems of diffusion, integration, and kinetics, which are introduced here, Monte Carlo methods have also been used to treat such problems as liquid structures, chromatographic processes, reaction dynamics, and scattering of atoms and ions from surfaces.

3.8.1 The Random Walk Problem

The random walk process may be visualized as follows: Assume that you are standing on a sidewalk and are completely undecided as to whether you would like to go north or south. If you just take a random step, you have a 50–50 chance of going in either direction. After taking one step, you take another, but if this is also chosen at random you may end up back where you started, or at two paces from your origin. Imagine that you continue this process of taking random steps. Now then, let us examine the question: How many random steps does it take, on the average, to advance N paces in either direction from the origin?

The following BASIC program will provide a visualization of the process:

```
10 INPUT N,M
20 KEY OFF
30 RANDOMIZE TIMER
40 SCREEN 1
50 CLS
60 DIM A(319)
70 FOR J=1 TO M
80 X=0 : I=0
90 R=INT(RND*2)
100 IF R=0 THEN X=X-1 ELSE X=X+1
110 I=I+1
120 IF ABS(X)<=N THEN 90
130 IF I>319 THEN 160
140 A(I)=A(I)+1
150 PSET(I,199-A(I))
160 NEXT J
```

Many additional features could be added to improve the presentation, such as the construction of axes and labels, but the program has purposely been kept very simple to illustrate the essential features.

The program begins with the input of an integer N representing the number of paces from the origin, and a second integer M representing the number of trials. The random numbers generated in statement 90 are either 0's or 1's. We take 0 to represent a step in the negative direction and 1 to represent a step in the positive direction (see statement number 100). If the distance from the origin has not yet reached N, then the program loops back from statement 120 to statement 90, where another random step is generated. Once the goal of N paces from the origin has been achieved, a new point is indicated at the appropriate location along the abscissa. It is unlikely that very high numbers will occur, but statement 130 is included in case a very large number that would be off the screen should occur; it is simply rejected.

It is fascinating to watch the figure develop for various values of N and M. Figure 3.17 shows the distribution for the preceding program as written with $N = 5$ and $M = 2000$.

The analogy between the above and a molecule in solution that may diffuse from a given location by random encounters should be obvious. A graphical illustration such as that of Figure 3.17 is instructive, but there are instances where we might want to have explicit formulas to describe the results of such random walk problems.

Suppose that we have a molecule in solution which undergoes some $2N$ random encounters with neighboring molecules, each of which displaces it a distance D from the point of encounter. We once again consider only one dimension; the results can readily be extended to three dimensions if desired. Our question is now phrased as

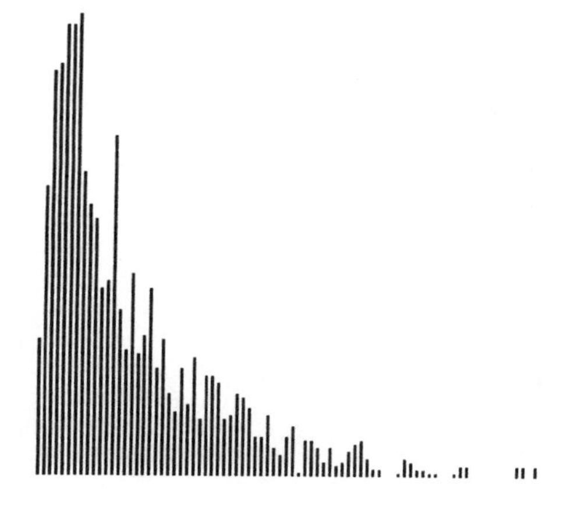

Figure 3.17 Appearance of the screen after execution of the random walk program described in the text for $M = 2000$.

follows: What is the probability that the molecule will have moved a distance $2RD$ from the origin, where $-N \le R \le N$? This implies that there have been $N + R$ displacements to the right and $N - R$ displacements to the left:

$$(N + R)(+D) + (N - R)(-D) = 2RD$$

The probability distribution for the molecule as a function of R after the $2N$ encounters is found to be

$$P(R) = C_{N+R}^{2N}\left(\frac{1}{2}\right)^{N+R}\left(\frac{1}{2}\right)^{N-R} \tag{3.99}$$

where

$$C_y^x = \frac{x!}{y!(x - y)!} \tag{3.100}$$

is the number of combinations of x objects taken y at a time.

The distribution function given by Eq. 3.99 is plotted for three different values of N in Figure 3.18. The result would look the same if, rather than considering the probability distribution for a single molecule, we started with a large number of molecules and simply tabulated the number of them at various locations after each had experienced a certain number of collisions. Since the average number of collisions for a given molecule is determined by the elapsed time, the curves of Figure 3.18 are closely related to the one-dimensional diffusion profiles of a solute as a function of time.

3.8.2 Monte Carlo Integration

Monte Carlo techniques are occasionally used for numerical quadrature when dealing with integrands that are particularly cumbersome. To illustrate the procedure we will consider a rather simple integral. Once the principles are understood, they can be easily extended to more complex cases.

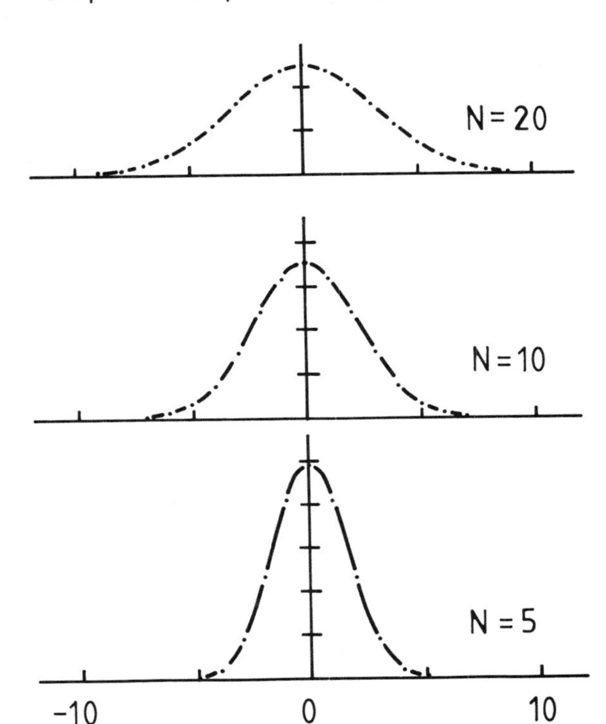

Figure 3.18 Plot of $P(R)$ versus R according to Eq. 3.99 for the three values of N indicated.

We will use the integral

$$I = \int_0^1 \sqrt{1 - x^2}\, dx \tag{3.101}$$

as our example. Note that the integrand $f(x) = \sqrt{1 - x^2}$ goes from $f(x) = 1$ to $f(x) = 0$ as x goes from 0 to 1, and this is in fact just the first quadrant of a circle whose radius is unity, as illustrated in Figure 3.19. We can calculate this integral with a Monte Carlo method using the following program:

```
10 LPRINT"   N","INTEGRAL"
20 INPUT N
30 RANDOMIZE TIMER
40 I=0
50 FOR J=1 TO N
60 X=RND
70 Y=RND
80 FX=SQR(1-X*X)
90 IF Y<=FX THEN I=I+1
100 NEXT J
110 LPRINT : LPRINT N,I/N
120 GOTO 20
```

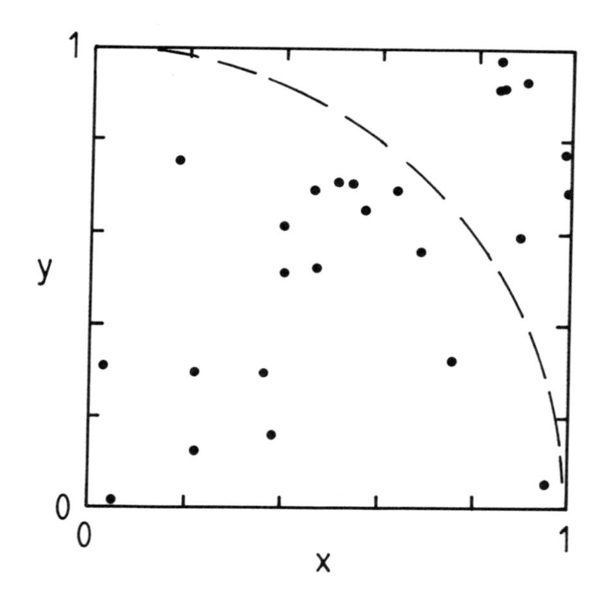

Figure 3.19 Integration of the function $f(x) = \sqrt{1 - x^2}$ according to the Monte Carlo method. The dashed curve is a plot of the integrand, and the dots show the coordinates for 25 random points within the range $0 \leq x \leq 1$ and $0 \leq y \leq 1$.

The procedure is as follows: We first input N, which is used to specify the number of random points for which the integrand is to be evaluated. For each cycle a pair of random numbers is generated (statements 60 and 70). These numbers are in the range 0 to 1, and they can be used to specify random coordinates within the unit square indicated in Figure 3.19. We count all the points that lie within the curve and divide this by the total number of points. This represents the fraction of the area that is enclosed by the integrand:

$$I = \frac{N_{\text{int}}}{N_{\text{tot}}} A_{\text{tot}}$$

In our particular case the total area is unity since we have a unit square, and for the case illustrated in Figure 3.19, where 25 points were used, we see that the approximate value of the integral is $18/25 = 0.72$.

This process can be extended to include any number of random coordinates, and we would expect that the integral would become exact if an infinite number of points were taken. In practice we must limit ourselves to some large finite number that is determined by the nature of the integrand.

We now return to a consideration of the program above, which carries out this Monte Carlo integration for an arbitrary number of random points. In statement 80 we use the random value of X to evaluate the integrand, and then we use this in statement 90 to see whether the random coordinates fall within the area bounded by the integrand or not. If so, we augment I, which serves as a counter for the number of random coordinates that fall within the area. Execution of the program for a few values of N leads to the following results:

N	INTEGRAL
10	.8
50	.82
100	.83
500	.79
1000	.775
5000	.79

In this case the precise value of the integral is readily obtained for comparison. The area in question is just one-fourth of a unit circle, and since $A_{tot} = \pi R^2/4$, the exact result is $\pi/4 = 0.785398. . . .$ We see that the Monte Carlo integration is converging toward this result but that a very large number of points will apparently be required if a precise result is desired.

3.8.3 Monte Carlo Kinetics Calculations

As a final example of the application of Monte Carlo techniques, we will consider how they can be applied to the treatment of complex reaction schemes. The basic concept is very simple. Suppose that we have a solution of molecules A_1 which undergo unimolecular decay to molecules A_2. We might construct a model for this reaction in which we set up a grid with a very large number of points. We randomly distribute a set of 1's representing the reactant molecules on these points. Now we generate a large but fixed number of random numbers spanning the entire set of points defined by our grid. If we find a 1 at that point, it is changed to a 2, indicating that reaction of that molecule has occurred; if the random number defines a point with a 2 or a 0 (which may represent a solvent molecule), no action is taken. It should be obvious that if this process is repeated over and over, a smaller number of 1's will be located each time the fixed number of random points is generated, and the number of "hits" may be taken as an indication of the course of the reaction. The time coordinate will be arbitrary, but the form of the concentration versus time curve should be correct except for an appropriate scale factor.

We can easily generalize the procedure described above for complex reaction schemes. If there are many reactants, we would use a set of numbers 1, 2, 3, . . . to represent them. Each can be changed into any other depending on the reactions involved, and the number of trials can be adjusted to reflect the relative sizes of the rate constants. If a process is bimolecular, pairs of random numbers are generated and only if a "hit" is scored by both are the digits changed to indicate that reaction has occurred.

As an illustration of the process, the following program has been constructed to treat the general reaction scheme represented by

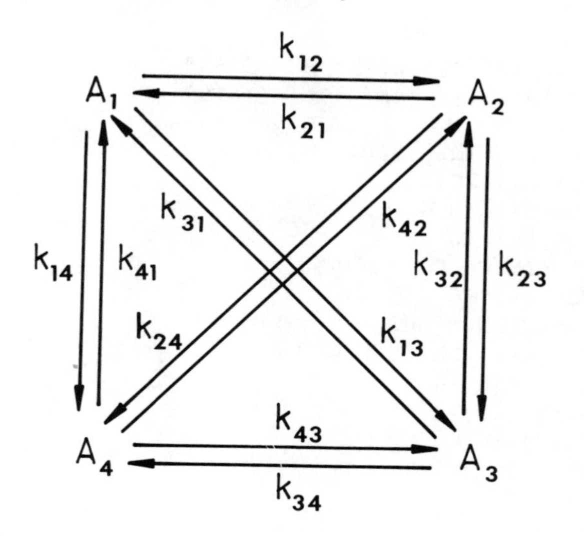


```
10 DIM MO(1000)
20 DATA 520,150,0,0
30 DATA 0,5,0,0
40 DATA 0,0,3,4
50 DATA 0,0,0,0
60 DATA 0,2,0,0
70 DATA 25,50
80 RANDOMIZE TIMER
90 FOR I=1 TO 4 : READ A(I) : NEXT I        'READ INITIAL CONCNS
100 FOR J=1 TO 4                               'READ RATE CONSTS
110 FOR L=1 TO 4 : READ K(J,L) : NEXT L
120 NEXT J
130 READ IM,IL                                 'READ LOOP PARAMS
140 LPRINT"MONTE CARLO CALCULATION WITH RATE CONSTANT MATRIX:"
150 FOR J=1 TO 4
160 FOR L=1 TO 4 : LPRINT K(J,L), : NEXT L
170 LPRINT
180 NEXT J
190 LPRINT
200 L=0                                        'GRID CONSTRUCTION
210 FOR I=1 TO 4
220 N=A(I)
230 IF N=0 THEN 280
240 FOR J=1 TO N
250 MO(L+J)=I
```

```
260 NEXT J
270 L=L+N
280 NEXT I
290 I=0
300 LPRINT" I"," A1"," A2"," A3"," A4"
310 LPRINT I,A(1),A(2),A(3),A(4)
320 I=I+1                                          'OUTER LOOP
330 IF I>IM THEN STOP
340 FOR L=1 TO IL                                  'INNER LOOP
350 FOR M=1 TO 4
360 FOR N=1 TO 4
370 IF K(M,N)=0 THEN 450
380 FOR J=1 TO K(M,N)                              'REACTION LOOP
390 R=INT(RND*1001)
400 IF MO(R)<>M THEN 440
410 A(M)=A(M)-1
420 A(N)=A(N)+1
430 MO(R)=N
440 NEXT J
450 NEXT N
460 NEXT M
470 NEXT L
480 LPRINT I,A(1),A(2),A(3),A(4)
490 GOTO 320
```

In statement number 10 the grid is defined to contain 1000 points, and statement 20 contains the initial concentrations of A_1, A_2, A_3, and A_4. They are treated here as integers, but they may be multiplied by an appropriate scale factor to convert them to corresponding concentrations. These initial concentrations are read in statement 90. The rate constants are contained in statements 30 to 60 in matrix form; the diagonal terms have no physical meaning and are required to be zero as this particular program is constructed. The off-diagonal terms are integers representing the relative sizes of the various rate constants. They are read in statements 100 to 120. IM and IL, which are read in statement 130, are parameters that define the number of cycles. The first is the number of times the concentrations are printed along the time axis, and the second is the number of times the program cycles through the process of generating changes in the concentrations for each specification of the time. Unless IL is a fairly large number, the statistical fluctuations will be severe and the actual course of the reaction will not be faithfully represented.

The loop over statements 200 to 280 is for the construction of the grid. It will be observed that the digits are not put into the MO array in the random manner indicated above, but this is of no consequence. Since the numbers are to be generated in a purely random fashion, their actual location within the grid does not matter. What is important is that the requisite number of digits for each component be inserted in MO, and this is accomplished by the loop indicated. We require here only that the sum of the components not exceed the number of grid points; if the sum is less, the extra points are zeros, which may be regarded as solvent molecules, as indicated above.

Statement 290 initializes the loop for the IM time values, and statements 300 and 310 print initial concentrations. The main iterative loop is contained in statements 340 to 470. A careful study of these statements will show that they carry out the process described above in a very general way as long as all the reactions are unimolecular.

As a specific example, the program above has been executed for the reaction scheme

$$A_1 \xrightarrow{k_1} A_2 \xrightarrow{k_2} A_3 \qquad A_2 \underset{k_4}{\overset{k_3}{\rightleftharpoons}} A_4$$

with $k_1 = 5$, $k_2 = 3$, $k_3 = 4$, and $k_4 = 2$, with the initial concentrations $[A_1]_0 = 520$, $[A_2]_0 = 150$, and $[A_3]_0 = [A_4]_0 = 0$. The result is

```
MONTE CARLO CALCULATION WITH RATE CONSTANT MATRIX:
 0          5          0          0
 0          0          3          4
 0          0          0          0
 0          2          0          0

 I          A1         A2         A3         A4
 0          520        150        0          0
 1          420        182        27         41
 2          329        209        56         76
 3          263        208        90         109
 4          201        217        116        136
 5          165        199        139        167
 6          131        180        173        186
 7          105        170        202        193
 8          72         153        238        207
 9          54         138        260        218
 10         43         129        273        225
 11         36         125        286        223
 12         27         104        311        228
 13         22         94         321        233
 14         16         89         337        228
 15         10         91         351        218
 16         8          84         367        211
 17         6          83         374        207
 18         5          82         387        196
 19         2          77         399        192
 20         2          78         410        180
 21         1          73         415        181
```

22	1	61	425	183
23	1	58	429	182
24	1	47	444	178
25	1	48	451	170

Example 3.16

Show that the spread of molecules from a point source by a diffusion-controlled process is proportional to $t^{1/2}$, where t is the time.

Solution. We can use the one-dimensional diffusion profiles given in Figure 3.18 to estimate the time dependence of the spreading process. We assume that the collision rate is constant, so the number of collisions is directly proportional to the time. This indicates that the widths of the curves of Figure 3.18 should be in the "theoretical" ratio $(5)^{1/2}:(10)^{1/2}:(20)^{1/2} = 1.00:1.41:2.00$ if the process is proportional to $t^{1/2}$.

We will take the width at half-height for the distribution curves as a measure of the spread of molecules in the diffusion process. Careful measurement of the curves shows that the widths are 3.92, 5.48, and 7.52 for the $N = 5$, $N = 10$, and $N = 20$ curves, respectively. These occur in "experimental" ratios of 1.00:1.40:1.92, which agree quite closely with the predicted ratios and confirms a $t^{1/2}$ dependence for the diffusion rate.

SUGGESTED ADDITIONAL READING

1. F. S. Acton, *Numerical Methods That Work*, Harper & Row, Publishers, Inc., New York, 1970.
2. R. Butler and E. Kerr, *An Introduction to Numerical Methods*, Sir Isaac Pitman & Sons Ltd., London, 1962.
3. Richard Hamming, *Numerical Methods for Scientists and Engineers*, McGraw-Hill Book Company, New York, 1973.
4. A. S. Householder, *Principles of Numerical Analysis*, McGraw-Hill Book Company, New York, 1953.
5. K. Jeffrey Johnson, *Numerical Methods in Chemistry*, Marcel Dekker, Inc., New York, 1980.
6. A. C. Norris, *Computational Chemistry—An Introduction to Numerical Methods*, John Wiley & Sons, Inc., New York, 1981.

EXERCISES

3.1. The hydrogen ion concentration of a weak monoprotic acid in aqueous solution satisfies the equation

$$[H^+]^3 + K_a[H^+]^2 - (K_w + K_aC_a)[H^+] - K_aK_w = 0$$

Determine $[H^+]$ for each of the following by graphical solution.

(a) 1.00×10^{-5} *M* HCN ($pK_a = 9.216$)
(b) 1.00×10^{-4} *M* phenol ($pK_a = 9.998$)
(c) 1.00×10^{-4} *M* NH_4Cl ($pK_a = 9.245$)
(d) 1.00×10^{-5} *M* NH_3OHBr ($pK_a = 5.96$)

3.2. The hydrogen ion concentration of a weak diprotic acid in aqueous solution satisfies the equation

$$[H^+]^4 + K_1[H^+]^3 + (K_1K_2 - K_w - K_1C_a)[H^+]^2 - (K_1K_w + 2K_1K_2C_a)[H^+] - K_1K_2K_w = 0$$

Determine $[H^+]$ for each of the following by graphical solution of the equation above.

(a) 1.00×10^{-5} *M* succinic acid ($pK_1 = 4.207$, $pK_2 = 5.635$)
(b) 1.00×10^{-6} *M* succinic acid
(c) 1.00×10^{-5} *M* carbonic acid ($pK_1 = 6.352$, $pK_2 = 10.33$)
(d) 1.00×10^{-7} *M* carbonic acid

3.3. Graph the logarithmic concentrations of each of the following pairs of metal ions as functions of $\log[S^{2-}]$, and state whether a quantitative separation (99.9% complete) is possible by sulfide precipitation.

(a) 0.10 *M* Ag^+ and 0.10 *M* Hg^{2+} ($pK_{sp} = 49.2$ for Ag_2S and 52.4 for HgS)
(b) 1.0×10^{-3} *M* Pb^{2+} and 1.0×10^{-2} *M* Zn^{2+} ($pK_{sp} = 27.9$ for PbS and 23.8 for ZnS)
(c) 1.0×10^{-3} *M* Fe^{2+} and 0.10 *M* Mn^{2+} ($pK_{sp} = 17.2$ for FeS and 12.6 for MnS)
(d) 1.0×10^{-2} *M* Co^{2+} and 0.10 *M* Fe^{2+} ($pK_{sp} = 20.4$ for CoS)

3.4. Write the proton condition for aqueous solutions of each of the following.

(a) Sodium formate
(b) Ammonium carbonate
(c) Sodium hydrogen oxalate
(d) Trimethylammonium chloride

3.5. Construct a logarithmic concentration diagram for a 0.010 *M* aqueous solution of benzoic acid ($pK_a = 4.212$) as a function of pH, and determine the following.

(a) The benzoic acid and benzoate concentrations at pH = 3
(b) The benzoic acid and benzoate concentrations at pH = 5
(c) The pH of a 0.010 *M* solution of benzoic acid
(d) The pH of a 0.010 *M* solution of sodium benzoate

3.6. Construct a logarithmic concentration diagram for a 1.0×10^{-3} *M* aqueous solution of oxalic acid ($pK_1 = 1.271$, $pK_2 = 4.266$) as a function of pH, and determine the following.

(a) The $H_2C_2O_4$ concentration if the pH is adjusted to 5.0
(b) The pH of a 1.0×10^{-3} *M* $H_2C_2O_4$ solution
(c) The pH of a 1.0×10^{-3} *M* $NaHC_2O_4$ solution
(d) The pH of a 1.0×10^{-3} *M* $Na_2C_2O_4$ solution

3.7. Solve the equation given in Exercise 3.1 by the method of successive approximations for each of the solutions indicated, using the graphical result as the starting point.

3.8. Solve the equation given in Exercise 3.2 by the method of successive approximations for each of the solutions indicated, using the graphical result as the starting point.

3.9. For an aqueous solution of a weak acid, it may generally be assumed that $[OH^-] \ll [H^+]$, in which case the following approximate relationship is valid:

$$K_a = \frac{[H^+]^2}{C_a - [H^+]}$$

Solve this equation by hand using the method of successive approximations for each of the following, using $[H^+] \approx \sqrt{K_a C_a}$ for the initial estimate of $[H^+]$.

(a) 0.100 M hydrofluoric acid ($pK_a = 3.14$)
(b) 1.00×10^{-3} M hydrocyanic acid ($pK_a = 9.216$)
(c) 0.0100 M formic acid ($pK_a = 3.752$)
(d) 1.00×10^{-4} M benzoic acid ($pK_a = 4.212$)

3.10. Write a BASIC computer program to carry out the calculations of Exercise 3.9, and run it for each case to check the hand calculations.

3.11. For an aqueous solution of the salt BA, the hydrogen ion concentration is given by

$$[H^+]^4 + (K_a + K_a' + C)[H^+]^3 + (K_a K_a' - K_w)[H^+]^2 - (K_a K_a' C + K_w(K_a + K_a'))[H^+] - K_a K_a' K_w = 0$$

where B is the acid form of a weak base with dissociation constant K_a', and A is the anion of a weak acid with dissociation constant K_a. Solve this equation by hand for each of the following cases by the method of successive approximations.

(a) 0.100 M methylammonium acetate ($pK_a = 10.624$ for the methylammonium ion and 4.757 for acetic acid)
(b) 0.0100 M trimethylammonium acetate ($pK_a = 9.796$ for the trimethylammonium ion and 4.757 for acetic acid)
(c) 0.200 M pyridinium formate ($pK_a = 5.22$ for the pyridinium ion and 3.752 for formic acid)
(d) 1.00×10^{-4} M ammonium benzoate ($pK_a = 9.245$ for the ammonium ion and 4.212 for benzoic acid)

3.12. Write a BASIC computer program that will carry out the calculations of Exercise 3.11, and run it to check the hand calculations for each case.

3.13. Evaluate each of the following using the Newton–Raphson method.

(a) $\sqrt{5}$
(b) $\sqrt{6}$
(c) $\sqrt[3]{7}$
(d) $\sqrt{e}$

3.14. If the dissociation of water is neglected, the hydrogen ion concentration of an aqueous solution of a weak acid is given by

$$[H^+]^2 + K_a[H^+] - K_a C_a = 0$$

Solve this equation by hand for each of the following cases using the Newton–Raphson method.

(a) 1.00×10^{-4} M acetic acid ($pK_a = 4.757$)
(b) 1.00×10^{-4} M formic acid ($pK_a = 3.752$)
(c) 1.00×10^{-5} M benzoic acid ($pK_a = 4.212$)
(d) 1.00×10^{-3} M hydrocyanic acid ($pK_a = 9.216$)

3.15. Write a BASIC computer program to carry out the calculations of Exercise 3.14, and use it to check the accuracy of the hand calculations.

3.16. Solve the equation given in Exercise 3.2 for each of the following aqueous solutions using a BASIC program based on the secant method.

(a) 0.0100 M carbonic acid ($pK_1 = 6.352$, $pK_2 = 10.33$)
(b) 1.00×10^{-5} M carbonic acid
(c) 1.00×10^{-6} M oxalic acid ($pK_1 = 1.271$, $pK_2 = 4.266$)
(d) 1.00×10^{-5} M succinic acid ($pK_1 = 4.207$, $pK_2 = 5.635$)

3.17. Given the data in the table below, find $f(7.7)$ with:

(a) A three-point Lagrange interpolation using the first three data points only
(b) A four-point Lagrange interpolation

x	$f(x)$
3	5.3
6	4.7
9	6.8
12	8.3

3.18. Compare the values of $f(11.5)$ calculated with the two interpolation formulas used in Exercise 3.17. Which is apt to be the more accurate, and why?

3.19. Evaluate $f'(7.7)$ using the three-point and four-point Lagrange differentiation formulas and the data of Exercise 3.17.

3.20. Find the value of x for which $f(x)$ is a minimum according to each of the formulas developed in Exercise 3.19.

3.21. Compare the values of the integral

$$\int_1^{10} \frac{dy}{y}$$

obtained by:

(a) Exact integration
(b) Five-point quadrature using the trapezoidal rule
(c) Five-point quadrature using the extended Simpson's rule
(d) Five-point Gaussian quadrature

3.22. Compare the percent error obtained in evaluating the integral

$$\frac{1}{\sqrt{2\pi}} \int_0^2 e^{-z^2/2}\, dz$$

for three and five quadrature points by each of the following methods (see Appendix A.8 for a table of values of this integral).

(a) The trapezoidal rule
(b) The (extended) Simpson's rule
(c) Gaussian quadrature

3.23. Derive $P_3(z)$ using Eq. 3.61, and solve for the values of z for which it is zero according to Eq. 3.64.

3.24. Using the results of Exercise 3.23, calculate the weighting factor for the $z_i = 0$ term of a three-point Gaussian quadrature.

3.25. Solve each of the following differential equations using a BASIC program similar to that given in Example 3.15. Check the solution to be sure that the interval is small enough to give a solution that is accurate to ± 0.001.

(a) $dy/dx = x - y$ in the range $0 \le x \le 4$ with $y = 3$ at $x = 0$

(b) $dy/dx = x^2 - y^2$ in the range $0 \le x \le 2$ with $y = 5$ at $x = 0$

(c) $dz/dt = 6z - 3zt$ in the range $5 \le t \le 10$ with $z = 2$ at $t = 5$

(d) $dx/dt = x - 4t$ in the range $2 \le t \le 4$ with $x = 1$ at $t = 2$

3.26. Solve the following systems of linear equations using the Gauss elimination technique.

(a)
$$4x + 8y + z = 2$$
$$x + 7y - 3z = -14$$
$$2x - 3y + 2z = 3$$

(b)
$$2x - 8y - z = 11$$
$$3x + 7y - z = 0$$
$$x - 3y - 5z = -18$$

(c) (*Note:* Requires pivoting.)
$$3x - 6y + z = -9$$
$$-6x + 12y - 5z = 21$$
$$x + 2y + 6z = -6$$

(d)
$$2w + 3x - y + 4z = -7$$
$$w - 3x - 4y - 8z = -2$$
$$w + x + y + z = 6$$
$$3w + 5x + 2y + 4z = 8$$

3.27. Write a BASIC program for the evaluation of each of the following integrals by a Monte Carlo technique.

(a) $$\int_0^2 e^x\,dx$$

(b) $$\int_0^1 \ln(1 + x)\,dx$$

(c) $$\int_0^1 \frac{\ln(1 + x)}{1 + x^2}\,dx$$

(d) $$\int_0^1 x \ln(1 + x)\,dx$$

3.28. Modify the BASIC Monte Carlo program given in the text for reaction kinetics to treat the scheme

$$\begin{array}{ccccc} A_1 & \xrightarrow{k_1} & A_2 & \underset{k_3}{\overset{k_2}{\rightleftharpoons}} & A_3 \\ & & k_4 \downarrow\uparrow k_5 & & \\ & & A_4 & & \end{array}$$

with $k_1 = k_2 = k_5 = 2$, $k_3 = k_4 = 3$, $[A_1]_0 = [A_4]_0 = 450$, and $[A_2]_0 = [A_3]_0 = 0$.

3.29. Write a BASIC program for a Monte Carlo calculation of the reaction scheme

$$A + B \xrightarrow{k_1} C + D$$

$$A + C \xrightarrow{k_2} D + E$$

with $k_1 = 2$, $k_2 = 5$, $[A]_0 = [B]_0 = 400$, and $[C]_0 = [D]_0 = [E]_0 = 0$.

3.30. Show that insertion of the three points $(-h, f_{-1})$, $(0, f_0)$, and (h, f_1) into the Lagrange interpolation formula (Eq. 3.14), followed by integration term by term, leads to Simpson's rule (Eq. 3.49).

3.31. Even for the very simple kinetic mechanism

$$A + B \underset{k_2}{\overset{k_1}{\rightleftharpoons}} C \xrightarrow{k_3} D$$

an exact solution by integration is impossible. Solve this problem numerically using the program RKC of Figure 3.16 for the initial concentrations $[A]_0 = 0.5$, $[B]_0 = 0.7$, $[C]_0 = [D]_0 = 0$ and rate constants $k_1 = 1.1$, $k_2 = 0.4$, $k_3 = 0.75$, and plot the concentrations versus time. How does changing the rate constant k_2 affect the concentration curve of D?

3.32. Modify the program RKC of Figure 3.16 to solve for the concentrations of A, B, C, D, and E as functions of time for the reaction scheme

$$\begin{array}{ccccc} A & \xrightarrow{k_1} & C & \underset{k_3}{\overset{k_2}{\rightleftharpoons}} & B \\ & & k_4 \downarrow \uparrow k_5 & & \\ & & D & \xrightarrow{k_6} & E \end{array}$$

Assume the rate constants $k_1 = 1.0$, $k_2 = 0.9$, $k_3 = 0.7$, $k_4 = 0.8$, $k_5 = 0.6$, $k_6 = 1.2$, and the initial concentrations $[A]_0 = 0.4$, $[B]_0 = 0.6$, $[C]_0 = [D]_0 = [E]_0 = 0$. Plot the concentrations of the components as functions of time.

3.33. Perform an analysis like that given in Example 3.14 for the molecule 1,3,5-hexatriene, $CH_2{=}CH{-}CH{=}CH{-}CH{=}CH_2$.

chapter four

STATISTICS

Statistics is an area of applied mathematics that is concerned with sampling procedures, design of experiments, analysis of data, and the drawing of inferences based on appropriate measurements. A proper statistical analysis can lead to an increased depth of understanding and to improved efficiencies for many processes. It does, however, require a careful analysis of the reliability of the various inferences as our subsequent discussion will show.

4.1 DETERMINATE AND INDETERMINATE ERRORS

There is no such thing as the perfect experiment. Even the best chemist in the world (whoever that may be!) is incapable of repeated experiments that yield only ideal results. There are always errors of one kind or another associated with a given measurement. In this chapter we introduce some of the techniques used to deal with them.

Before we discuss experimental errors, it is important to note that there are various kinds of data, and the statistical techniques that have been developed to treat them are quite varied. *Nominal* and *ordinal* data arise from discrete, noncontinuous observations. Only whole numbers are involved. Nominal data consist of a tabulation of the frequencies of occurrence in various categories, where there is no ranking or ordering of the categories. Some examples are the distribution of chemists among the subdisciplines physical, organic, inorganic, analytical, . . . , or the number of molecules in each of several possible isomeric forms that may result from a complex synthesis. Ordinal data are also discrete and noncontinuous, but specify a distribution in ranked or ordered categories. Here we might cite as examples the ranking of chemists according to their educational levels (number with B.S. degrees, M.S. degrees, etc.), or the number of molecules in the ground state, first excited state, second excited state, . . . under a given set of conditions.

Most of the measurements with which the chemist is concerned yield continuous quantitative data; that is, the results may be expressed in terms of fractional parts of a gram, milliliter, and so forth. One may distinguish between *interval* and *ratio* data, but the difference is not an important one for most purposes. Both are based on observations of continuous variables. For interval data our zero or reference point is at some intermediate point on the measuring scale, while ratio data are structured such that they are referred to an absolute zero. For example, temperatures measured on the Celsius scale may form a set of interval data; they can easily be converted to the Kelvin scale and treated as ratio data by a simple linear transformation. Some other examples include the equilibrium concentration of a particular reactant as a function of the initial amount taken, or the moles of reactant remaining as a function of the time of reaction.

Interval and ratio data, which will be our principal concern, are subject to errors of two types. *Determinate errors* are of identifiable origin. Perhaps we have a certain color-blind chemist who always overruns his endpoints. If all chemists in the laboratory use the same stock solution, it will appear that the samples of the color-blind chemist always contain more analyte than those of the other chemists. We might never recognize the problem unless many comparisons are made using different chemists and/or analytical methods. Determinate errors lead to a certain *bias* in the results. They can be identified and eliminated, although this can sometimes be very difficult in practice. Determinate errors are also called *systematic errors*.

Indeterminate errors are also known as *random errors*, and as the names imply, they are of unknown origin. They have a physical basis just as determinate errors do; the essential difference is that they are totally unpredictable. For example, our experiment might require the reading of an electronic device such as a spectrophotometer. All such devices have certain "noise levels" associated with them, and our readings will vary within these noise limits, but in a random fashion. No two instruments behave identically, and no two chemists work with precisely the same level of reproducibility. We all have our own built-in "noise level." It is possible to improve the reproducibility of our measurements with practice and concentration, even though the origin of the indeterminate errors is not always clearly understood.

We often use the terms *precision* and *accuracy* in describing chemical data. We may be able to reduce indeterminate errors to a very low level for a given method. Then, if we make several replicate determinations using the method, the values all lie within a narrow range. We describe such a determination as *precise*. A very precise value may be highly inaccurate if there is a large bias, or systematic error, in the determination. Therefore, precision is a measure of how tightly grouped the results are, whereas *accuracy* is a measure of how near the grouping is to the true value. Generally, an accurate result is also precise, but the converse is not necessarily true. These concepts will be put on a more concrete footing as our discussion proceeds.

In general, our goal is to characterize a certain entity fully with respect to some chosen property. The entity with which we deal may be large (e.g., the ocean, or the earth's crust), or it may be small (e.g., a gallon of gasoline, or 2 g of phthalic acid). The term used by statisticians to describe the entity as a whole is the *population,* and in order to have a complete characterization of it, we would have to analyze it all. This is usually impractical, and we generally select certain quantities that are chosen to be as representative of the whole as possible. On the basis of analysis of the smaller set, we attempt to describe the properties of the whole. This smaller set chosen for analysis is called a *sample* by the statistician. This is a possible source of confusion to chemists, who generally refer to an individual quantity submitted for analysis as a sample. Throughout this chapter we will understand the term *sample* to mean the entire set of quantities chosen for analysis, and we will assume that proper consideration has been given to making the sample representative of the entire population.

Determinate or systematic errors will not be discussed in detail here. They are very specific to the type of measurement under consideration, and even though they may be the dominant source of error, it is difficult to give a complete account of them. In general, they are associated with a particular method of analysis, and their elucidation usually requires comparison of the work of different analysts, different instruments, different methods, and different standard samples. By appropriate analysis, an estimate of the magnitude of such systematic errors can be made, and they can be incorporated as appropriate in the equations we will derive later for the specification of experimental error. For the present we will concentrate on the treatment of random errors.

4.2 DISTRIBUTION CURVES

Let us say that we have taken a sample consisting of the following 19 values from a very large population: 9.81, 7.82, 10.77, 8.55, 8.67, 11.14, 7.05, 10.17, 6.43, 9.33, 5.63, 9.06, 8.98, 9.49, 7.46, 8.23, 8.52, 9.27, 8.17. If there has been no bias in the sampling process itself, we can assume that this distribution of values is representative of the entire population and we need to consider how we can present these data in an easily interpreted, meaningful way.

A technique in common use for the visualization of such data is the construction of a *histogram*. In order to construct a histogram, we must divide the data into *classes*. We will take 1.00 as our *class interval,* and group all the values together that fall in the range 5.00–5.99, 6.00–6.99, 7.00–7.99, and so forth. Then we have one value that falls in the class whose midpoint is 5.50, one that falls in the class whose midpoint is 6.50, three in the class whose midpoint is 7.50, and so forth. The entire *frequency distribution* is 1:1:3:6:5:2:1, and it can be represented by the histogram shown in Figure 4.1. We recognize this as a type of bar graph with heights equal to the frequency of occurrence of values within each of the respective classes.

It is generally not convenient to display an entire frequency distribution each time we wish to report data, so we use certain numerical quantities to represent the characteristics of the distribution curve. Two properties are of special interest; they may be called the central tendency and the dispersion. By *central tendency* we mean

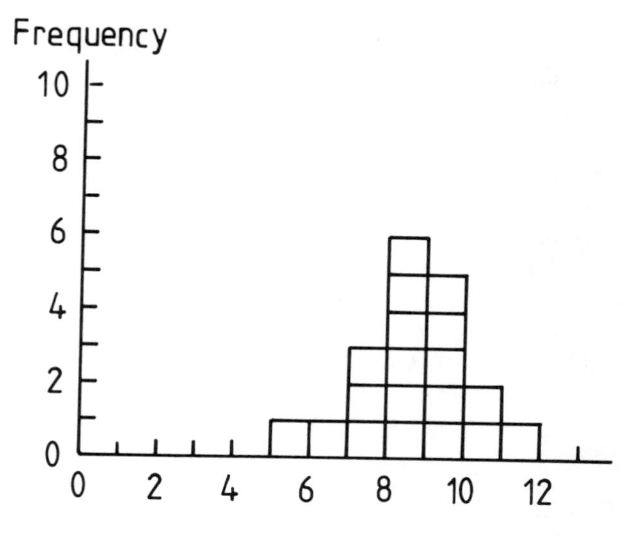

Figure 4.1 Frequency distribution representing the data described in the text.

the one number we might quote to best represent a typical value for the sample, whereas the term *dispersion* signifies the magnitude of the spread of actual values about that represented by the central value.

The *mode* is sometimes used as a measure of central tendency. It is defined as the value with the greatest frequency of occurrence. This is useful for dealing with nominal or ordinal data where only whole numbers occur, if they are likely to occur repeatedly. The concept is less useful here, although we could give an approximate mode by citing the midpoint of the class of most frequent occurrence, which in this case would be 8.50.

Another measure of central tendency that is useful in some instances is the *median*. The median is chosen such that there are equal numbers of values less than and greater than this value. In the case of a sample with an odd number of values, the middle value is chosen as the median; if the number of values is even, an average of the two middle values is taken. For our example, the median is 8.67.

Still another method used to represent the central tendency is the *mean*, which is just another word for the average. It is constructed by adding together all values in the sample and dividing by the total number of values. In mathematical terms we represent the mean $\bar{x}$ by

$$\bar{x} = \frac{\Sigma x_i}{N} \tag{4.1}$$

where N is the number of values. Throughout this chapter, unless designated otherwise, it is understood that the summation is to be carried out over the entire sample; that is, the summation index runs from 1 to N. It may readily be verified that for our example the mean is 8.66.

The mean is usually the method of choice for specification of the central tendency, although the median is sometimes used when dealing with very small samples, and the reason for it can be stated as follows. If we have a sample consisting of but three values let us say, and one value differs greatly from the other two, there is good reason to believe that the one value is spurious. If we compute an average, this one spurious value produces a large shift from the true central value of the population, but the median is not so sensitive to such an anomalous value.

The dispersion of a frequency distribution can also be specified in quite a number of different ways. A measure sometimes used is the *range*, which is the difference between the smallest and largest values. For the data set given above, the range is $11.14 - 5.63 = 5.51$. The range is generally not a very good measure of dispersion; it is obviously very sensitive to the occurrence of abnormally large or small values.

If the mean is properly constructed, it is obvious that the sum of all the deviations from the mean $\Sigma\, d_i = \Sigma\, (x_i - \bar{x})$ is equal to zero, and this cannot serve as a measure of dispersion. The sum of the *absolute values* of the deviations is nonzero, however, and the *average deviation,*

$$\bar{d} = \frac{\Sigma\, |d_i|}{N} = \frac{\Sigma\, |x_i - \bar{x}|}{N} \tag{4.2}$$

is sometimes used. The average deviation for the data above is 1.06. Occasionally, the *quartile deviation* is given for a data set. It is defined to include one-half of the data points, with one-fourth of the points being too small and one-fourth being too large for inclusion. For our case, the quartile deviation is 1.67.

None of the measures of dispersion thus far mentioned is entirely satisfactory. A suitable measure should be as independent of the size of the sample as possible, and it should be defined in such a way that comparisons between independent data sets or groupings of related data sets can readily be made. In the older literature the *root-mean-square,* or RMS, deviation was often quoted. As the name implies, one squares the deviations, divides by N to construct the mean, and then takes the square root of the result:

$$\text{RMS} = \sqrt{\frac{\Sigma\, d_i^2}{N}} = \sqrt{\frac{\Sigma\, (x_i - \bar{x})^2}{N}}$$

It is now recognized that the RMS deviation is a reliable measure of dispersion when dealing with very large samples, but it is not valid for small ones. The reason is that we have in fact overspecified the data set by using this together with the mean. For a given number of data points N, we can completely characterize the set by giving the mean and $N - 1$ deviations from the mean, since we can solve for the Nth deviation using the mean and all the other $N - 1$ deviations. Thus there are really only $N - 1$ *degrees of freedom,* and we should use this as a divisor in place of N. When this is done our measure of dispersion is a quantity known as the *standard deviation:*

$$s = \sqrt{\frac{\Sigma\, d_i^2}{N - 1}} = \sqrt{\frac{\Sigma\, (x_i - \bar{x})^2}{N - 1}} \tag{4.3}$$

Obviously, the difference between the RMS deviation and the standard deviation becomes insignificant as the sample size becomes very large.

Another quantity in common use by statisticians is the *variance*. The variance is the square of the standard deviation, and rather than introduce a new symbol, we will simply represent it by s^2. It is a more useful measure of dispersion in the sense that it is s^2 rather than s that occurs in most of the equations where these quantities are to be combined or compared. Chemists are more prone to quote the standard deviation of a data set, however, because this quantity has the same units as the data itself and the spread of the data points is more readily visualized. For the data given above, the standard deviation is 1.40 and the variance is 1.96.

There is an alternative method of calculating the standard deviation which is more useful for hand calculation. We see that the application of Eq. 4.3 requires that the mean already be known. It is of course readily calculated by Eq. 4.1, but this necessitates a reentry of the data as a two-step process, which can be time consuming. As long as there are at least two storage registers available to accumulate the sums, the calculation of s can be carried out using an alternative formula which is readily derived by expanding Eq. 4.3 and inserting Eq. 4.1 as follows:

$$s = \sqrt{\frac{\Sigma x_i^2 - \Sigma 2\bar{x}x_i + \Sigma \bar{x}^2}{N - 1}}$$

$$= \sqrt{\frac{\Sigma x_i^2 - 2\dfrac{\Sigma x_i}{N}\Sigma x_i + N\left(\dfrac{\Sigma x_i}{N}\right)^2}{N - 1}}$$

$$= \sqrt{\frac{\Sigma x_i^2 - \dfrac{(\Sigma x_i)^2}{N}}{N - 1}} \tag{4.4}$$

Many modern calculators have a statistics function that automatically accumulates the quantities necessary to carry out the calculation of s, but if this is not available, calculation according to Eq. 4.4 is much more convenient.

In certain cases one needs to calculate the mean and standard deviation from a frequency distribution rather than from raw data. Some small discrepancies might be expected, but an approximate value of the mean may be calculated from

$$\bar{x} = \frac{\Sigma f_i X_i}{\Sigma f_i} \tag{4.5}$$

and the corresponding value of the standard deviation is given by

$$s = \sqrt{\frac{\Sigma f_i (X_i - \bar{x})^2}{\Sigma f_i - 1}} \tag{4.6}$$

Here we use X_i to represent the midpoint of the class i, and f_i is the frequency of occurrence within that class. Applying these formulas to the frequency distribution given above gives $\bar{x} = 8.71$ and $s = 1.44$, and we see that these values are close approximations of those calculated directly from the raw data.

In summary we will generally use the mean and the standard deviation as the best characterization of a frequency distribution, although alternative measures are used in some special cases. We hope that our sample is representative of the entire population. In actuality, however, we expect that even in the best of cases there will be some small deviations from the true values, and the deviations are apt to be greater if the sample size is small. We use the Greek letters μ and σ to represent the true mean and standard deviation of the entire population, to distinguish them from the corresponding values $\bar{x}$ and s calculated for a sample taken from the population.

4.2.1 The Binomial Distribution

Even though our principal interest is with interval and ratio data, it will help us understand the nature of distribution functions if we first consider a simple case where given events are classified only as successful or unsuccessful. There are many illustrations that can be cited, such as the flipping of a coin, the rolling of a die, and so forth. We will assume two important restrictions in the discussion that follows: namely, that there is a well-defined probability of occurrence for a single event, and that each event is independent of all the others.

Let us use the letter p to represent a successful event and the letter q to represent an unsuccessful event. For example, if the event consists of flipping a coin, and we are hopeful of getting a head, we say that p is equal to the probability that a head will be obtained on a single toss of the coin, and q is the probability that a head will not be obtained on a single toss. Assuming that we have a normal coin, we can take $p = 1/2$ and $q = 1/2$.

Now suppose that we roll a die and that we desire to obtain a 3 on a single roll. If we assume that we are dealing with a normal die which is free of any bias, the probability of success on a given roll is $p = 1/6$. The likelihood that we will get any one of five other equally probable numbers is $q = 5/6$. Note that by definition it is required that $p + q = 1$ in every case.

We will continue to explore the latter case, namely the rolling of a die with the hope of getting a 3. First let us consider the probability of getting two 3's on two rolls of the die. Since the probability of getting a 3 on the second roll does not depend on the result of the first roll, it has exactly the same probability of occurrence; that is, $p = 1/6$. The probability of 3 occurring on both rolls is therefore $1/6 \times 1/6 = 1/36$. In the same way, the probability that a 3 would fail to occur on either the first or the second roll is $5/6 \times 5/6 = 25/36$. The probability of getting just one 3 on two rolls of the die requires that we consider two possibilities. We could have a success on the first roll and a failure on the second, in which case the probability is $1/6 \times 5/6 = 5/36$, or we could have a failure on the first roll and a success on the second roll, for which the probability is $5/6 \times 1/6 = 5/36$. Since either of these two cases satisfies the requirement that we get just one 3, the total probability is the sum of the two: namely, $5/36 + 5/36 = 10/36$. We can summarize these results as follows:

two 3's: $\quad p^2 = 1/6 \times 1/6 = 1/36$

one 3: $pq + qp = 1/6 \times 5/6 + 5/6 \times 1/6 = 10/36$

no 3's: $q^2 = 5/6 \times 5/6 = 25/36$

Note that in this case also the sum of the individual probabilities is 1. That is, the probability of getting two 3's plus the probability of getting one 3 plus the probability of getting no 3's is $1/36 + 10/36 + 25/36 = 1$.

It will be recognized that the result above can conveniently be represented in the simple form

$$(p + q)^2 = p^2 + 2pq + q^2$$

where p^2 represents the probability of two successful events, $2pq$ represents the probability of one successful event, and q^2 represents the probability that no successful event will be observed in two trials.

The analysis above can readily be extended to include three trials, in which case the result is

$$(p + q)^3 = p^3 + 3p^2q + 3pq^2 + q^3$$

and p^3 represents the probability of success in each of three trials, $3p^2q$ is the probability of two successes and one failure, $3pq^2$ is the probability of one success and two failures, and q^3 is the probability of no success in any of the three trials.

This process can be continued indefinitely; the general result obtained in n trials is

$$(p + q)^n \tag{4.7}$$

Fortunately, it is not necessary that we multiply Eq. 4.7 out for the large values of n that may occur in practice. The general form may be written

$$(p + q)^n = C_0^n q^n + C_1^n pq^{n-1} + C_2^n p^2 q^{n-2} + \cdots + C_r^n p^r q^{n-r} + \cdots + C_n^n p^n \tag{4.8}$$

with the expansion coefficients given according to the *binomial expansion theorem* by

$$C_r^n = \frac{n!}{r!(n - r)!}$$

These same binomial coefficients were used previously (see Eq. 3.100) in a somewhat different context. Equation 4.8 defines the *binomial distribution*, and our discussion leads directly to a statement of *Bernoulli's theorem:* If we designate the probability of success and failure for a given trial by p and q, respectively, the probability of r successes after n independent trials is

$$P_r = C_r^n p^r q^{n-r} \tag{4.9}$$

We conclude this section by calculating the various probabilities that in eight rolls of a die, no 3 is obtained, one 3 is obtained, two 3's are obtained, and so forth. The individual probabilities are as follows:

$$
\begin{array}{rll}
r = 0 & P_0 = C_0^8(1/6)^0(5/6)^8 & = 0.2326 \\
1 & P_1 = C_1^8(1/6)^1(5/6)^7 & = 0.3721 \\
2 & P_2 = C_2^8(1/6)^2(5/6)^6 & = 0.2605 \\
3 & P_3 = C_3^8(1/6)^3(5/6)^5 & = 0.1042 \\
4 & P_4 = C_4^8(1/6)^4(5/6)^4 & = 0.0260 \\
5 & P_5 = C_5^8(1/6)^5(5/6)^3 & = 0.0042 \\
6 & P_6 = C_6^8(1/6)^6(5/6)^2 & = 0.0004 \\
7 & P_7 = C_7^8(1/6)^7(5/6)^1 & = 0.0000 \\
8 & P_8 = C_8^8(1/6)^8(5/6)^0 & = \underline{0.0000} \\
 & \Sigma P_i & = 1.0000
\end{array}
$$

A histogram representing the binomial distribution for this particular example is shown in Figure 4.2.

We again note that the sum of all the probabilities is equal to 1. Use is made of this fact not only to check the accuracy of the results, but to simplify certain calculations as well. For example, if we wish to know the probability of getting one or more 3's on 8 rolls of a die, it is simpler to evaluate $1 - P_0 = 0.7674$ than $P_1 + P_2 + P_3 + P_4 + P_5 + P_6 + P_7 + P_8 = 0.7674$.

4.2.2 The Normal Distribution

Let us for the moment assume a hypothetical situation. Suppose that we are attempting to determine some physical quantity with a well-defined value, such as Avogadro's

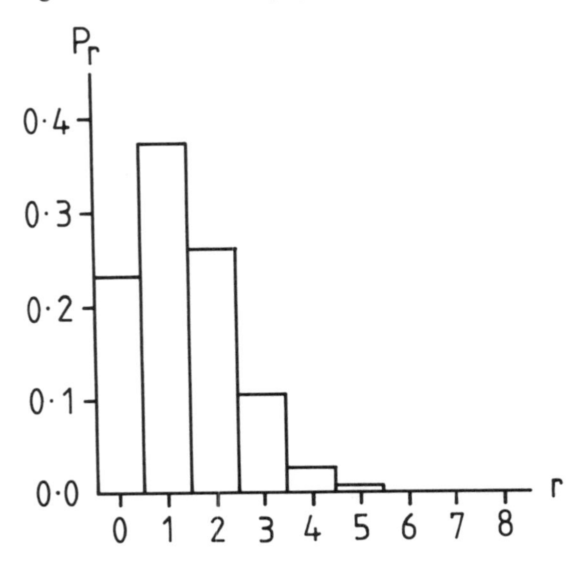

Figure 4.2 Binomial distribution showing the probability P_r of getting r 3's on 8 rolls of a die.

number or the charge of an electron. Let us also suppose that our measurement is free of determinate errors, but that it is subject to indeterminate errors. For the sake of argument we will make the naive assumption that there are six different sources of error, and that each may contribute a small error of magnitude e to the measured result with positive and negative contributions occurring entirely at random with equal probability from each of the six sources.

With but a slight difference of interpretation, we can use the binomial distribution discussed in the preceding section to describe the current situation. We will use p to represent an error $+e$ from any one of the six independent sources, and q to represent a corresponding error of $-e$. Then a binomial expansion can be used to represent the distribution of errors about the true value as follows:

$$\begin{array}{lllllllll} (p+q)^6 = & p^6 & + 6p^5q & + 15p^4q^2 & + 20p^3q^3 & + 15p^2q^4 & + 6pq^5 & + & q^6 \\ \text{error} = & +6e & +4e & +2e & 0 & -2e & -4e & & -6e \end{array}$$

This situation may be depicted by a histogram as shown in Figure 4.3. This histogram is a representation of the frequency distribution of measured values for this particular case. It shows that if a large number of determinations were made, the most probable value would be the true value. There is also quite a large probability for the occurrence of small errors, but their likelihood becomes smaller as the magnitude of the total error increases. Alternatively stated, it is highly unlikely that all six sources of error make contributions in the same direction for a given measurement if each is truly random.

Of course we will not find such simplistic situations in the real world, but the model helps us appreciate how individual errors can be compounded to give a distribu-

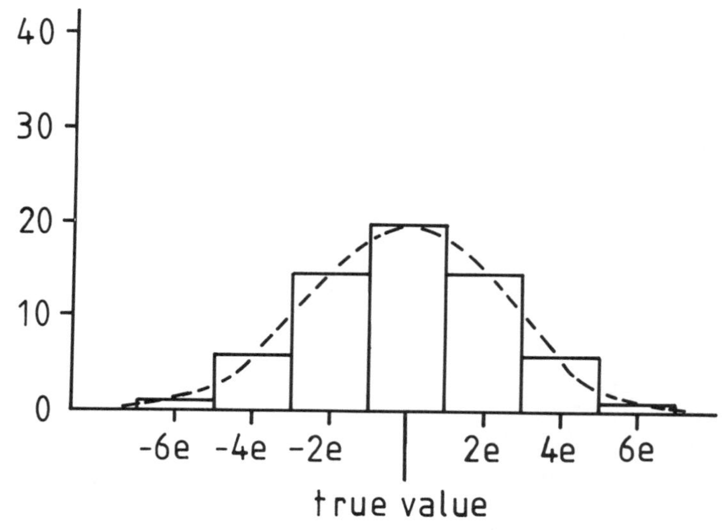

Figure 4.3 Probability distribution for the determination of a quantity that is subject to six random errors of equal magnitude.

tion about the true value. In actuality, a given measurement may be subject to many more than six sources of error, and certainly most random errors do not occur in simple discrete steps of equal magnitude. Both of these considerations would lead to a smoothing of the frequency distribution, and what we might expect as a result is a continuous curve having the general shape of that illustrated in Figure 4.3. A curve that has these characteristics is the *Gaussian distribution:*

$$p = \frac{f}{N} = \frac{1}{\sqrt{2\pi s^2}} e^{-\frac{1}{2}\left(\frac{x-\bar{x}}{s}\right)^2} \tag{4.10}$$

Here $\bar{x}$ represents the value of x about which the curve is centered, and s is a measure of its width. Note that the frequency f is divided by the total number of observations N, so the equation as written defines the probability distribution p. The pre-exponential factor $1/\sqrt{2\pi s^2}$ is chosen such that the total area under the curve is equal to 1, which is required if it is to be interpreted in terms of probabilities. With this choice, the curve is said to be *normalized*.

Equation 4.10 gives rise to what is usually described as a bell-shaped curve. It is shown plotted in Figure 4.4 for arbitrarily chosen values of $\bar{x}$ and s. As noted, the curve is much broader for a larger value of s, but the total area under the curve in either case is unity. The Gaussian function represents the theoretical distribution of a large number of observations about the true value, assuming that the results are subject only to random errors. The calculation of appropriate values of $\bar{x}$ and s from a given set of data has already been outlined.

We see that the standard deviation provides a direct indication of how tightly

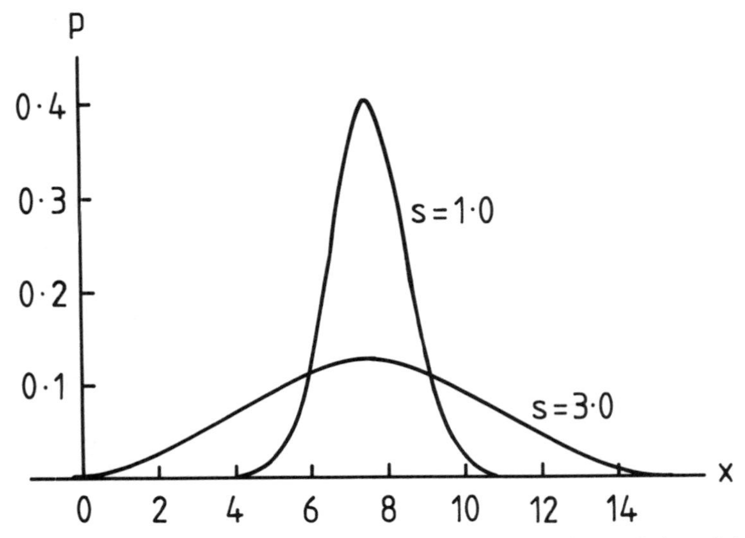

Figure 4.4 Gaussian distribution curve (Eq. 4.10) for two different choices of the standard deviation s, with $\bar{x} = 7.5$.

grouped the observations are about the mean, and as such it may be used as a quantitative measure of precision. A small value of s corresponds with a precise measurement. The values are grouped about the mean $\bar{x}$, and if this is near the true value μ, the measurement is also accurate. Unfortunately, for many applications the values of μ and σ of the population are not known; we will consider in a later section with what certainty they may be specified using the values $\bar{x}$ and s determined for a sample taken from the population.

There are as many Gaussian distribution curves as there are values of $\bar{x}$ and s (i.e., an infinite number). Fortunately, it is not necessary to plot a new curve for each situation we may encounter. We can interpret our data in terms of the *normal distribution function,* also sometimes called the *error function:*

$$p = \frac{1}{\sqrt{2\pi}} e^{-z^2/2} \tag{4.11}$$

A graph of this function is shown in Figure 4.5. Equation 4.11 is obtained from Eq. 4.10 by the substitution

$$z = \frac{x - \bar{x}}{s} \tag{4.12}$$

This substitution has two obvious effects: A transformation along the abscissa is made so that the figure is centered about $z = 0$, and s has been incorporated in the defini-

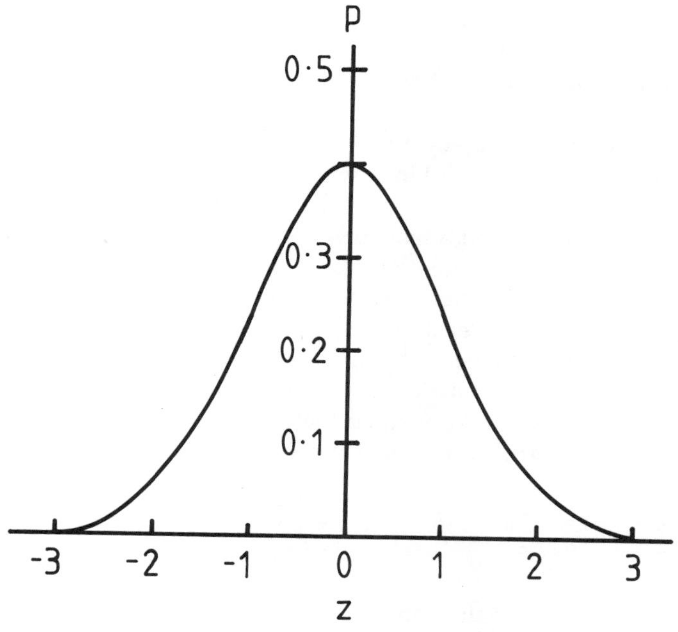

Figure 4.5 The normal error curve (Eq. 4.11).

tion of z so that the normal distribution curve has an apparent standard deviation $s = 1.00$. By doing this we obtain a single standard distribution function which can be used for the analysis of any number of individual cases.

An integration of Eq. 4.12 can be carried out to determine the probability that a given measurement will fall within a certain range of values. Because the integrand is an even function, it is generally evaluated in the form

$$\frac{1}{\sqrt{2\pi}} \int_0^{z_1} e^{-z^2/2}\, dz \tag{4.13}$$

It is not possible to carry out the integration of the normal error function analytically; a numerical quadrature must be performed. A table of values of the integral as a function of z_1 is given in Appendix A.8.

It will be observed that the values of p listed in Appendix A.8 tend toward the limiting value 1/2 as $z = x/s$ becomes very large. This is consistent with the observation made above; namely, only half the integral is tabulated because the remainder is readily obtained by symmetry. These values allow us to make a clear statement of the precise meaning of the standard deviation of a given sample. We observe, for example, that the value of p tabulated for a value of z which is 1.0 unit from the origin is 0.3413. This indicates that approximately 34% of the curve lies between the origin and one standard deviation unit to the right. We conclude that if a large number of measurements were made, approximately 68% of them would lie within $\bar{x} - s$ and $\bar{x} + s$. Similarly, about 95% of them would lie in the range $\bar{x} - 2s$ to $\bar{x} + 2s$, and so forth.

4.2.3 The Distribution of Means

Assume that several samples are taken from a certain population, and a mean is calculated for each. We wish to address the question: How are these means distributed? We intuitively expect that their average (the mean of the means) should be the same as the population mean they represent. It is not so clear how the individual means would be distributed about this population mean, although we predict that the larger the size of the samples, the closer each sample mean would lie to the true mean. In the limit of infinitely large samples, each represents the entire population exactly and there would be no spread in the values. This situation is illustrated graphically in Figure 4.6; we will now derive a mathematical relationship that defines this result quantitatively.

As a first step, let us consider the addition of two independent variables x and y to obtain a third variable which we will call z. We find that the mean value of z is given by

$$\bar{z} = \frac{\Sigma\, z_i}{N} = \frac{\Sigma\,(x_i + y_i)}{N} = \frac{\Sigma\, x_i}{N} + \frac{\Sigma\, y_i}{N} = \bar{x} + \bar{y} \tag{4.14}$$

The variance of the z values obtained by addition of x and y is obtained using Eq. 4.14 as follows:

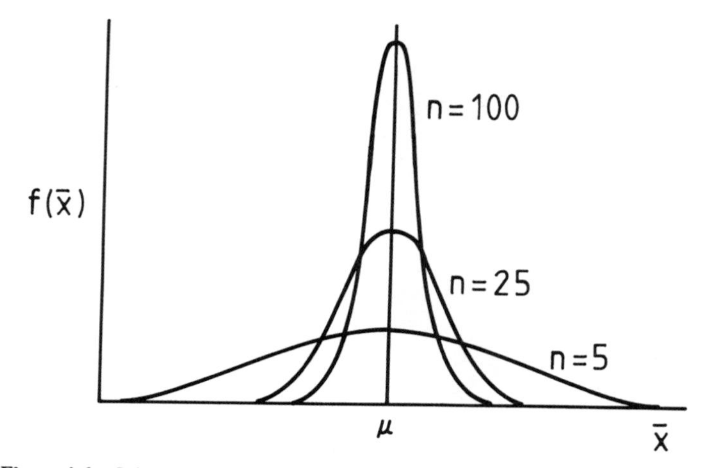

Figure 4.6 Schematic representation of the distribution of sample means about the population mean as a function of the sample size.

$$s_z^2 = \frac{\Sigma\,(z_i - \bar{z})^2}{N-1} = \frac{\Sigma\,(x_i + y_i - \bar{x} - \bar{y})^2}{N-1}$$

$$= \frac{\Sigma\,[(x_i - \bar{x}) + (y_i - \bar{y})]^2}{N-1}$$

$$= \frac{\Sigma\,(x_i - \bar{x})^2}{N-1} + \frac{\Sigma\,(y_i - \bar{y})^2}{N-1} + 2\,\frac{\Sigma\,(x_i - \bar{x})(y_i - \bar{y})}{N-1}$$

Since x and y are independent variables we have

$$\Sigma\,(x_i - \bar{x})(y_i - \bar{y}) = 0$$

and so

$$s_z^2 = s_x^2 + s_y^2 \tag{4.15}$$

Equations 4.14 and 4.15 can readily be generalized to include addition of any number of independent variables.

We now consider the variable z to be defined by $z = ax$, where a is a constant and x is again an independent variable. In this case the mean of the z values is

$$\bar{z} = \frac{\Sigma\,z_i}{N} = \frac{\Sigma\,ax_i}{N} = a\,\frac{\Sigma\,x_i}{N} = a\bar{x} \tag{4.16}$$

with variance given by

$$s_z^2 = \frac{\Sigma\,(z_i - \bar{z})^2}{N-1} = \frac{\Sigma\,(ax_i - a\bar{x})^2}{N-1}$$

$$= a^2\,\frac{\Sigma\,(x_i - \bar{x})^2}{N-1} = a^2 s_x^2 \tag{4.17}$$

We use the results derived in Eqs. 4.14 to 4.17 to write the quantitative relationships for distribution of means. We will assume that N samples are taken from a population with mean μ and standard deviation σ. Let us further assume that the sample size is sufficiently large so that the mean and standard deviation of each are nearly equal to μ and σ, respectively. Then the mean of the means may be written

$$\bar{x} = \frac{\bar{x}_1 + \bar{x}_2 + \bar{x}_3 + \cdots}{N} \tag{4.18}$$

and since we have $\bar{x}_1 = \bar{x}_2 = \bar{x}_3 = \cdots = \mu$, we obtain

$$\bar{x} = \frac{\mu + \mu + \mu + \cdots}{N} = \frac{N\mu}{N} = \mu \tag{4.19}$$

as expected.

With the individual means combined as indicated in Eq. 4.18, we can apply Eqs. 4.15 and 4.17 to calculate the variance of the means:

$$\sigma_{\bar{x}}^2 = \frac{1}{N^2}\sigma_{x1}^2 + \frac{1}{N^2}\sigma_{x2}^2 + \frac{1}{N^2}\sigma_{x3}^2 + \cdots$$

But $\sigma_{x1}^2 = \sigma_{x2}^2 = \sigma_{x3}^2 + \cdots = \sigma^2$, and we have

$$\sigma_{\bar{x}}^2 = \frac{1}{N^2}(\sigma^2 + \sigma^2 + \sigma^2 + \cdots) = \frac{N\sigma^2}{N^2} = \frac{\sigma^2}{N}$$

By taking the square root we obtain the *standard deviation of means*, also known as the *standard error:*

$$\sigma_{\bar{x}} = \frac{\sigma}{\sqrt{N}} \tag{4.20}$$

This is a highly important statistical relationship. From it we can predict the distribution of means based on the mean and standard deviation determined for a single sample.

We see that the narrowing of the distribution of means predicted above and illustrated qualitatively in Figure 4.6 is borne out by a quantitative derivation. It is important to note that the narrowing is not linear, however, and this has some important implications for practical analysis. We may take a sample for analysis that we assume to be representative of the entire population. We can apparently lower the range of uncertainty of the result by the factor 1.414 by averaging the results of two samples, or by the factor 1.732 by averaging three samples, and so forth. Obviously, it is possible to improve the result to any desired degree. However, since the narrowing is proportional to the square root of N, increasingly large numbers of samples are required to bring about a given level of reduction in the uncertainty. For example, the first reduction by a factor of 2 requires 4 samples, a second factor of 2 requires 16, a third factor of 2 requires 64, and so forth. As a result, the chemist usually chooses to run a few samples in parallel for analysis, say 3 to 5, to take advantage of the reduction in uncertainty which can be brought about by relatively small numbers of samples. It is often impractical to go beyond that point, however.

An important observation concerning the distribution of means is that they usually follow a normal distribution, even in those cases where the population itself may show some deviation from the normal. Because of this, analysis of problems based on the distribution of means is also made using Eq. 4.11, with z defined for this case by

$$z = \frac{\bar{x} - \mu}{\sigma_{\bar{x}}} = \frac{\bar{x} - \mu}{\sigma/\sqrt{N}} \tag{4.21}$$

4.2.4 Student's t-Distribution

We have seen that with just a single sample, we can calculate the form of the distribution of means using Eq. 4.11 with z defined by Eq. 4.21. It is required, however, that the sample be large so that the values of μ and σ are well determined. This is not always possible, and we must frequently base our analysis on a very small sample of say, four or five values. Does the above still apply?

We can approach the problem in much the same way, but there are certain modifications that must be made. The difficulty stems from the uncertainties in both the mean and the standard deviation of the population as estimated from a small sample. Now the mean and standard deviation are independent quantities, so we do not expect their variations to be correlated. In other words, if the mean were abnormally small it is just as probable that the standard deviation would be abnormally large as small, or vice versa. When we evaluate the ratio of these terms as in Eq. 4.21 to define the distribution, we expect more variability in the parameter calculated from a small sample than for a large one that gives reliable values of both μ and σ.

These considerations were placed on a firm quantitative basis through studies made about the turn of the century by an Irish chemist named W. S. Gosset. Gosset expressed his results in terms of a certain quantity which he called t, defined by analogy with Eq. 4.21 as

$$t = \frac{\bar{x} - \mu}{s_{\bar{x}}} = \frac{\bar{x} - \mu}{s/\sqrt{N}} \tag{4.22}$$

Note that the essential difference is that the values of μ and σ must be replaced by those inferred from a small sample consisting of n observations.

For some unknown reason Gosset published his results in 1908 under the pseudonym "Student," so we refer to the distribution of means for small samples as *Student's t-distribution*. The greater variability noted above is reflected in the distribution curve tending to be broader for small samples than for large ones. Of course, as the sample size gets larger, the characterization of the population becomes more reliable, so we expect that in the limit where n becomes very large, the t-distribution should become the equivalent of the normal distribution.

Figure 4.7 shows a comparison of the normal distribution and the t-distribution

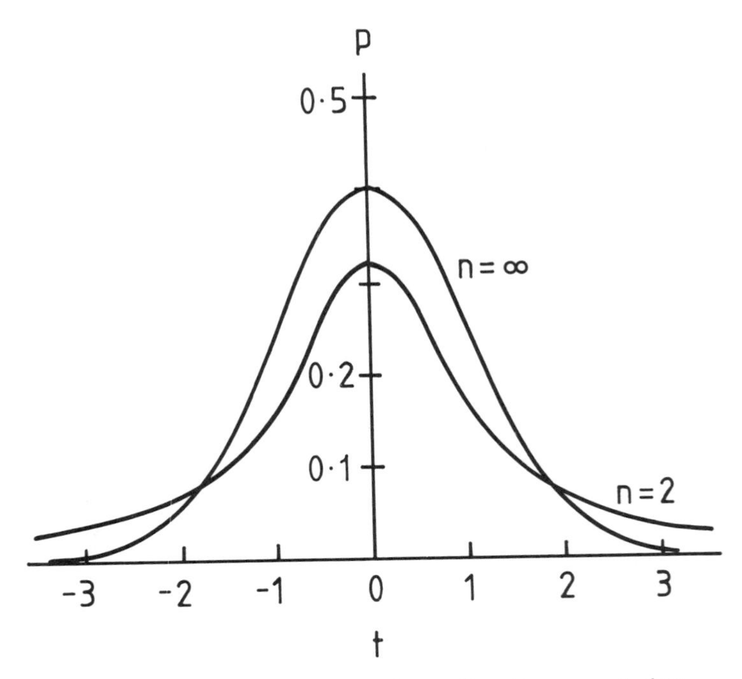

Figure 4.7 Student's t-distribution for small and large sample sizes.

for the rather extreme case of but two values in the sample. As n increases, the t-distribution curve approaches that for the normal distribution, labeled in Figure 4.7 as the $n = \infty$ curve. Actually, the curves become almost identical beyond about $n = 30$, so this is usually taken as a crude limit to distinguish between large and small samples.

Because the curve is distinct for each value of n, at least up to values of about $n = 30$, it is important in using the t-distribution to state the size of the sample under consideration. This is usually done by specifying the degrees of freedom $n - 1$, it being again noted that once the mean is computed, there can be only $n - 1$ other conditions required to specify the set completely.

The actual form of the t-distribution function is rather complex, and it is seldom used directly. For practical purposes, what is needed are the *critical values of t,* which, like z, indicate the cutoff point such that the area under the curve is a certain fraction of the total. The meaning of this statement can be clarified by a comparison of Appendix A.8, which gives areas under the normal distribution curve, with Appendix A.9, which gives critical values of t as a function of the number of degrees of freedom. Let us assume, for example, that we wish to select a value of z for the normal distribution such that only 5% of the total area lies to the right. The value of p corresponding to this case is 0.45, since this combined with the value 0.50 for negative values of z accounts for 0.95, or 95% of the total area. We see that an interpolation is required; the value of z $(= x/s)$ being 1.645. If we now examine Appendix A.9, we see that this value is listed as the critical t value for what is known as a "one-sided t-test" for a large number of degrees of freedom. The one-sided t-test will be defined in Section 4.4; it is sufficient for our present purposes to note that this is the value of t such that 5% of the distribution curve lies to the right. As already stated, the curve is broader for fewer degrees of freedom. We see that we must accordingly integrate to greater distances from the origin to include 95% of the area as the degrees of freedom decreases.

There are applications where we may want to specify that a certain fraction of the total distribution curve lies within a given range rather than placing it all to one side of the critical value as was done above. Suppose, for example, that we are again interested in the inclusion of 95% of the area, but this time we wish to distribute the remaining 5% equally in the wings to the right and left. We must then find the value of z in Appendix A.8 corresponding to $p = 0.475$; then we will have 0.025 or 2.5% of the total area to the right and a corresponding 2.5% in the tail to the left of $-z$. The interpolation in this case gives the value 1.960, so we can state that 95% of the normal distribution curve lies in the range $-1.960 \leq z \leq +1.960$.

If we look again at Appendix A.9, we see that this value is listed as the critical t value for a "two-sided t-test" for a large number of degrees of freedom. We will once again defer a discussion of the manner in which the test is carried out until Section 4.4, but note that the t-distribution for a large sample contains 2.5% of the area to the right of $+1.960$, and 2.5% to the left of -1.960. Once again we see that the critical values of t become much larger as the degrees of freedom decrease since the distribution curve is flatter, having a larger proportion of the total area in the wings.

Example 4.1

Consider a sample composed of the values 34.72, 36.84, 33.25, 37.78, 39.16, 34.54, and 35.06. Assume that these values define a normal population and calculate the probability that in a large sample (a) a value greater than 37.00 will be obtained, and (b) a value equal to or less than 36.00 will be obtained.

Solution. Using Eq. 4.1 and either Eq. 4.3 or 4.4, the mean and standard deviation are calculated to be

$$\bar{x} = 35.91 \qquad s = 2.08$$

(a) If it is assumed that a normal distribution is defined by these parameters, then according to Eq. 4.12, the z value corresponding to a value of $x = 37.00$ is

$$z = \frac{37.00 - 35.91}{2.08} = 0.525$$

The portion of the probability distribution curve required is represented schematically in the following figure:

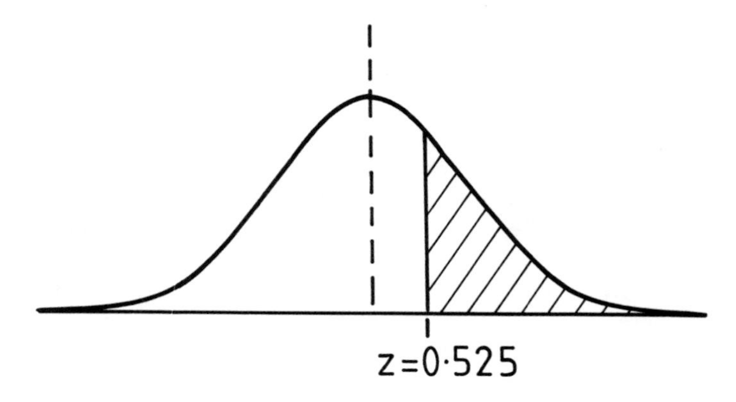

By interpolation of Appendix A.8 we find

$$0.1915 + \frac{0.525 - 0.5}{0.6 - 0.5}(0.2257 - 0.1915) = 0.2001$$

Thus the probability of all values up to and including 37.00 is $0.5000 + 0.2001 = 0.7001$, so the probability of values greater than 37.00 is $1.0000 - 0.7001 = 0.2999$.

(b) The z value in this case is

$$z = \frac{36.00 - 35.91}{2.08} = 0.0433$$

and the desired probability is represented by the shaded area of the following figure:

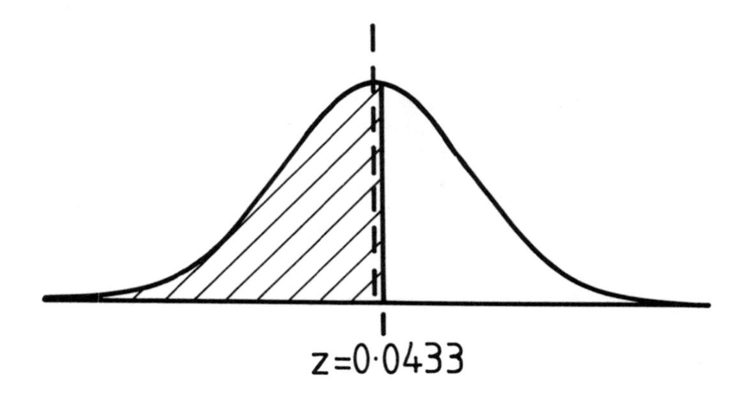

Interpolation of Appendix A.8 now gives

$$0.0000 + \frac{0.0433 - 0.0}{0.1 - 0.0}(0.0398 - 0.0000) = 0.0172$$

and the total probability of all values up to 36.00 is therefore $0.5000 + 0.0172 = 0.5172$.

Example 4.2

If it is again assumed that the data of Example 4.1 define a normal distribution, find the range of x values that includes 95% of the values of a large sample.

Solution. The question implies that we place 2.5% of the distribution in each of the tails. We have seen that the corresponding z values are $-1.960 \leq z \leq +1.960$. Thus the lower limit is given by

$$-1.960 = \frac{x_l - \bar{x}}{s} = \frac{x_l - 35.91}{2.08}$$

$$x_l = 35.91 - 2.08(1.960) = 31.83$$

and the upper limit by

$$+1.960 = \frac{x_u - \bar{x}}{s} = \frac{x_u - 35.91}{2.08}$$

$$x_u = 35.91 + 2.08(1.960) = 39.99$$

The range that encompasses 95% of the x values is therefore 31.83 to 39.99.

Example 4.3

If repeated samples of seven values are taken from the same population as that of Example 4.1, what would be the standard deviation of the means about the true mean of the population?

Solution. By analogy to Eq. 4.20, the standard deviation of means for a small sample is given by

$$s_{\bar{x}} = \frac{s}{\sqrt{n}}$$

For the data of Example 4.1 this is

$$s_{\bar{x}} = \frac{2.08}{\sqrt{7}} = 0.79$$

4.3 THE CONFIDENCE INTERVAL

We can represent the results of an analysis in a rather complete way for a given sample by reporting both the mean and standard deviation. However, we have just seen that the reliability of the values is dependent on the sample size, and it is desirable to have a quantity that specifies just how reliable our result is apt to be. The *confidence interval* provides such a measure.

Let us assume that we are dealing with a population whose mean μ and standard deviation σ are well defined. If we take a sample of size N from the population, we now know that the sample mean provides an approximation of the true population mean, subject to deviations that follow the normal distribution obtained by inserting Eq. 4.21 in Eq. 4.11. We recognize that only for a sample of infinite size can we be assured that the sample mean will be an exact representation of the population mean. In any practical situation, we will have to content ourselves with a certain degree of uncertainty in the result.

In principle a deviation of any magnitude is possible. We recognize, however, that deviations of small magnitude are much more probable than large ones (see Section 4.2.2), so we must establish some limit of the uncertainty which we will consider acceptable. The actual value may vary depending on the circumstances. Some workers choose to define a probability greater than 99% as highly significant, greater than 95% as significant, greater than 90% as marginal, and so forth. Extensive tables are available in other sources to treat these various cases, but for purposes of illustration we will confine ourselves to a 95% level, which is adequate for many cases. In other words, we will determine a range such that if we were to examine a large number of samples, we would be assured that 95% of the time the mean we determine would fall within a certain range of values, known as the *confidence interval*. We will call α the complement of this value, so that $\alpha/2$ lies in each of the two wings of the normal distribution curve (see Figure 4.8). For $\alpha = 0.05$, we of course require the value of z such that 2.5% of the curve lies to the right. We have seen that the value of z for this case is 1.960, and we write $z_{.025} = 1.960$.

By insertion of $z_{\alpha/2}$ in Eq. 4.21, we can solve for the cutoff value of $\bar{x}$ such that the fraction $\alpha/2$ of the values lies in the upper wing of the distribution curve. We call this the *upper confidence limit* $\bar{x}_u$:

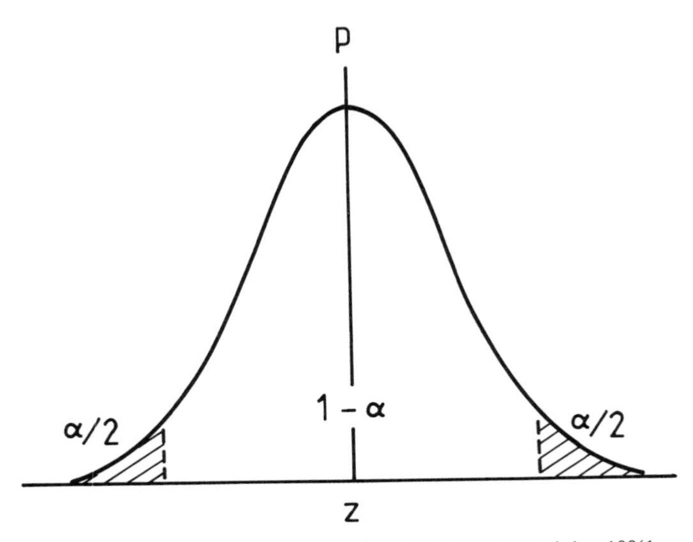

Figure 4.8 Normal distribution curve showing the interval containing $100(1 - \alpha)\%$ of the values. The remainder is distributed equally in the two wings.

$$z_{\alpha/2} = \frac{\bar{x}_u - \mu}{\sigma/\sqrt{N}}$$

$$\bar{x}_u = \mu + z_{\alpha/2} \frac{\sigma}{\sqrt{N}}$$

Similarly, the fraction $\alpha/2$ of the x values lies below the *lower confidence limit* $\bar{x}_l$ given by

$$-z_{\alpha/2} = \frac{\bar{x}_l - \mu}{\sigma/\sqrt{N}}$$

$$\bar{x}_l = \mu - z_{\alpha/2} \frac{\sigma}{\sqrt{N}}$$

These two limits bracket the range in which a determination of $\bar{x}$ for given samples of size N will fall $100(1 - \alpha)\%$ of the time, and we call this the $100(1 - \alpha)\%$ confidence interval. In most cases we must use the sample mean $\bar{x}$ as our best approximation of μ, and we write the confidence interval as

$$\bar{x} \pm z_{\alpha/2} \frac{\sigma}{\sqrt{N}} \tag{4.23}$$

For the 95% confidence interval that we use most frequently, we have

$$\bar{x} \pm 1.960 \frac{\sigma}{\sqrt{N}} \tag{4.24}$$

We assumed in the discussion above that the standard deviation of the population σ was well known, either through previous analysis or by use of a sample of large size (say, at least 20 to 30). Frequently, the use of such large samples is impossible or at least impractical, and we wish to determine the confidence interval for a small sample. From our discussion of the preceding section it is clear that our analysis should be based on the t-distribution rather than on the normal distribution. Otherwise, the argument is the same. Thus from Eq. 4.22 we see that for a small sample, the confidence interval is written

$$\bar{x} \pm t_{\alpha/2} \frac{s}{\sqrt{n}} \tag{4.25}$$

This equation is used and interpreted in the same way as Eq. 4.23. The only difference is that the standard deviation s of the small sample of size n is used in place of σ, and the critical value of $t_{\alpha/2}$ that defines the cutoff enclosing $100(1 - \alpha)\%$ of the curve depends on the degrees of freedom. The values are read from Appendix A.9 for the 95% confidence level. Tables are available in various reference works for other levels. It is noted that the critical t values for the 95% confidence level approach 1.960 for large values of n, but increase for small values of n. In other words, the confidence interval for a small sample is greater, and this is a reflection of the fact that there are uncertainties in both μ and σ which must be accounted for.

Example 4.4

Compute the 95% confidence interval for the mean of the data given in Example 4.1.

Solution. The critical value $t_{\alpha/2}$ for 6 degrees of freedom according to Appendix A.9 is 2.447, so by Eq. 4.25 we have

$$\bar{x} \pm t_{\alpha/2}\frac{s}{\sqrt{n}} = 35.91 \pm (2.447)\left(\frac{2.08}{\sqrt{7}}\right)$$

$$= 35.91 \pm 1.92$$

4.4 TESTS OF SIGNIFICANCE

There are many situations where we must make decisions regarding the significance of differences that may be observed. We may wish to compare analytical methods to determine which is the most accurate or precise, various preparations to see which gives the best yield, various samples to find which has the greatest purity, and so forth. If the difference is large, the choice may be clear, but we require general methods that will allow us to make such decisions without bias in marginal cases. We recognize that in any case there will be uncertainties in the parameters we determine using samples of limited size, so we must rely on arguments based on the probability that an observed difference might be real. Thus we must once again adopt an arbitrary *significance level* α, and determine whether the difference is likely to be real for that particular level.

We will again assume that our significance level is $\alpha = 0.05$ for purposes of illustration. This is the most commonly used level, but once again, other levels may be appropriate in certain instances. The standard reference works may be consulted for more extensive tables as required.

The rationale for a test of significance is as follows: We define a certain *test statistic* that defines the difference in which we are interested. If the difference that we have defined is likely to occur with a probability of at least α, we do not consider the occurrence to be unusual and we label the difference as insignificant. If the probability of occurrence is less than α, however, we regard the difference as real, recognizing of course that there is a certain small probability (i.e., α) that we are in error and that the result is a normal statistical fluctuation. At the 0.05 level, we are saying that if a difference of a certain magnitude occurs at least one time in every 20 observations the result is not considered unusual, but an occurrence less than this is unlikely, so the difference is taken to be significant. There is a corresponding probability of 1/20 that our conclusion is in error. If this is unacceptable, we can use an alternative significance level, but there will always be an element of uncertainty as long as our samples are of finite size.

In statistical terminology, we are asking the question: Are the two samples we are comparing drawn from the same population? A test of significance is based on what is called the *null hypothesis,* which is generally represented by H_0. The null hypothesis states that no significant difference exists. We may be comparing means, standard deviations, or any other statistically defined quantity, which we will represent by S. For a test of significance between two samples 1 and 2, we may represent the null hypothesis symbolically by

$$H_0\colon S_1 = S_2$$

The test statistic is compared with the values in the standard tables, and if the occurrence is unusual at the significance level *that has been selected in advance,* the null hypothesis is rejected and an *alternative hypothesis,* H_a, is accepted. Note that if the occurrence is not unusual, we *do not* accept the null hypothesis. We merely state that there is insufficient data to reject the null hypothesis. The possibility always exists that a real difference will be revealed by further sampling, so technically we can never accept the null hypothesis.

The alternative hypothesis may be stated in various ways depending on the nature of the circumstances. For example, we may be interested in whether a difference exists, regardless of which of the two samples has the larger value of S. Then the alternative hypothesis may be written

$$H_a\colon S_1 \neq S_2$$

Since we are unconcerned in this case whether the difference is positive or negative, we spread the probability α into the two wings of the distribution curve and make a *two-sided test*. If, on the other hand, we wish to know whether S_1 is significantly larger (or smaller) than S_2, our alternative hypothesis reads

$$H_a\colon S_1 > S_2$$

or

$$H_a\colon S_1 < S_2$$

Now the entire probability α is in the upper (or lower) wing of the distribution curve, and we make a *one-sided test*.

4.4.1 The Comparison of Variances (F-Test)

A comparison of the dispersion or relative variabilities of two samples is under consideration here. There are various examples that might be cited where the test is appropriate, including whether one analytical method is more precise than another, whether one instrument delivers results with greater variability than another, and so forth.

Let us assume that we have two samples, whose standard deviations are s_1 and s_2. If they represent the same population, the variances should be equal (i.e., $s_1^2 = s_2^2$). However, we expect s_1 and s_2 to be subject to random errors for finite sample sizes, so they will not be identically equal even if they represent the same population. If the difference is sufficiently small, we might conclude that it is due to random statistical fluctuation and it is likely to be insignificant.

The test statistic that is used in this case is the ratio of the two variances, which is designated by the letter F:

$$F = \frac{s_1^2}{s_2^2} \tag{4.26}$$

Obviously, the value of F should be 1 in the ideal case of two samples taken from a given population. In a real case one of the variances will be larger than the other, and by convention we always construct the F value with the larger of the two variances in the numerator. This procedure does not limit the generality of the F-test but merely simplifies the tabulation of test values, since the critical values of the distribution are always greater than or equal to 1.

The F-distribution is, like the t-distribution, very complex and is seldom used directly. As we might expect, it also depends on sample size. In this case there are two samples involved, and we must specify the degrees of freedom for each. A table of critical values of F at the 0.05 level is given in Appendix A.10. These are the cutoff values of the F-distribution such that 5% of the area lies in the tail beyond that point. As we see, the value must be read along a given row that specifies the degrees of freedom for the denominator (corresponding to the sample with the smaller variance), and down a given column that specifies the degrees of freedom for the numerator (corresponding to the sample with the larger variance). Note that with the exception of extremely small samples, the critical F values become smaller as we move to the right along a given row or down a given column; this is indicative of decreasing variability in the determination of the variance with increasing sample size.

To carry out the F-test we must first construct the null hypothesis, which states that the two variances represent the same population. We represent the null hypothesis by

$$H_0\colon \sigma_1^2 = \sigma_2^2$$

The table given in Appendix A.10 is constructed for a one-sided test. If our question is whether the variance of sample 1 is larger than the variance of sample 2, the alternative hypothesis is

$$H_a\colon \sigma_1^2 > \sigma_2^2$$

and the table may be used directly. That is, we construct the F-statistic (Eq. 4.26) and look up the critical F value from the table using the appropriate degrees of freedom for both numerator and denominator. If our calculated F value exceeds the critical value, we reject the null hypothesis and conclude that sample 1 does show greater variability than sample 2 at the 0.05 level. If our calculated F value is smaller than the critical value, we cannot reject the null hypothesis at the 0.05 level without additional data.

If we had posed the question differently and asked only if there were a significant difference in the variances of samples 1 and 2, without specifying which might be the larger, the alternative hypothesis is written in the form

$$H_a: \sigma_1^2 \neq \sigma_2^2$$

and our test must be applied somewhat differently. If either sample could have the larger variance, the 5% probability should be distributed in two wings and we should apply a two-sided test. We have in fact stipulated that the larger variance be placed in the numerator. Appendix A.10 can still be applied if the test is two-sided, but the significance level will then be $\alpha = 0.10$ since $\alpha/2$ must occur in each tail.

Example 4.5

You have just been appointed laboratory director, and you have to make a decision whether you should continue to do your protein determinations by the traditional Kjeldahl method (tedious acid digestion and all) or by the new near-infrared reflectance technique, which, it is claimed, is very quick and easy. You carry out analyses by both methods on standard samples with the following results:

	Percent N
Kjeldahl method	1.56, 1.78, 1.46, 1.32, 1.49
Infrared method	1.38, 1.76, 1.59, 1.83, 1.29, 1.40

Should you conclude that there is a significant difference in the variability of the results produced by the two methods?

Solution. It is found that the standard deviation of the results obtained by the Kjeldahl method is 0.169 and that obtained for the infrared method is 0.220. From these we calculate the F value:

$$F = \frac{(0.220)^2}{(0.169)^2} = 1.71$$

We find the critical F value from Appendix A.10 under 4 degrees of freedom for the denominator and 5 degrees of freedom for the numerator to be 6.26. Since this is larger than our calculated F value, we cannot reject the null hypothesis; that is, the difference in the variability (precision) of the two methods is regarded as insignificant at the 0.05 level.

4.4.2 The Comparison of Means (t-Test)

There are several different comparisons that may be made involving means of samples. We will discuss three important tests here: the comparison of a mean to a known or assumed value, the comparison of two means, and the comparison of two means for paired data. In each case the comparison is based on the t-distribution, so these are all referred to as *t-tests*. Some authors also define similar z-tests based on the normal distribution. We have seen that the t-distribution becomes the equivalent of the z-

distribution for large sample sizes. Thus we view the normal distribution as just a special case of the t-distribution for large samples, so we will not regard it as a different entity as far as tests of significance are concerned.

The Comparison of a Mean with a Known or Assumed Value

We refer once again to the fact that the distribution of means about the population mean is represented by Student's t-distribution as indicated in Section 4.2 (see Eq. 4.22). This suggests that if we wish to test whether the sample mean represents the population mean μ_0, which is known or assumed, we should use

$$t = \frac{\bar{x} - \mu_0}{s/\sqrt{n}} \tag{4.27}$$

as our test statistic. Since we take $\bar{x}$ to be a representation of the population mean, we may write the null hypothesis in the form

$$H_0: \mu = \mu_0$$

What we write for the alternative hypothesis will depend on the way the question is phrased. We may have any one of the following:

$$H_a: \mu \neq \mu_0 \qquad H_a: \mu > \mu_0 \qquad H_a: \mu < \mu_0$$

In the first case a two-sided test would be applied; either of the latter two forms implies a one-sided test.

The test is made by computing the t-statistic as indicated in Eq. 4.27. This is compared with the critical t value for the appropriate degrees of freedom as given in Appendix A.9. If the computed t value is smaller than that listed in the table, this result would occur at least one time in 20 by normal statistical fluctuation and is not regarded as unusual. If the computed t value exceeds the critical value found in the table, the occurrence is unlikely and the null hypothesis is rejected.

Example 4.6

A new method of analysis for bromide ion has been developed. To test it, a standard sample with a certified analysis 22.34% bromide was obtained. Five determinations of the bromide concentration in the standard sample were made using the new method, and the results were 23.26, 22.43, 23.37, 22.44, and 22.61. Is there reason to suspect that the new method gives values which are too large?

Solution. The mean and standard deviation of the set of five determinations are

$$\bar{x} = 22.82 \qquad s = 0.457$$

Insertion of these values in Eq. 4.27 gives

$$t = \frac{22.82 - 22.34}{0.457/\sqrt{5}} = 2.36$$

The null hypothesis for this case may be written H_0: $\mu = \mu_0$, and the alternative hypothesis is H_a: $\mu > \mu_0$. In other words, we will perform a one-sided t-test with $5 - 1 = 4$

degrees of freedom. The critical value of $t_{.05}$ found in Appendix A.9 is 2.132. Since our calculated value is greater than this, we reject the null hypothesis. It is concluded that there is a bias in the new method and it can be expected to yield results which are significantly larger than the true values. The 95% confidence level of the result obtained by the new method is

$$\bar{x} \pm t_{.05/2} \frac{s}{\sqrt{n}} = 22.82 \pm (2.776)\left(\frac{0.457}{\sqrt{5}}\right) = 22.82 \pm 0.57$$

The Comparison of Two Means

Consider two populations with means and standard deviations μ_1,σ_1 and μ_2,σ_2. Suppose that we draw a random sample of size n_1 from population 1, and a random sample of size n_2 from population 2, and that we compute $\bar{x}_1,s_1$ and $\bar{x}_2,s_2$ for these samples. The question we need to answer is: How do the means $\bar{x}_1$ and $\bar{x}_2$ differ?

Now the two populations have a difference of means that is equal to $\mu_1 - \mu_2$, and since $\bar{x}_1$ is an approximation of μ_1 and $\bar{x}_2$ is an approximation of μ_2, we would expect that $\bar{x}_1 - \bar{x}_2$ should reflect this same difference. In other words, with repeated sampling we should find that values of $\bar{x}_1 - \bar{x}_2$ would be centered about the true difference of means $\mu_1 - \mu_2$. The nature of the distribution follows from our discussion of Section 4.2. An examination of the argument leading up to Eq. 4.15 shows that the same result would be obtained for combination of two variables as either a sum or a difference. Applying this to the present case gives

$$s_{x_1-x_2}^2 = s_{x_1}^2 + s_{x_2}^2$$

It follows that

$$s_{\bar{x}_1-\bar{x}_2}^2 = s_{\bar{x}_1}^2 + s_{\bar{x}_2}^2 = \frac{s_1^2}{n_1} + \frac{s_2^2}{n_2}$$

or

$$s_{\bar{x}_1-\bar{x}_2} = \sqrt{\frac{s_1^2}{n_1} + \frac{s_2^2}{n_2}} \tag{4.28}$$

Thus we conclude that the differences in sample means $\bar{x}_1 - \bar{x}_2$ are distributed about the true difference in population means $\mu_1 - \mu_2$ with a distribution that is characterized by the standard deviation given by Eq. 4.28. This is a normal distribution for large numbers of sample pairs, but follows the t-distribution for small numbers. Based on the above, we might expect

$$t = \frac{\bar{x}_1 - \bar{x}_2}{\sqrt{s_1^2/n_1 + s_2^2/n_2}} \tag{4.29}$$

to serve as the general test statistic for the comparison of two means.

Now if the two means $\bar{x}_1$ and $\bar{x}_2$ represent similar populations, we would have $\sigma_1^2 = \sigma_2^2$ and $s_1^2 \approx s_2^2$. This suggests that the comparison of two sample means should only be made if the variances are the same or nearly so. If the variances are quite different, there could be large differences in sample means even when the population means are identical, and this would invalidate the t-test. It may be necessary to apply an F-test in some instances to see if we are really justified in attempting a comparison. Let us assume that the variances may be taken as equal, in which case we may write $s_1^2 = s_2^2 \equiv s^2$, and our test statistic becomes

$$t = \frac{\bar{x}_1 - \bar{x}_2}{s\sqrt{1/n_1 + 1/n_2}} \tag{4.30}$$

We may use either s_1 or s_2 to approximate s in Eq. 4.30, but it would obviously be preferable to use a weighted average of the two to get the best approximation of the true standard deviation σ of the population. Thus we write

$$s = \sqrt{\frac{(n_1 - 1)s_1^2 + (n_2 - 1)s_2^2}{n_1 + n_2 - 2}} \tag{4.31}$$

for the *pooled standard deviation*. Note that s_1^2 and s_2^2 are each weighted by their respective degrees of freedom, and the total degrees of freedom for the pooled standard deviation is $n_1 + n_2 - 2$.

Finally, we note that the arguments above may be generalized to include a hypothetical difference between the two means. The comparison is still valid as long as the two populations have similar variances. We represent the hypothesized difference of means by D_0 (which will be zero in many cases), and then the test statistic becomes

$$t = \frac{\bar{x}_1 - \bar{x}_2 - D_0}{s\sqrt{1/n_1 + 1/n_2}} \tag{4.32}$$

The null hypothesis is written H_0: $(\mu_1 - \mu_2) = D_0$. The alternative hypothesis is H_a: $(\mu_1 - \mu_2) \neq D_0$ for a two-sided t-test; this implies that rejection of the null hypothesis is made for $t < -t_{\alpha/2}$ or $t > t_{\alpha/2}$, where $t_{\alpha/2}$ is the critical t value found in the table (Appendix A.9) for $n_1 + n_2 - 2$ degrees of freedom. Obviously, if our alternative hypothesis is H_a: $(\mu_1 - \mu_2) > D_0$ or H_a: $(\mu_1 - \mu_2) < D_0$, we use $t > t_\alpha$ or $t < -t_\alpha$ as the rejection criterion in a one-sided test.

Example 4.7

Assume that we have two carloads of iron ore, and we wish to know if their iron content is the same. Analyses are performed on samples withdrawn at random from the two cars with the following results:

Car	Percent Fe
1	38.65, 36.24, 36.37, 35.41, 37.92
2	38.86, 39.01, 36.22, 38.13, 37.79, 38.67

Are we justified in assuming that any significant difference in the percent iron of the two cars exists at the 0.05 level?

Solution. The means and standard deviations for the two sets of data are first computed with the following results:

$$\text{Car 1:} \quad \bar{x}_1 = 36.92 \qquad s_1 = 1.33$$

$$\text{Car 2:} \quad \bar{x}_2 = 38.11 \qquad s_2 = 1.04$$

The standard deviations are sufficiently different to suggest that we should compare them with an F-test to see if we are justified in assuming that they represent the same population. We find

$$F = \frac{s_1^2}{s_2^2} = \frac{1.76}{1.07} = 1.64$$

and the critical F value for 5 degrees of freedom in the denominator and 4 in the numerator according to Appendix A.10 is 5.19. We conclude that the comparison of means by the t-test is justified. We calculate the pooled standard deviation according to Eq. 4.31:

$$s = \sqrt{\frac{(5 - 1)1.76 + (6 - 1)1.07}{5 + 6 - 2}} = 1.17$$

The t value that we calculate by Eq. 4.30 is

$$t = \frac{36.92 - 38.11}{1.17\sqrt{1/5 + 1/6}} = -1.68$$

The null hypothesis may be written H_0: $(\mu_1 - \mu_2) = 0$ and the alternative hypothesis is H_a: $(\mu_1 - \mu_2) \neq 0$. We see that a two-sided t-test is appropriate, and we find that the critical value from Appendix A.9 for $5 + 6 - 2 = 9$ degrees of freedom is 2.262. Now we are concerned only with the magnitude of the difference, not the sign of t. We see that the magnitude of our calculated value is less than the critical value, so the difference in the means of the two cars is not regarded as significant at the 0.05 level.

The Comparison of Means for Paired Data

We will illustrate what is involved in the comparison of means for paired data by consideration of a specific example. Let us assume that we have developed a new bleach for fabrics, and we wish to determine whether the bleaching process has a significant effect on the strength of the fibers. Now we recognize that there is a certain degree of nonuniformity between the fibers, so we take five short lengths of fiber from five different spools and cut each into two shorter lengths. Even though there is quite a lot of variation in the fibers, it seems plausible that we should be able to compare the means of the sets of shorter fibers before and after bleaching to see if the strength has been materially affected. The details of the strength test need not concern us here; presumably a measure of the force required to break the fibers could easily be devised with the strengths tabulated in some appropriate units.

Assume that such a series of tests has been completed, with the following results:

Fiber	Strength before Treatment (1)	Strength after Treatment (2)
1	11.7	11.3
2	9.9	9.5
3	10.2	9.8
4	9.7	9.2
5	8.9	8.6
	$\bar{x}_1 = 10.08$	$\bar{x}_2 = 9.68$
	$s_1^2 = 1.05$	$s_2^2 = 1.02$

We will attempt a comparison of these two means with the t-test outlined in the preceding section. Using Eq. 4.31, we have

$$s = \sqrt{\frac{(5-1)1.05 + (5-1)1.02}{5+5-2}} = 1.02$$

and the t value according to Eq. 4.30 is

$$t = \frac{10.08 - 9.68}{1.02\sqrt{1/5 + 1/5}} = 0.62$$

From Appendix A.9 we find that the critical t value $t_{.05/2}$ for 8 degrees of freedom for a two-sided t-test is 2.306. Our computed t value is far less than the critical value at the 0.05 level, and we cannot reject the null hypothesis; that is, the difference in the strength of the fibers is not regarded as significant.

A little reflection will show that this result is really not very reasonable. As we examine the data above we see that for *every* pair of fibers, there was a decrease in the strength after bleaching. How then can we conclude that the effect is without significance? The answer to this apparent paradox lies in the fact that we have made an improper application of the comparison of means. In the preceding section we considered two *random* samples withdrawn from the two populations. That condition is not met here, and this is the essence of the problem. There is greater variability in the strength of the fibers themselves than in the effect we are attempting to measure, and this is the reason for the failure of the test.

Our test was specifically designed to take account of the fact that the fibers themselves showed considerable nonuniformity, but this feature was not included in the comparison above. When dealing with *paired data* (i.e., when there is a one-to-one correspondence between values in the two groups) they must be compared in a pairwise fashion. We construct the average of the differences,

$$\bar{d} = \frac{1}{n} \Sigma\, d_i = \frac{1}{n} \Sigma\, (x_{1_i} - x_{2_i}) \tag{4.33}$$

and the standard deviation of the differences,

$$s_d = \sqrt{\frac{\Sigma\, d_i^2}{n-1}} = \sqrt{\frac{\Sigma\,(x_{1_i} - x_{2_i})^2}{n-1}} \tag{4.34}$$

The test statistic is defined in terms of these as follows:

$$t = \frac{\bar{d} - 0}{s_d/\sqrt{n}} \tag{4.35}$$

The null hypothesis for the example above may be written H_0: $\mu_d = 0$, where $\mu_d = \mu_1 - \mu_2$, and the alternative hypothesis is H_a: $\mu_d \neq 0$. The calculations in this case give the following:

Fiber	d_i
1	0.4
2	0.4
3	0.4
4	0.5
5	0.3
$\bar{d} = 0.40$	$s_d = 0.0707$

so

$$t = \frac{0.40 - 0}{0.0707/\sqrt{5}} = 12.65$$

Now the critical value of $t_{.05/2}$ for a two-sided t-test with 4 degrees of freedom is 2.776. We should clearly reject the null hypothesis and conclude that the treatment of the fibers with bleach leads to a significant decrease in strength.

The magnitude of the effect is probably best represented in terms of the confidence interval as defined in Section 4.3. The 95% confidence interval for the decrease in strength is (see Eq. 4.25)

$$\bar{d} \pm t_{.05/2}\frac{s_d}{\sqrt{n}} = 0.40 \pm (2.776)\left(\frac{0.0707}{\sqrt{5}}\right) = 0.40 \pm 0.09$$

Example 4.8

Professor X has developed a new procedure for treating catalyst surfaces which he claims will result in a significant enhancement in the number of active sites. The number of active sites can be determined by adsorption of H_2 gas. It is recognized that the native catalysts before treatment may show considerable variation, so Professor X tested each sample before and after the treatment and obtained the following for H_2 uptake in terms of mmol g^{-1}:

Sample Number	Before Treatment	After Treatment
1	165	172
2	146	189
3	174	168
4	186	176
5	147	198
6	153	184
7	132	188
8	175	197

Has the treatment resulted in a significant increase in the number of active sites on the catalyst surfaces?

Solution. Because there is a one-to-one correspondence of the data before and after the treatment, a paired difference test is appropriate. We find

Sample Number	Difference d_i
1	7
2	43
3	−6
4	−10
5	51
6	31
7	56
8	22
$\bar{d} = \frac{\Sigma d_i}{8} = 24.25$	$s_d = \sqrt{\frac{\Sigma d_i^2}{8-1}} = 25.39$

and the calculated t value is

$$t = \frac{24.25 - 0}{25.39/\sqrt{8}} = 2.70$$

The critical t value from Appendix A.9 for a one-sided test (i.e., for the alternative hypothesis H_a: $\mu_2 > \mu_1$) for $8 - 1 = 7$ degrees of freedom is 1.895. It appears that Professor X is on to something and should see a patent attorney right away. In other words, we reject the null hypothesis and conclude that there is a significant increase in the number of active sites at the 0.05 significance level.

4.4.3 The Comparison of Frequencies (χ^2-Test)

In Section 4.2 we considered a simple case in which events could be classified as either successful or unsuccessful, and this led us to the binomial distribution. In a similar fashion, we can consider that we have a large number N of identical events, but that the outcome of any given event can fall in any one of k classes. The probability of occurrence for a given class is the same for each event, and the sum of these probabilities is unity:

$$p_1 + p_2 + p_3 + \cdots + p_k = 1$$

Suppose that we now carry out the process N times, and note the number of times the result occurs in each of the k classes. We represent these frequencies of occurrence by $N_1, N_2, \ldots$, and note that

$$N_1 + N_2 + N_3 + \cdots + N_k = N$$

We have just described a *multinomial experiment*, and the problem we wish to examine is whether the observed frequencies correspond with those that would be expected based on a set of known or assumed probabilities $p_1, p_2, \ldots$. Now if the probability of occurrence for the ith class for a single event is p_i, the expected frequency of occurrence in the ith class after N independent trials is

$$E_i = Np_i$$

We can construct the expected values for each of the classes in this way and compare them with the observed frequencies N_i. It seems clear that our test statistic in this case should involve the terms

$$N_i - Np_i$$

In 1900 the British statistician Karl Pearson showed that in repeated sampling the sum of the squares of these deviations, each divided by their expected values,

$$\sum_{i=1}^{k} \frac{(N_i - Np_i)^2}{Np_i}$$

follows what is known as a *chi-square probability distribution*. We will again disregard the mathematical form of the distribution function and concern ourselves only with the critical values of χ^2, which includes $100(1 - \alpha)\%$ of the area under the curve. For purposes of illustration we again quote only the values for $\alpha = 0.05$. They are given in Appendix A.11. This is suitable for many cases but there are instances where other levels may be more appropriate; reference works should be consulted for more extensive compilations.

The calculated test statistic for a χ^2-test is usually written in the form

$$\chi^2 = \sum_{i=1}^{k} \frac{(O_i - E_i)^2}{E_i} \tag{4.36}$$

where E_i is the expected frequency and O_i is the observed frequency of the ith class. The test is made by comparison of this calculated value of χ^2 with the critical value from the table. As usual, the critical values of χ^2 depend on the degrees of freedom. This will usually be equal to the number of classes minus one $(k - 1)$, because of the constraint

$$N_1 + N_2 + N_3 + \cdots + N_k = N$$

In some special cases there may be additional constraints on the distribution, in which case the degrees of freedom must be correspondingly reduced.

The null hypothesis is of the general form

$$H_0: p_1 = a_1, \quad p_2 = a_2, \quad p_3 = a_3, \quad \ldots$$

and the alternative hypothesis

$$H_a: p_1 \neq a_1, \quad p_2 \neq a_2, \quad p_3 \neq a_3, \ldots$$

Note that because the difference of observed and expected frequencies is squared in Eq. 4.36, the value of χ^2 is always positive regardless of whether the actual difference is positive or negative. Thus the χ^2-test is inherently a two-sided test, and the tables are constructed accordingly.

Example 4.9

You have just been appointed regional director for five analytical laboratories operating in five different locations. You wish to impress the vice president, to whom you report, so you decide to set up an incentive program to stimulate greater productivity. A blue ribbon laboratory will be selected and all employees given a bonus, provided that it reports a significantly greater number of analyses per employee than its competitor laboratories.

A check of the records of the past year provides the following data:

Laboratory	Number of Employees	Analyses
A	10	9,261
B	15	13,954
C	20	19,226
D	7	6,527
E	9	8,535
	61	57,503

Is one or more of the laboratories worthy of the blue ribbon designation?

Solution. We are dealing here with frequency distributions, so the results are compared with a χ^2-test. The average number of analyses per employee for all five laboratories is

$$\frac{57{,}503}{61} = 942.67$$

We can then readily calculate the expected number of analyses for each laboratory based on the number of employees, and construct the following table:

Laboratory	E_i	$(O_i - E_i)^2/E_i$
A	9,426.7	2.91
B	14,140.1	2.45
C	18,853.4	7.36
D	6,598.7	0.78
E	8,484.1	0.31
		$\chi^2 = 13.81$

Under $\chi^2_{.05}$ for 4 degrees of freedom in Appendix A.11 we find the critical value 9.48. Thus we reject the null hypothesis and conclude that there is a significant difference in productivity between laboratories.

It is obvious that laboratory C has made the largest contribution to χ^2. If we omit it and repeat the analysis using only laboratories A, B, D, and E, we find that $\chi^2 = 2.89$. The critical value of $\chi^2_{.05}$ for 3 degrees of freedom is 7.81, and the deviation from the expected values is not regarded as significant. Thus laboratory C is the only one deserving of recognition, so you pass out the bonuses and begin listening to all the reasons why the other laboratories were placed at a disadvantage!

4.5 DEALING WITH OUTLIERS IN SMALL SAMPLES

The term *outliers* is used to represent values that are far removed from most of the others. They can be very difficult to deal with, particularly in small samples. Suppose that you have performed four separate determinations of the amount of chloride in a material by a gravimetric method in which the AgCl precipitate is weighed. If three of the values show good consistency but one is much larger than the rest, the various possible reasons should be examined. Maybe a visitor to the laboratory dropped some cigar ashes in one of your beakers, or possibly you transposed two digits when you recorded the weight. Many kinds of gross errors or blunders are possible. It is also conceivable that there was a sampling problem and one of the determinations really represented a different population. Some populations inherently contain an unusually large proportion of extreme cases, and it is possible that the sample just happens to include one. How can these cases be distinguished?

Unfortunately, there is no simple answer that covers all situations. Certainly the first thing to do with a small sample that contains an outlier is to examine carefully all the data to satisfy yourself that there is no obvious source of error. Second, it is generally useful to give some consideration to the question of the method itself. If deviations of the observed magnitude are routinely to be expected for the method used, there is probably no justification for regarding the outlier as spurious.

If, after the foregoing considerations, there is still reason to question whether the unusual value should be included in the analysis or discarded, a simple test called the q-test may be applied. We assume that we are dealing with a small sample of, say, 3 to 10 values. The values are written in ascending order, so that the outlier in question is labeled either x_1 or x_n. We then determine the q value from either

$$q = \frac{x_2 - x_1}{w} \tag{4.37}$$

or

$$q = \frac{x_n - x_{n-1}}{w} \tag{4.38}$$

depending on whether the outlier is abnormally small or large. In these expressions w represents the range $w = x_n - x_1$. The test is usually made against critical values of q

for a normal distribution at the 90% confidence level. A table of such values as a function of the sample size in the range 3 to 10 is given in Appendix A.12. If the calculated value of q exceeds the critical value, the outlier may be rejected at the 90% confidence level; otherwise, it should be retained and used in the analysis.

It will be recalled that a suggestion was made previously that the median should be given consideration as a representation of the central tendency when dealing with very small samples. This becomes particularly important if the data set contains an outlier that cannot be excluded on the basis of the q-test. The median is far less sensitive to the presence of the outlier, so it can be included in the data set without exerting an inordinate effect.

It is found that for very small samples, a reliable estimate of the dispersion can often be obtained using the range rather than a direct calculation of the standard deviation. There is a relationship between the two parameters for small samples that may be represented by

$$s = K_w w \tag{4.39}$$

where K_w is a proportionality factor that depends on the size of the sample. Values of K_w for small samples in the range $n = 2$ to $n = 10$ are given in Appendix A.12.

In the same vein, one may choose to estimate the confidence interval of a very small sample using the range. By insertion of Eq. 4.39 in Eq. 4.25, we have

$$\bar{x} \pm t_{\alpha/2} \frac{K_w w}{\sqrt{n}} = \bar{x} \pm t_w w \tag{4.40}$$

with

$$t_w = t_{\alpha/2} \frac{K_w}{\sqrt{n}}$$

Values of t_w are listed in Appendix A.12 for the estimation of the 95% confidence interval for small samples using the range. These shortcut methods can save quite a lot of time, and they generally yield parameters that are as reliable as the more involved calculations for very small samples.

Example 4.10

Assume that the values 6.75, 8.61, 11.24, and 7.23 have been determined. The value 11.24 may be regarded as an outlier.

(a) Should the value 11.24 be rejected?

(b) Estimate the standard deviation using the range, and compare with the value obtained by direct calculation.

(c) Estimate the 95% confidence interval using the range, and compare with the value obtained by direct calculation.

Solution. (a) According to the definition of Eq. 4.38, the q value for testing whether 11.24 should be rejected is

$$q = \frac{11.24 - 8.61}{11.24 - 6.75} = 0.59$$

The critical q value for a sample composed of four values is 0.74 (Appendix A.12). Since our calculated value is less than the critical value, we cannot reject the outlier at the 0.10 level. We note that the means calculated with and without the outlier are 8.46 and 7.53, while the medians are 7.92 and 7.23, respectively. Because the median is less sensitive to the inclusion of the outlier, it is chosen as the measure of central tendency.

(b) The standard deviation is estimated using the range according to Eq. 4.39:

$$s = (0.49)(4.49) = 2.20$$

This may be compared with the value $s = 2.02$ calculated by direct application of Eq. 4.3 or 4.4.

(c) Using Eq. 4.40, we estimate the 95% confidence interval to be

$$7.92 \pm (0.72)(4.49) = 7.92 \pm 3.23$$

The corresponding value by direct calculation using Eq. 4.25 is

$$7.92 \pm 3.21$$

4.6 ANALYSIS OF VARIANCE

The tests of significance for means described in Section 4.4.2 were designed to make a pairwise comparison for two values. A general method that can be applied to the effect of any number of variables on a given result is based on an analysis of the factors that contribute to the total variance of the population. This is called an *analysis of variance,* and it is often abbreviated ANOVA. We may wish, for example, to compare several methods of analysis, the effect of various physical conditions on the product of a reaction, and so forth. An extension of the simple t-test of Section 4.4.2 for comparison of an arbitrary number of factors is called the *multiple group experiment;* similarly, an extension of the t-test for paired data of Section 4.4.2 for comparison of an arbitrary number of factors is called the *randomized block experiment.*

4.6.1 The Multiple Group Experiment

We assume that we have a certain number of factors k which we wish to examine. The members of the population x_{ij} may be represented by

$$x_{ij} = \mu + f_i + \epsilon_{ij}$$

where μ is the true population mean, ϵ_{ij} is a random error normally distributed about the value zero, and f_i is the effect due to factor i, defined such that the average effect of all the factors is zero: $\sum_{i=1}^{k} f_i = 0$. The question we wish to examine is whether or not the values of x_{ij} corresponding to the various factors i belong to the same population. Thus the null hypothesis is written

$$H_0\colon \mu_1 = \mu_2 = \cdots = \mu_i = \cdots \mu_k$$

The alternative hypothesis states that one or more of the means is different at the particular significance level for which the test is applied.

We will describe the techniques involved by carrying out the ANOVA for a hypothetical example of a multiple group experiment. Let us assume that a given analyst wishes to compare the results of chloride analysis by several methods. Those chosen are titrimetric determinations by the Mohr, clear point, and Volhard methods, and gravimetric analysis. Six analysis ($n = 6$) were carried out on a given standard sample by each of the four ($k = 4$) methods, with the following results:

n \ k	1 (Mohr)	2 (clear point)	3 (Volhard)	4 (gravimetric)
1	14.003	14.031	14.012	14.005
2	14.025	14.042	14.022	14.028
3	14.013	14.035	14.023	14.012
4	14.017	14.027	14.030	14.018
5	14.015	14.039	14.007	14.019
6	14.031	14.034	14.024	14.024
$\bar{x}$	14.0173	14.0347	14.0197	14.0177
s^2	9.507×10^{-5}	2.907×10^{-5}	7.227×10^{-5}	6.827×10^{-5}

We can calculate a pooled variance for the entire population by combining those given above according to an extension of Eq. 4.31; the result is

$$s_p^2 = \frac{5(9.507 \times 10^{-5}) + 5(2.907 \times 10^{-5}) + 5(7.227 \times 10^{-5}) + 5(6.827 \times 10^{-5})}{4(6-1)}$$

$$= 6.617 \times 10^{-5}$$

A second expression for the variance is obtained from the standard error or standard deviation of means $s_{\bar{x}}$. Using the data above we find that the mean of the means $\bar{\bar{x}}$ is 14.0224, and by Eq. 4.3

$$s_{\bar{x}} = \sqrt{\frac{(14.0173 - 14.0224)^2 + (14.0347 - 14.0224)^2 + (14.0197 - 14.0224)^2 + (14.0177 - 14.0224)^2}{4-1}}$$

$$= 8.301 \times 10^{-3}$$

This is related to the standard deviation for a given method through $s_{\bar{x}}^2 = s^2/n$ (see Eq. 4.20), so the variance based on the standard deviation of means is $s_m^2 = ns_{\bar{x}}^2 = 6(8.301 \times 10^{-3})^2 = 4.134 \times 10^{-4}$. Now if there were no difference in the means corresponding to the various methods, s_m^2 would be equal to zero. Thus the ratio $F = s_m^2/s_p^2$ serves as a measure of the relative contribution to the total variance by the effect of the different methods. In our particular case $F = 6.25$. The critical value of F at the 0.05 level for $k - 1 = 3$ degrees of freedom in the numerator and $k(n - 1) = 20$ degrees of freedom in the denominator, according to Appendix A.10, is 3.10. Thus the F-test shows that there is a significant difference in the means of the various methods at the 0.05 level, and the null hypothesis is rejected.

If the F-test reveals that there is a difference in the means corresponding to the various methods, as in the present case, further analysis is called for to determine which difference(s) may be significant. This is done by means of a *D-test*. D represents a range of differences between means which may be considered significant at the level used for comparison. They are defined by

$$D = Qs'_{\bar{x}} \tag{4.41}$$

where the values of Q are given in Appendix A.13 for the 0.05 level, and $s'_{\bar{x}}$ is the mean standard error of a given method. This can be calculated from the pooled variance of the entire population by $s'_{\bar{x}} = \sqrt{s_p^2/n} = 3.321 \times 10^{-3}$.

We rank the values of $\bar{x}$ given above and compute differences between them to complete the analysis. Thus $r_1 = 14.0347$, $r_2 = 14.0197$, $r_3 = 14.0177$, and $r_4 = 14.0173$. The values of Q listed in Appendix A.13 depend on differences in rank; the values of k' for which the tabulation is made are one unit greater than the differences in rank. Thus in our case $r_1 - r_4 = 0.0173$ is three ranks apart, and we see that the value of Q for $k' = 4$ and $k(n - 1) = 20$ degrees of freedom is 3.96. This gives a value for the range of differences of $D = (3.96)(3.321 \times 10^{-3}) = 0.0132$. Similarly, $r_1 - r_3 = 0.0170$ is two ranks apart and $D = (3.58)(3.321 \times 10^{-3}) = 0.0119$. By continuation of this process the following difference table is constructed:

	$\bar{x}$	$r - r_4$	$r - r_3$	$r - r_2$
r_1	14.0347	0.0173 ± 0.0132	0.0170 ± 0.0119	0.0150 ± 0.0098
r_2	14.0197	0.0023 ± 0.0119	0.0020 ± 0.0098	
r_3	14.0177	0.0003 ± 0.0098		
r_4	14.0173			

It will be observed that the range of differences for r_1 compared with each of the other means is greater than zero, so we conclude that r_1 (corresponding to $\bar{x}_2$, or the clear point method) differs significantly from the others at the 0.05 level. However, the ranges between r_2, r_3, and r_4 [corresponding to $\bar{x}_3$ (Volhard method), $\bar{x}_4$ (gravimetric method), and $\bar{x}_1$ (Mohr method), respectively] include zero, so their differences are not considered significant at the 0.05 level.

4.6.2 The Randomized Block Experiment

We may have samples that are known or suspected to have inherent differences in them; they may be used as a basis for analysis using the randomized block experiment. It should be noted that there must be some clear rationale for separation into distinct *blocks*, in the same sense that the samples were grouped in pairs for a t-test as described in Section 4.4.2. Our purpose is to assess differences in the processes to which the blocked data may be subjected, such as different methods of analysis, different conditions of compound synthesis, different temperatures of catalyst activation, and so forth.

The experimental design may be illustrated as follows. Consider material from five sources that we wish to subject to three separate processes. We will assume that each of the five sources is sampled three times; the samples from a given source constitute a block. For convenience we may label the 15 samples 1.1, 1.2, 1.3, 2.1, 2.2, . . . , 5.3. We use a process of *randomization* to assign the samples of a given block for treatment by each of the three processes. This may be done using the random number generator of a calculator or computer, or by use of random number tables that are available in various sources, such as the *CRC Standard Mathematical Tables*. We may represent the result in a given case as follows:

		Block				
	$k \backslash n$	1	2	3	4	5
	1	1.3	2.2	3.1	4.3	5.1
Process	2	1.2	2.1	3.2	4.2	5.3
	3	1.1	2.3	3.3	4.1	5.2

Note that the samples are randomized among the processes but *not* among the blocks; we assume inherent differences in the latter which require that the samples be grouped accordingly.

The members of the population in this case may be described by

$$x_{ij} = \mu + b_i + p_j + \epsilon_{ij}$$

where μ is the mean of the entire population and ϵ_{ij} is a random error normally distributed about zero. b_i represents the differences due to the various blocks, and p_j represents the differences due to the various processes. The sums of the block and process effects are zero: $\sum_{i=1}^{n} b_i = \sum_{j=1}^{k} p_j = 0$. We wish to determine whether there are significant differences in the x_{ij} values due to the processes. Our null hypothesis states that the x_{ij} belong to the same population for various values of j, and it may be written

$$H_0: \mu_1 = \mu_2 = \cdots = \mu_j = \cdots = \mu_k$$

The alternative hypothesis states that one or more of the means is different from the population mean at the significance level used for the test.

As before, we will use a specific example to illustrate the process of ANOVA for a randomized block experiment. Assume that we have five silver wires which we wish to analyze for their silver content using three different methods. Three short lengths are taken from each of the five wires and randomly assigned for analysis by the three methods according to the foregoing experimental design. The results are:

k \ n	1	2	3	4	5	Sum
1	96.22	94.72	93.65	92.89	98.14	475.62
2	96.15	94.60	93.54	92.72	98.04	475.05
3	96.21	94.73	93.61	92.88	98.12	475.55
Sum	288.58	284.05	280.80	278.49	294.30	1426.22

The first step of the ANOVA is to calculate the sum of each row (process) and column (block), and a general sum for the entire experiment. These results have been included in the table. The second step is to calculate a general correction factor (see Eq. 4.4)

$$C = \frac{\left(\sum_{i=1}^{n} \sum_{j=1}^{k} x_{ij}\right)^2}{k_n} = \frac{(1426.22)^2}{15} = 135{,}606.8992$$

which is required for the calculation of the variances. We next calculate the total sum of squares,

$$SS_T = \sum_{i=1}^{n} \sum_{j=1}^{k} x_{ij}^2 - C = [(96.22)^2 + (96.15)^2 + \cdots] - C = 53.3538$$

the block sum of squares,

$$SS_B = \frac{1}{k} \sum_{j=1}^{n} x_{ij}^2 - C = \frac{(288.58)^2 + (284.05)^2 + \cdots}{3} - C = 53.3105$$

and the process sum of squares,

$$SS_P = \frac{1}{n} \sum_{i=1}^{k} x_{ij}^2 - C = \frac{(475.62)^2 + (475.05)^2 + (475.55)^2}{5} - C = 0.0387$$

The error sum of squares is the difference in the total sum of squares and the sum of the block and process sums of squares:

$$SS_E = SS_T - (SS_B + SS_P) = 0.0046$$

The test statistic used is the F value or ratio of the process variance and the error variance. The former has $n - 1 = 4$ degrees of freedom, while the latter has $kn - k - n + 1 = 8$ degrees of freedom. Thus

$$F = \frac{0.0387/4}{0.0046/8} = 16.8$$

The critical value of F at the 0.05 level with 4 degrees of freedom in the numerator and 8 degrees of freedom in the denominator, from Appendix A.10, is 3.84. Therefore, we reject the null hypothesis and conclude that there are significant differences in the analyses of silver content by the three methods.

Since we have found differences in the methods, we again use the D-test to determine which are significant. We first compute variances for each of the blocks (1.433×10^{-3}, 5.233×10^{-3}, 3.100×10^{-3}, 9.100×10^{-3}, 2.800×10^{-3}), which may be pooled to estimate the variance of an entire population if there were no differences in the blocks (4.333×10^{-3}). The standard deviation of means for five analyses taken from such a uniform population would be $s_{\bar{x}} = \sqrt{(4.333 \times 10^{-3})/5} = 2.944 \times 10^{-2}$. Under $k(n - 1) = 12$ degrees of freedom we find that $Q = 3.08$ for $k' = 2$ (one rank apart) and $Q = 3.77$ for $k' = 3$ (two ranks apart), and the corresponding values of D according to Eq. 4.41 are 0.091 and 0.111. The mean values $\bar{x}$ for the three methods are 95.124, 95.010, and 95.110, respectively, so the difference table is

	$\bar{x}$	$r - r_3$	$r - r_2$
r_1	95.124	0.114 ± 0.111	0.014 ± 0.091
r_2	95.110	0.100 ± 0.091	
r_3	95.010		

We see that the ranges of differences between r_1 and r_3, and between r_2 and r_3, do not include zero, so r_3 (method II) gives results that are less than the other two at the 0.05 level. The range of difference between r_1 and r_2 (methods I and III) includes zero, so these methods do not give results that are considered significantly different at the 0.05 level.

SUGGESTED ADDITIONAL READING

1. Philip R. Bevington, *Data Reduction and Error Analysis for the Physical Sciences*, McGraw-Hill Book Company, New York, 1969.
2. George E. P. Box, William G. Hunter, and J. Stuart Hunter, *Statistics for Experimenters: An Introduction to Design, Data Analysis & Model Building*, John Wiley & Sons, Inc., New York, 1978.
3. C. J. Brookes, I. G. Betteley, and S. M. Loxston, *Fundamentals of Mathematics and Statistics for Students of Chemistry and Allied Subjects*, John Wiley & Sons, Inc., New York, 1979.
4. Walter Clark Hamilton, *Statistics in Physical Science*, The Ronald Press Company, New York, 1964.
5. P. MacDonald, *Mathematics and Statistics for Scientists and Engineers*, Van Nostrand Reinhold Company, New York, 1966.
6. William Mendenhall, *Introduction to Probability and Statistics*, 5th ed., Duxbury Press, North Scituate, Mass., 1979.

7. Emerson M. Pugh and George H. Winslow, *The Analysis of Physical Measurements,* Addison-Wesley Publishing Co., Inc., Reading, Mass., 1966.
8. Hubert L. Youmans, *Statistics for Chemistry,* Charles E. Merrill Publishing Company, Columbus, Ohio, 1973.
9. Hugh D. Young, *Statistical Treatment of Experimental Data,* McGraw-Hill Book Company, New York, 1962.

EXERCISES

4.1. A number of iron ore samples from a given location were analyzed with the following results (% Fe): 26.08, 24.16, 25.93, 28.62, 26.33, 32.21, 27.12, 21.44, 27.05, 25.22, 27.01, 29.88, 22.84, 31.55, 26.66, 30.70, 24.47, 27.59, 25.30, 29.43. Calculate the mean $\bar{x}$ and the standard deviation s for this set of values.

4.2. Plot the data of Exercise 4.1 as a frequency distribution using the class intervals 20–21, 21–22, 22–23, ...

4.3. Using the values of $\bar{x}$ and s found in Exercise 4.1, plot the frequency f given by Eq. 4.10 on the figure constructed in Exercise 4.2.

4.4. Calculate the values of $\bar{x}$ and s from the frequency distribution using Eqs. 4.5 and 4.6, and compare them with the values obtained directly in Exercise 4.1.

4.5. Calculate the following for the iron ore samples of Exercise 4.1, assuming that they follow a normal distribution.

(a) The probability that the percent Fe will be greater than 30 in a given sample
(b) The probability that the percent Fe will be greater than 25 in a given sample
(c) The probability that the percent Fe will be less than 20 in a given sample
(d) The probability that the percent Fe will lie between 22 and 28 in a given sample

4.6. The discharge of suspended solids from a phosphate mine follows a normal distribution, with a mean daily discharge of 27 mg L^{-1} and a standard deviation of 14 mg L^{-1}. During how many days of a typical year might one expect to find a discharge that exceeds 50 mg L^{-1}, assuming that the mine is in operation every day?

4.7. A number of determinations were made with the following results: 24.25, 25.37, 23.85, 26.31, 28.17, 25.01, 28.35, 26.73, 26.70, 28.40, 24.84, 26.27, 27.53, 25.24, 29.76, 27.30, 27.13, 22.46, 25.36, 27.82, 28.75, 24.42, 25.99, 24.56, 22.94, 21.33, 23.24, 27.21, 27.77, 25.37, 22.53, 30.97.

(a) Calculate the mean and standard deviation of the set.
(b) Group the results pairwise, and calculate the mean and the standard deviation of the 16 pairwise mean values.
(c) Group the results by fours, and calculate the mean and the standard deviation of the eight mean values.
(d) Calculate the theoretical standard errors for cases (b) and (c) using $s_{\bar{x}} = s/\sqrt{n}$, and compare them with the results obtained by direct calculation.

4.8. A certain student reported the following for the percent sulfate in an unknown: 18.63, 19.24, 19.22, 18.46. The actual value is 18.31.

(a) Calculate the mean $\bar{x}$ and the standard deviation s for the four student results.
(b) What is the standard error $s_{\bar{x}}$ for the four student results?

(c) Calculate the t value.

(d) Is it likely that the student had some systematic error(s) in his/her work?

4.9. A new method of chloride analysis has been devised. Use of the method on a standard sample, known to contain 25.76% Cl, gave the following for five replicate determinations: 25.63, 25.42, 25.67, 25.38, 25.54% Cl. Can one conclude that the method is free of systematic errors on the basis of these data?

4.10. Farmer Brown is trying to decide if he should plant his pinto beans in the north forty and his sugar beets in the south forty, or vice versa. He has fairly high levels of soluble salts in his soils, and is aware of the fact that the yield of pinto beans is quite susceptible to high salt levels, being reduced by about 50% if the electrical conductivity is 4 mmho cm^{-1}. He brings in five samples from the north forty, and the conductances are found to be 3.18, 3.47, 3.38, 3.59, 3.79. Four comparable samples from the south forty give conductances of 4.22, 3.56, 3.71, 3.96. Would you conclude that there is a significant difference in the salinity of the two fields at the 0.05 level?

4.11. What if Farmer Brown (see Exercise 4.10) brought in five samples from each field with conductances 3.18, 3.47, 3.38, 3.59, and 3.79 mmho cm^{-1} from the north forty and 4.22, 3.56, 3.71, 3.96, and 3.86 mmho cm^{-1} from the south forty. Would this change the conclusion reached in Exercise 4.10?

4.12. "Happy Tummy" pills are said to consume 87 times their weight in excess stomach acid. The new improved product is supposed to be even better. A set of five of the original pills gave the values 85.27, 89.43, 88.17, 90.06, and 87.49, while a set of five new improved pills gave 88.26, 92.07, 87.66, 90.72, and 89.36. Is the new improved product significantly better at the 0.05 level?

4.13. Seven test strips were painted with the usual formulation, and half of each was then painted with an overcoat that is supposed to help maintain a higher reflectance value. After six months of exposure to normal weathering conditions, the reflectance values of the painted strips were 79.3, 64.8, 66.7, 82.2, 77.1, 73.5, and 68.8%, while the values of the corresponding strips also treated with the overcoat were 80.4, 75.9, 84.6, 69.4, 78.3, 83.3, and 76.9%. Would you conclude that the overcoat has a significant effect at the 0.05 level?

4.14. Ten different analytical laboratories agreed to run replicate determinations on a given standard sample to check one another's results. A standard deviation of 0.32 was found for all 50 results combined. The values found by laboratory A were 18.63, 18.47, 18.73, 18.22, and 18.38%. Is there a significant difference in the standard deviation of laboratory A results compared with those of all laboratories combined? (Assume that the degrees of freedom is infinite for the combined results.)

4.15. A set of six analyses on a given sample by method A had a standard deviation of 0.94, while a set of five analyses of the same sample by method B had a standard deviation of 2.31. Can one conclude that the precision of method A is superior at the 0.05 level?

4.16. Many people have a personal bias in estimating the terminal digit when reading a buret. In checking back through data recorded over a period of time, a certain chemist found the following occurrences of the terminal digit for buret readings:

Terminal Digit	Number of Occurrences
0	21
1	22
2	21
3	16
4	12
5	10
6	11
7	8
8	12
9	11

Is there an indication of bias in the buret readings of this person at the 0.05 level?

4.17. An organic preparation was carried out five times in each of three different solvents with the following results: solvent A: 24.3, 27.2, 23.6, 22.9, 28.5; solvent B: 28.1, 26.4, 23.9, 29.2, 25.7; solvent C: 24.8, 22.2, 23.4, 22.4, 25.6. Perform an ANOVA of these data as a multiple group experiment to determine whether the solvent effect is significant at the 0.05 level.

4.18. Design a randomized block experiment for ANOVA of four blocks that are to be treated by four processes.

4.19. Catalysts from four different sources have been treated by four different processes. Assume that they were sampled and analyzed for catalytic activity according to the experimental design developed in Exercise 4.18, and that the results in terms of an arbitrary measure of activity are as follows:

Source	Process 1	Process 2	Process 3	Process 4
1	31.33	33.47	31.16	32.21
2	46.28	47.93	45.82	46.40
3	38.19	39.89	38.07	37.94
4	34.56	36.17	34.61	34.77

Complete an ANOVA for this randomized block experiment to determine which, if any, of the processes show significant differences at the 0.05 level.

chapter five

SOME SPECIAL TOPICS IN CHEMOMETRICS

Having reviewed the more important basic mathematical and computational techniques, we now consider some specialized applications to chemical problems. Chemistry has been traditionally regarded as an exact science that relies in general on precise measurements. In many areas we have been able to make major advances in understanding chemical behavior making use of only rather elementary mathematical and/or statistical analyses. As we expand our knowledge base further, we must use increasingly complex mathematical approaches which optimize the useful information that can be extracted from a given set of measurements. This is the aim of *chemometrics*.

In this chapter we briefly outline some of the major topics that are generally dealt with in modern chemometrics. In addition to the reference books listed at the end of the chapter, the reader may wish to consult *Analytical Chemistry Reviews* for the even years beginning with 1980, where comprehensive reviews with extensive references to the original literature may be found.

5.1 OPTIMIZATION AND THE CONCEPT OF RESPONSE SURFACES

Let us consider a general function r which depends on several independent variables:

$$r = f(x_1, x_2, x_3, \ldots x_n)$$

For our present discussion we will call r a *response function*. The response function may represent the performance of an instrument that depends on a number of control settings, the yield of a chemical reaction that depends on several physical or chemical parameters, and so forth. We assume that there is a definite relationship such that a change in any one of the x's results in an observable response of the system. The process of *optimization* refers to finding that set of conditions for which the response is either a maximum (product of a reaction, signal-to-noise ratio of an instrument, etc.) or a minimum (yield of a side product, deviation of an observed quantity from the ideal, etc.). The functional relationship between r and the x's may not always be easily defined in analytical terms, but we assume that the maximum or minimum value sought in the optimization process is some clearly defined quantity, and that r is a *unique* (single-valued) function of the x's.

The response may be viewed as a surface in the $(n + 1)$-dimensional space defined by the x's and r. We may have difficulty visualizing such a surface if n is large, but conceptually it is no different than a three-dimensional surface such as one of those sketched in Figure 5.1.

Various complications can arise in the search for the maximum or minimum on a response surface. Often there are small *local maxima* or *minima,* and they must be distinguished from the one highest maximum or lowest minimum corresponding to the true optimum. The latter is generally referred to as a *global maximum* or *minimum*. In complex cases one must sample a large part of the response surface to distinguish between global and false local maxima or minima. We base our subsequent discussion for the most part on the assumption that we are dealing with a moderately simple surface, or at least with a limited part of a surface with a single clearly defined optimum.

The search for an optimum on a response surface can be one of two types. The first is called a *nonsequential search*. It is one in which all the data points are available at the outset, and there is no possibility of using some known data points to construct new ones. This is generally not the method of choice, but it may be the only possibility if the data must be collected simultaneously. Faced with this situation, one can attempt a characterization of the response surface by either of two methods. The first is a *random search;* as the name implies, a random search is done by generating sets of random x variables, and determining the response at each of these points. This is generally less efficient than a *factorial search,* in which a grid of equally spaced x values defines the points at which the response is to be evaluated. The random search is sometimes preferable to a factorial search even though it is less efficient; it is completely unbiased, and it may reveal the occurrence of local optima on complex surfaces which might be overlooked otherwise.

In a *sequential search,* we use the local characteristics of the surface to guide us to new sampling points in the direction of the optimum. Two sequential search methods are described below. The first is called the *simplex method* and it is an outgrowth of studies in *linear programming,* a special branch of optimization theory in which the response is a linear function of each of the x's. The method is in fact quite general; it is not particularly efficient, but it is nevertheless very useful because of its simplicity and generality. The second method we will describe is a *gradient search method,* which makes use of the derivatives of the response function r to define the gradient or slope of

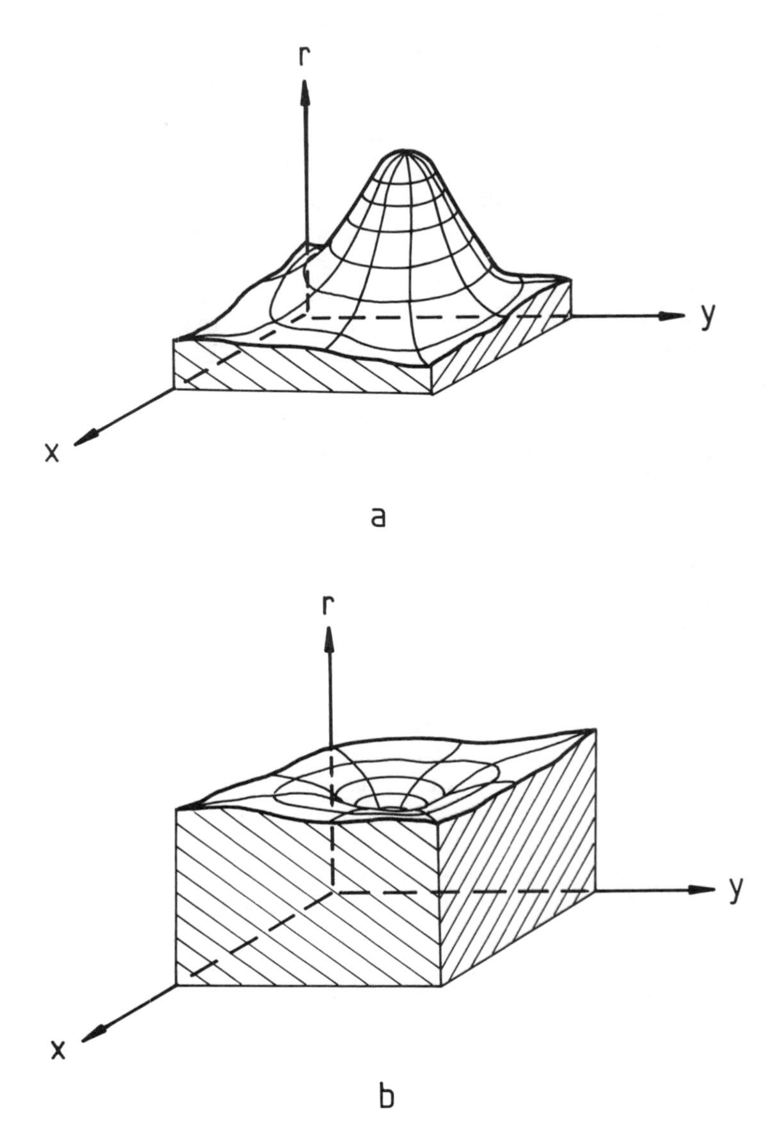

Figure 5.1 Schematic diagrams of response surfaces displaying (a) a maximum and (b) a minimum.

the surface at a particular point. Such gradients can be used to guide the search toward the optimum in a more efficient manner.

5.1.1 The Simplex Method

One technique for sequential optimization involves the construction of an $(n + 1)$-sided polygon on the response surface, known as a *simplex*. The response is calculated at each of the vertices of the simplex, and the responses are ranked. The least desirable vertex is then replaced by one obtained by reflection from the least desirable vertex through the centroid of the simplex. In this way the simplex moves about on the response surface in the direction of more desirable responses.

The *simple simplex method* can be used with any number of factors that might influence the response, but we will assume only two factors in the subsequent discussion for the sake of simplicity. The arguments can readily be extended if required. Let us assume that the optimum on the response surface we wish to find is a minimum, and that we have tested the response for three pairs of values of the two factors. The three responses are ranked and labeled H for high, M for medium, and L for low as illustrated on the schematic response surface of Figure 5.2a. We replace the point H by the point R, which is obtained by reflection through the centroid at point C. If the response at point R is better than that at M or L, the points can be relabeled and the process continued since the simplex is moving in a desirable direction. However, if the response at R is worse than at M or L, the procedure would result in oscillation back and forth between simplexes HML and RML. To prevent this occurrence, we reflect from M (the-next-to-least-desirable vertex in the general case) rather than point H if the response is worse than M or L. This prevents simple oscillation, but it does not exclude the possibility of return to the original simplex in a cyclic process.

Because of these complications the simple simplex method that we have just described is seldom used in practice. A more powerful technique allows for expansion and contraction of the simplex as it moves over the response surface, depending on the nature of the response. According to this *modified simplex method,* the response at R is tested and if it is better than M or L, an expansion coordinate E at twice the distance from the centroid is calculated (see Figure 5.2b). If the response at E is better than at R, EML is taken as the new simplex; if E is worse than R, RML is used. If the test of R shows that it is not better than M or L, the simplex is contracted. For a value of R intermediate between M and L a contraction to point I (halfway between R and C) is made and the new simplex is IML. However, if R is worse than H, the contraction is to the point K (halfway between H and C) and KML is the new simplex. Unless the response surface has some unusual features, the size of the simplex becomes smaller as the optimum is approached, so that the distance between the response at H and L can be used as a criterion of convergence.

The simplex procedure lends itself readily to control of factors that may be difficult to describe in analytical terms. It has been used for such chemical applications as optimization of yields in organic reactions, control of magnetic field homogeneity in

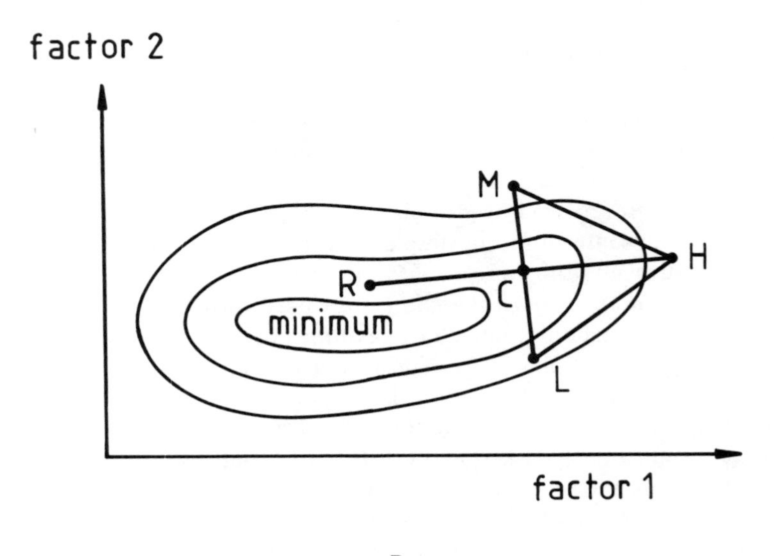

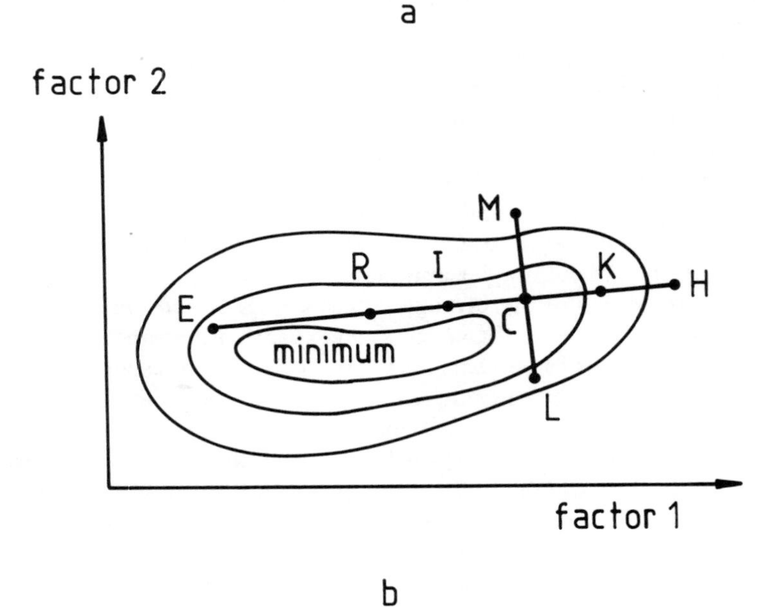

Figure 5.2 Schematic representation of a contour diagram for a response surface for two independent factors, illustrating (a) the simple simplex method and (b) the modified simplex method as described in the text.

NMR, and adjustment of parameters for chromatography and spectroscopy and other instrumental procedures.

The simplex method can also be used to fit equations to experimental data. Although alternative procedures are frequently used (see Section 5.2), this application provides a convenient illustration of the technique. In the fitting of an equation $f(x)$ to a set of data points (x_i, y_i), use is generally made of an error function of the form (this process will be discussed in detail in the following section):

$$E = \sum_{i=1}^{n} [y_i - f(x_i)]^2$$

We will consider this to be our response function, and optimization consists of finding the point where it is a minimum.

As a specific application, we will fit the van der Waals equation of a gas,

$$p = \frac{RT}{V_m - b} - \frac{a}{V_m^2}$$

to the experimental isotherm of CO_2 at 330.90 K. The necessary pVT data are found in the *International Critical Tables,* Vol. 3 (McGraw-Hill, New York, 1928), and we will use the values of the parameters a and b defined in terms of the critical constants

$$a = \frac{27R^2T_c^2}{64p_c} = 3.61 \text{ L}^2 \text{ atm mol}^{-2} \qquad b = \frac{RT_c}{8p_c} = 0.043 \text{ L mol}^{-1}$$

as our initial guesses. As is well known, the parameters determined at the critical point generally fail to give a good representation of pVT data under other conditions.

A BASIC program that carries out the fit of van der Waals equation to the experimental data is shown in Figure 5.3. The parameters read in statement 40 are the molar gas constant (R), the Kelvin temperature (T), the initial estimates of a and b (AI and BI), the error limit (EL), the maximum number of simplexes allowed (MNS), and the number of (p_i, V_i) data points (NP). The initial simplex is constructed by arbitrarily changing each of the initial estimates by 10% (see statement 70). The operation of the program should be clear from the discussion above and the internal documentation provided. The output is also shown in Figure 5.3, and a plot of the equation with optimized a and b parameters together with the experimental data is shown in Figure 5.4.

We often use the van der Waals equation in our discussion of real gases because the physical significance of the parameters a and b is apparent. It certainly does not give the best representation of the data, however. Another equation of state has been proposed by Dieterici,

$$p = \frac{RT}{V_m - b} e^{-a/RTV_m}$$

The program of Figure 5.3 was modified to fit the Dieterici equation to the CO_2 data at 330.90 K. A convergence to the parameters $a = 4.1485 \text{ L}^2 \text{ atm mol}^{-2}$ and $b = 0.04936$

```
10 REM--PROGRAM SMPX--OPTIMIZATION BY SIMPLEX METHOD
20 DIM P(20),V(20)
30 DEF FNP(C,D,F)=R*T/(F-D)-C/F/F                  'VAN DER WAALS EQUATION
40 READ R,T,AI,BI,EL,MSN,NP
50 FOR I=1 TO NP : READ V(I),P(I) : NEXT I
60 'CONSTRUCTION OF THE INITIAL SIMPLEX
70 X(1)=AI : Y(1)=BI : X(2)=1.1*AI : Y(2)=BI : X(3)=AI : Y(3)=1.1*BI
80 SN=1 : LPRINT "SIMPLEX","  A","  B","  ERROR"
90 FOR I=1 TO 3                                      'EVALUATE THE VERTICES
100 A=X(I) : B=Y(I) : GOSUB 400
110 E(I)=EE : NEXT I
120 L=1 : H=1 : FOR I=2 TO 3                    'FIND LOW AND HIGH VERTICES
130 IF E(I)<E(L) THEN L=I
140 IF E(I)>E(H) THEN H=I
150 NEXT I
160 FOR I=1 TO 3                                       'FIND MIDDLE VERTEX
170 IF I<>L AND I<>H THEN M=I
180 NEXT I
190 LPRINT SN,X(L),Y(L),E(L)
200 'CHECK THE TERMINATION CONDITIONS
210 IF SN=MSN OR (E(H)-E(L))/E(L)<EL THEN STOP ELSE SN=SN+1
220 XC=(X(L)+X(M))/2 : YC=(Y(L)+Y(M))/2              'CALCULATE CENTROID
230 XR=2*XC-X(H) : YR=2*YC-Y(H)                   'REFLECTION COORDINATES
240 A=XR : B=YR : GOSUB 400 : ER=EE
250 IF ER>E(L) THEN 320
260 XE=2*XR-XC : YE=2*YR-YC                        'EXPANSION COORDINATES
270 A=XE : B=YE : GOSUB 400
280 IF EE<ER THEN 300
290 X(H)=XR : Y(H)=YR : E(H)=ER : GOTO 310             'RETAIN R VERTEX
300 X(H)=XE : Y(H)=YE : E(H)=EE                        'RETAIN E VERTEX
310 N=H : H=M : M=L : L=N : GOTO 190
320 IF ER>E(M) THEN 350
330 X(H)=XR : Y(H)=YR : E(H)=ER                        'RETAIN R VERTEX
340 N=H : H=M : M=N : GOTO 190
350 IF ER>E(H) THEN 370
360 X(H)=(XC+XR)/2 : Y(H)=(YC+YR)/2 : GOTO 380 'CONTRACT TO I VERTEX
370 X(H)=(XC+X(H))/2 : Y(H)=(YC+Y(H))/2          'CONTRACT TO K VERTEX
380 A=X(H) : B=Y(H) : GOSUB 400
390 E(H)=EE : GOTO 120
400 EE=0                                        'ERROR FUNCTION SUBROUTINE
410 FOR J=1 TO NP
420 EE=EE+(P(J)-FNP(A,B,V(J)))^2
430 NEXT J
440 RETURN
450 DATA 8.20575E-2,330.90,3.61,.043,.001,50,14
460 DATA .29528,66.27,.27695,69.20,.25969,72.18,.24223,74.76
470 DATA .22434,78.99,.20780,82.49,.19012,86.62,.17187,91.16
480 DATA .15533,95.79,.13702,101.32,.12041,107.06,.10301,114.45
490 DATA .08506,126.10,.07668,135.81
```

(a)

SIMPLEX	A	B	ERROR
1	3.61	.043	3688.37
2	3.61	.043	3688.37
3	3.835625	.041925	3443.421
4	3.835625	.041925	3443.421
5	3.70025	.04085	2347.665
6	3.621281	4.165625E-02	1075.322
7	3.485906	4.058124E-02	729.8233
8	3.485906	4.058124E-02	729.8233
9	3.485906	4.058124E-02	729.8233
10	3.371684	3.913671E-02	299.5432
11	3.371684	3.913671E-02	299.5432
12	3.371684	3.913671E-02	299.5432
13	3.257462	3.769217E-02	125.2958
14	3.257462	3.769217E-02	125.2958
15	3.257462	3.769217E-02	125.2958
16	3.143239	3.624763E-02	106.6821
17	3.143239	3.624763E-02	106.6821
18	3.204537	3.707593E-02	102.7338
19	3.166937	3.666677E-02	102.2829
20	3.164488	3.655949E-02	100.2006
21	3.185125	3.684453E-02	100.0949
22	3.178741	3.671964E-02	99.40671
23	3.178741	3.671964E-02	99.40671
24	3.178741	3.671964E-02	99.40671

(b)

Figure 5.3 (a) (*Opposite page*) BASIC computer program for fitting van der Waals equation to pVT data for CO_2 at 330.90 K. The optimum values of the constants a and b are found by the modified simplex method. (b) Output from a run of the BASIC computer program shown in Figure 5.3a.

L mol^{-1} was readily obtained, with an error of 1.5629. This represents a considerable improvement over the fit with the van der Waals equation, as is apparent from an examination of Figure 5.4.

5.1.2 The Method of Steepest Ascent or Descent

One widely used gradient search method is the *method of steepest ascent* or *descent,* which was first introduced by Cauchy in 1847. It consists of moving over the surface in the direction defined by the gradient of largest magnitude until the maximum or minimum is found. It is the equivalent of climbing a hill or descending into a valley by always moving in the direction of maximum slope.

We will show how the method works by consideration of a response function of two independent variables x and y,

$$r = f(x, y)$$

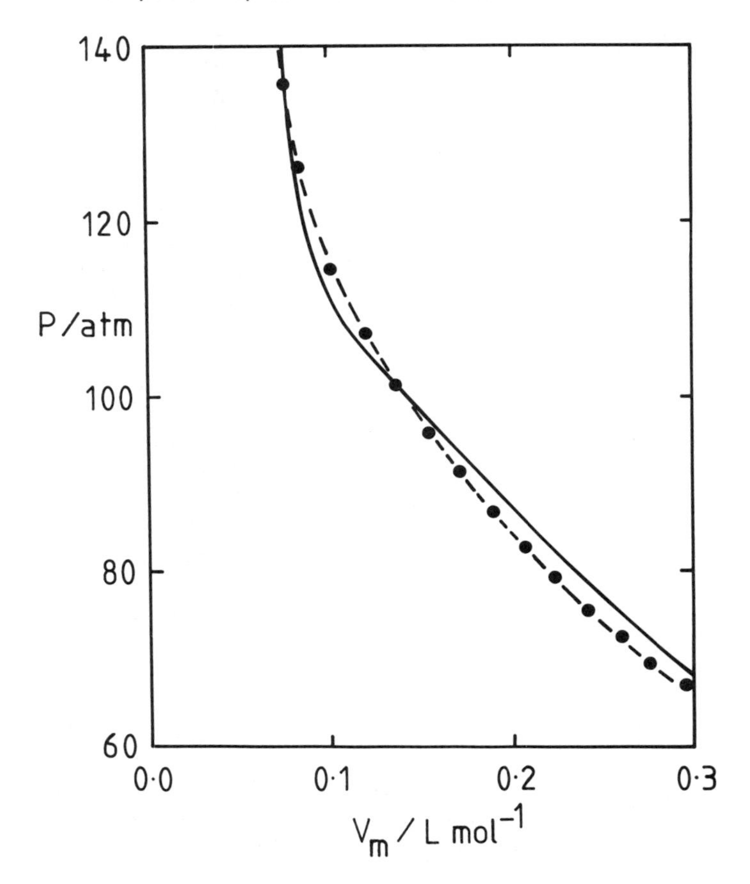

Figure 5.4 Experimental isotherm of CO_2 at 330.90 K compared with van der Waals and Dieterici equations fit by simplex optimization. Solid curve: van der Waals equation with a = 3.1787 L^2 atm mol^{-2} and b = 0.03672 L mol^{-1}. Dashed curve: Dieterici's equation with a = 4.1485 L^2 atm mol^{-2} and b = 0.4936 L mol^{-1}.

which can be schematically represented by a two-dimensional contour diagram like that shown in Figure 5.5. Starting at some arbitrary point P, we wish to find a line s of optimum gradient passing through the point. For an infinitesimal change in x and y,

$$dr = \left(\frac{\partial r}{\partial x}\right)_P dx + \left(\frac{\partial r}{\partial y}\right)_P dy$$

and the slope in the direction s is therefore

$$\frac{dr}{ds} = \left(\frac{\partial r}{\partial x}\right)_P \frac{dx}{ds} + \left(\frac{\partial r}{\partial y}\right)_P \frac{dy}{ds} = \left(\frac{\partial r}{\partial x}\right)_P \cos\theta + \left(\frac{\partial r}{\partial y}\right)_P \sin\theta \tag{5.1}$$

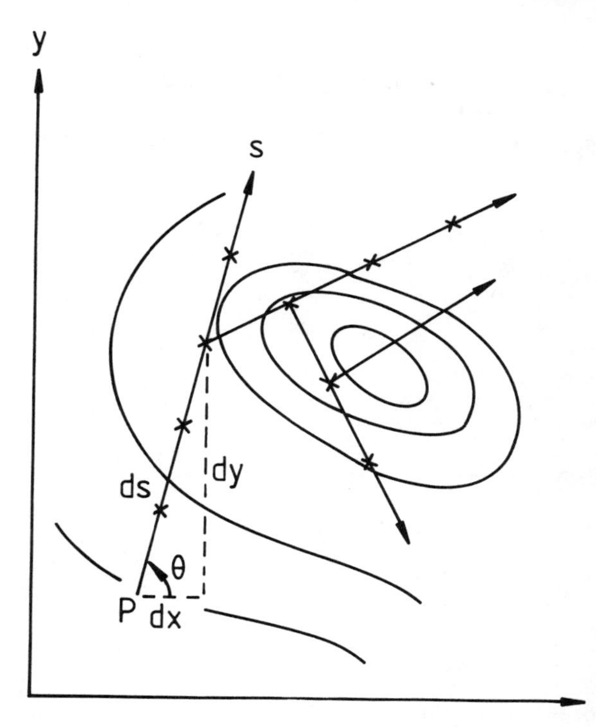

Figure 5.5 Schematic diagram of a response surface where the response function r is assumed to be a function of the two independent variables x and y. The direction of steepest ascent (descent) from the point P is along the line s.

The value of θ for which $dr/ds = 0$ is given by

$$\tan \theta_1 = -\frac{\left(\dfrac{\partial r}{\partial x}\right)_P}{\left(\dfrac{\partial r}{\partial y}\right)_P}$$

It is apparent that this angle θ_1 defines the tangent along the contour line at point P. There are actually two angles for which $dr/ds = 0$, corresponding to the two directions along the contour line; one value is θ_1 and the other is $\theta_1 + \pi$. There must be some intermediate values of θ such that dr/ds is a maximum or a minimum. They are defined by the condition

$$\frac{d(dr/ds)}{d\theta} = 0 = -\left(\frac{\partial r}{\partial x}\right)_P \sin \theta + \left(\frac{\partial r}{\partial y}\right)_P \cos \theta$$

or

$$\tan \theta_2 = \frac{\left(\dfrac{\partial r}{\partial y}\right)_P}{\left(\dfrac{\partial r}{\partial x}\right)_P} \tag{5.2}$$

Obviously, $\tan\theta_1 \tan\theta_2 = -1$, which is just the condition of perpendicularity of two lines in the x,y plane. Thus the direction defined by the gradient of maximum value is perpendicular to the contour line at point P, as we might have anticipated.

Inserting the value of $\tan\theta_2$ defined by Eq. 5.2 in Eq. 5.1, we have for the extremum slope

$$\frac{dr}{ds} = \cos\theta_2 \left[\left(\frac{\partial r}{\partial x}\right)_P + \left(\frac{\partial r}{\partial y}\right)_P \frac{\left(\frac{\partial r}{\partial y}\right)_P}{\left(\frac{\partial r}{\partial x}\right)_P} \right] \tag{5.3}$$

Now Eq. 5.2 can be solved for $\cos\theta_2$ to give

$$\cos\theta_2 = \pm \frac{\left(\frac{\partial r}{\partial x}\right)_P}{\sqrt{\left(\frac{\partial r}{\partial x}\right)_P^2 + \left(\frac{\partial r}{\partial y}\right)_P^2}} \tag{5.4}$$

and substituting this in Eq. 5.3 gives

$$\frac{dr}{ds} = \pm\sqrt{\left(\frac{\partial r}{\partial x}\right)_P^2 + \left(\frac{\partial r}{\partial y}\right)_P^2} \tag{5.5}$$

where the + sign is used for steepest ascent and the − sign for steepest descent. This result may be readily generalized to n dimensions and expressed in the form

$$m_j = \pm \frac{\frac{\partial r}{\partial x_j}}{\sqrt{\sum_{i=1}^{n} \left(\frac{\partial r}{\partial x_i}\right)^2}} \tag{5.6}$$

Here m_j is the relative change in the jth coordinate for a step in the direction of steepest ascent or descent.

As an example, assume that we have the function $r = 100(y - x^2)^2 + (1 - x)^2$ and that we wish to find the minimum by the method of steepest descent, beginning our search at the point (3, 2). We have

$$\frac{\partial r}{\partial x} = -400x(y - x^2) - 2(1 - x) = 8404 \qquad \frac{\partial r}{\partial y} = 200(y - x^2) = -1400$$

$$m_x = -0.9864 \qquad m_y = +0.1643$$

As we move along the line s defined by these numbers, the value of the function r changes. If we evaluate the function for several lengths (as indicated schematically by the x's along the line s of Figure 5.5), we should find that r goes through a minimum. For the example above, we have for unit lengths along s,

$$r(3, 2) = 4904$$
$$r(2.0136, 2.1643) = 358.35$$
$$r(1.0272, 2.3286) = 162.17$$
$$r(0.0408, 2.4929) = 621.55$$

At the minimum [$r(1.0272, 2.3286)$ in this particular case] we again evaluate the gradient, and this defines a new starting direction of steepest descent. From this point we move along the new line of steepest descent until a second minimum is found, and this process is repeated as many times as necessary until the minimum of the response surface is located (Figure 5.5).

Some care must be taken in choosing proper step sizes; obviously, they should decrease as the optimum is approached. There are other factors that may complicate the process in certain cases. For example, if there are long narrow ridges on the surface, the method of steepest ascent or descent tends to zigzag along the ridge in a rather inefficient manner, and it may be desirable to scale the variables so that the contours are more circular. Many other refinements too numerous to mention here have been introduced by various authors to circumvent complications of this kind. They are discussed in various standard works on optimization theory, together with other gradient search methods which are less prone to oscillation in the vicinity of the optimum (see, e.g., G. S. G. Beveridge and R. S. Schechter, *Optimization: Theory and Practice,* McGraw-Hill Book Company, New York, 1970).

5.2 CORRELATION AND LEAST SQUARES CURVE FITTING

The tests of significance discussed in Section 4.4 all have one thing in common: they tell us whether deviations from the norm are sufficiently great that they should be regarded as real effects rather than as routine statistical fluctuations. Sometimes we require additional information, such as which variables are actually related and the nature of their functional relationship. The techniques used to assess the extent of correlation between parameters and the optimum fitting of equations to experimental data are the topics addressed in this section.

5.2.1 The Fitting of a Line to a Set of Data Points

We will assume that we have n pairs of variables which we will label (x_i, y_i). Let us also assume that we have reason to believe x and y are related so that by changing one of the variables, one expects a corresponding change in the other. Either may presumably be regarded as the independent variable; for the following discussion we will take it to be x. We may suppose that the relationship between x and y is a simple linear equation of the form

$$y = b_0 + b_1 x \tag{5.7}$$

Once we have written an explicit functional relationship between the variables we have specified a *deterministic mathematical model*. In other words, we assume a precise relationship between the variables such that when one is specified, the other is determined. We deal with such deterministic models a great deal in chemistry and other natural sciences; the law of chemical equilibrium, Newton's laws of motion, and Einstein's mass-energy relationship are all familiar examples.

Because of random errors, the actual values of x and y that are determined from a laboratory study do not lie exactly on any given line, even if Eq. 5.7 is a faithful representation of the physical system. It is therefore appropriate that we define a *probabilistic mathematical model* to represent the real system. In the present case we write

$$y = b_0 + b_1 x + \epsilon \tag{5.8}$$

where ϵ is a random error whose *expected* value is zero if b_0 and b_1 are properly chosen, but which is actually distributed about the value 0 with a standard deviation σ.

Our task is to find the values of b_0 and b_1 such that Eq. 5.7 gives the best possible representation of the data. Let us assume for the sake of argument that values of the independent variable x (e.g., time, temperature, etc.) can be determined with high precision, but that the dependent variable y is subject to significant random error. This is shown schematically in Figure 5.6, where it will be seen that the fluctuations from ideal linear behavior are all indicated in the y direction. Our problem is to find the equation of a line such that the deviations represented by the solid line segments linking the various points to it are as small as possible. Line 1 has the appearance of being a good representation of the data as a whole; the question we must address is how one constructs the line without bias so that it is in some sense an optimum fit to the data.

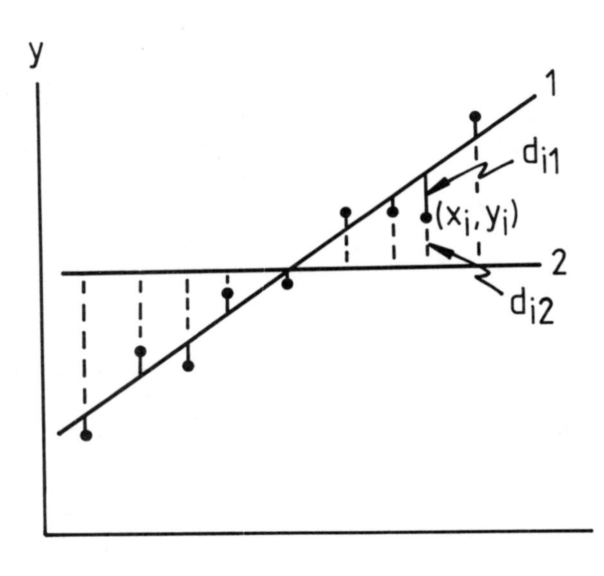

Figure 5.6 Two straight lines, 1 and 2, drawn through the set of data points (x_i, y_i). The deviations in the y-coordinate between each of the points and the two lines are indicated.

By Eq. 5.7 it is possible to calculate a value of y for any arbitrary value to x, but apparently it is only by rare coincidence that the calculated value would correspond exactly with the measured value of y for that value of x. The difference between the observed and calculated values of y for a particular value of x is just the length of the small vertical line segments of Figure 5.6; we represent these deviations by $d_i = y_i - b_0 - b_1x_i$. We wish to minimize the sum of these deviations, but this alone will not necessarily provide the best representation of the data. For example, line 2 of Figure 5.6 may minimize the sum of deviations just as well as line 1 because of the cancellation of positive and negative deviations, but we would definitely not choose this as the best representation of the data.

To avoid this cancellation of positive and negative deviations, we customarily minimize the sum of the squares of the deviations,

$$\text{SS} \equiv \Sigma\, d_i^2 = \Sigma\, (y_i - b_0 - b_1x_i)^2 \tag{5.9}$$

This provides the basis of the *least squares method* of analysis. Among statisticians, the technique we are about to describe is termed *linear regression*.

Now the b_0 and b_1 occurring in Eq. 5.7 are actually constants, but we may regard them as variables in order to determine their optimum values. In other words, we require that the sum of squares of the deviations as given by Eq. 5.9 be a minimum with respect to variations of either one of them. We find the minimum by differentiating with respect to each variable and setting the result equal to zero:

$$\begin{aligned} \frac{\partial\, \text{SS}}{\partial b_0} &= \Sigma\, 2(y_i - b_0 - b_1x_i)(-1) = 0 \\ \frac{\partial\, \text{SS}}{\partial b_1} &= \Sigma\, 2(y_i - b_0 - b_1x_i)(-x_i) = 0 \end{aligned} \tag{5.10}$$

We note that a second differentiation with respect to b_0 and b_1 gives $2n$ and $2\,\Sigma\, x_i^2$, respectively. Since both of these terms can only be positive, we are assured that we are dealing with a minimum in each case.

The first Eq. 5.10 may be written

$$b_0n + b_1\,\Sigma\, x_i = \Sigma\, y_i \tag{5.11}$$

while the second yields

$$b_0\,\Sigma\, x_i + b_1\,\Sigma\, x_i^2 = \Sigma\, x_iy_i \tag{5.12}$$

We may solve Eq. 5.11 for b_0,

$$b_0 = \frac{\Sigma\, y_i - b_1\,\Sigma\, x_i}{n} \tag{5.13}$$

and this result is inserted in Eq. 5.12 to give

$$\frac{\sum y_i - b_1 \sum x_i}{n} \sum x_i + b_1 \sum x_i^2 = \sum x_i y_i$$

$$b_1\left[\sum x_i^2 - \frac{(\sum x_i)^2}{n}\right] = \sum x_i y_i - \frac{\sum x_i \sum y_i}{n}$$

$$b_1 = \frac{\sum x_i y_i - \dfrac{\sum x_i \sum y_i}{n}}{\sum x_i^2 - \dfrac{(\sum x_i)^2}{n}} \tag{5.14}$$

Equation 5.14 allows us to calculate the value of the slope b_1 from the raw data. By comparison of Eqs. 4.3 and 4.4 we see that the denominator is just the sum of squares of the deviations in x from the mean value,

$$\sum x_i^2 - \frac{(\sum x_i)^2}{n} = \sum (x_i - \bar{x})^2 \equiv \mathrm{SS}_x$$

and similarly,

$$\sum x_i y_i - \frac{\sum x_i \sum y_i}{n} = \sum (x_i - \bar{x})(y_i - \bar{y}) \equiv \mathrm{SS}_{xy}$$

Thus Eq. 5.14 is often written in the alternative form

$$b_1 = \frac{\sum (x_i - \bar{x})(y_i - \bar{y})}{\sum (x_i - \bar{x})^2} = \frac{\mathrm{SS}_{xy}}{\mathrm{SS}_x} \tag{5.15}$$

Having determined the value of b_1, we can now complete the solution for b_0. We see that Eq. 5.13 is equivalent to $b_0 = \bar{y} - b_1\bar{x}$, and by use of Eq. 5.15 we may calculate b_0 from

$$b_0 = \bar{y} - \frac{\mathrm{SS}_{xy}}{\mathrm{SS}_x}\,\bar{x} \tag{5.16}$$

Now that b_0 and b_1 have been determined, we need to next consider how we can assess the quality of the fit of the model to the data. First we estimate the standard deviation of the random errors represented by ϵ in Eq. 5.8. We recall that the sum of squares of the deviations between the calculated and observed values of y is given by Eq. 5.9. By analogy with Eq. 4.3, we should divide this by the degrees of freedom and then take the square root to determine the standard deviation of the fit:

$$s = \sqrt{\frac{\mathrm{SS}}{n - 2}} \tag{5.17}$$

Note that the degrees of freedom is $n - 2$. This is because we have fixed values of two parameters by the least squares fitting process, and this has required the introduction of two constraints (Eqs. 5.11 and 5.12).

The calculation of SS by Eq. 5.9 is both tedious and subject to round-off errors, so its use is discouraged. We can obtain an alternative formula as follows:

$$\begin{aligned} SS &= \Sigma\,(y_i - b_0 - b_1 x_i)^2 = \Sigma\left[y_i - \left(\bar{y} - \frac{SS_{xy}}{SS_x}\,\bar{x}\right) - \frac{SS_{xy}}{SS_x}\,x_i\right]^2 \\ &= \Sigma\,(y_i - \bar{y})^2 - 2\,\frac{SS_{xy}}{SS_x}\,\Sigma\,(x_i - \bar{x})(y_i - \bar{y}) + \frac{SS_{xy}^2}{SS_x^2}\,\Sigma\,(x_i - \bar{x})^2 \\ &= SS_y - 2\,\frac{SS_{xy}}{SS_x}\,SS_{xy} + \frac{SS_{xy}^2}{SS_x^2}\,SS_x \\ &= SS_y - \frac{SS_{xy}^2}{SS_x} \end{aligned} \tag{5.18}$$

By use of Eq. 5.15 we have

$$SS = SS_y - b_1 SS_{xy} \tag{5.19}$$

The parameters b_0 and b_1 are of course subject to some uncertainty because they are determined from a sample of limited size n. In repeated sampling the values of b_1 follow a distribution for which the approximate standard deviation is

$$s_{b_1} = \sqrt{\frac{s^2}{SS_x}} \tag{5.20}$$

with s equal to the standard deviation of the fit as determined by Eq. 5.17. It can be shown that the values of b_0 follow a distribution with standard deviation approximated by

$$s_{b_0} = \sqrt{\frac{s^2\,\Sigma\,x_i^2}{nSS_x}} \tag{5.21}$$

Since the parameters b_0 and b_1 are subject to the uncertainties given by Eqs. 5.20 and 5.21, it follows that the expected values of y calculated by using them in Eq. 5.7 are likewise subject to uncertainties. In repeated sampling the expected values of y for a particular input value of x, which we will call x_p, are distributed with a standard deviation which may be represented in approximate form by

$$s_{y_p} = s\sqrt{\frac{1}{n} + \frac{(x_p - \bar{x})^2}{SS_x}} \tag{5.22}$$

To the uncertainty in this expected value of y must be added the random error represented by ϵ in Eq. 5.8, if we wish to estimate the total error in predicting a particular value of y. Thus the approximate standard deviation that would be observed in repeated sampling for the predicted value of y corresponding to the particular value of the independent variable x_p is

$$s_{\text{err}} = s\sqrt{1 + \frac{1}{n} + \frac{(x_p - \bar{x})^2}{SS_x}} \tag{5.23}$$

In presenting Eqs. 5.20 to 5.23 we have emphasized that these standard deviations are approximate; that is, we have assumed that a small sample is being used, so our characterization of the population is imperfect. As the size of the sample becomes very large, the s's should be replaced by σ's and the results become exact. In the limit of large n, we also note that the second and third terms of Eq. 5.23 become small and s_{err} approaches the standard deviation σ which is characteristic of the random errors ϵ.

5.2.2 Assessing the Extent of Correlation between Variables

On the basis of the discussion above we infer that when a large sample is taken so that σ is well determined, the values of b_1 for repeated sampling are normally distributed with a standard deviation (see Eq. 5.20)

$$\sigma_{b_1} = \sqrt{\frac{\sigma^2}{\mathrm{SS}_x}}$$

The corresponding z value that should be inserted in Eq. 4.11 to generate the error curve is

$$z = \frac{b_1 - \beta}{\sigma/\sqrt{\mathrm{SS}_x}}$$

where β is the true value of b_1 which is characteristic of the entire population.

Now for a small sample it is the t-distribution rather than the normal distribution that is followed, so the appropriate statistic for a test of significance should be

$$t = \frac{b_1 - \beta}{s/\sqrt{\mathrm{SS}_x}} \tag{5.24}$$

Here again β is the true value of b_1, which may be known or assumed. If the 95% confidence interval for b_1 is desired, it can be calculated from

$$b_1 \pm t_{.05/2} \frac{s}{\sqrt{\mathrm{SS}_x}} \tag{5.25}$$

We may wish to use a calculated b_1 value to determine whether there is sufficient evidence to conclude the existence of a correlation between the variables. If x and y are not correlated, we would expect that $\beta = 0$, and so we write the test statistic in the form

$$t = \frac{b_1}{s} \sqrt{\mathrm{SS}_x} \tag{5.26}$$

This value is compared with the critical value as listed in a t-table under $n - 2$ degrees of freedom, and if it exceeds the critical value the null hypothesis, H_0: $\beta = 0$, is rejected and we conclude that the correlation of the variables is significant. Depending on how the alternative hypothesis is stated, the test may be two-sided with rejection for either $t > t_{.05/2}$ or $t < -t_{.05/2}$, or it may be one-sided with rejection for $t > t_{.05}$ or $t < -t_{.05}$.

The correlation between variables is often described in terms of the *correlation coefficient*,

$$r \equiv \frac{\Sigma\,(x_i - \bar{x})(y_i - \bar{y})}{\sqrt{\Sigma\,(x_i - \bar{x})^2\,\Sigma\,(y_i - \bar{y})^2}} = \frac{SS_{xy}}{\sqrt{SS_x SS_y}} \tag{5.27}$$

From Eq. 4.3 we see that the alternative expression

$$r = \frac{SS_{xy}}{(n-1)s_x s_y} \tag{5.28}$$

is also valid, where s_x is the standard deviation in x values, and s_y is the standard deviation in y values.

Let us now examine the characteristics of this correlation coefficient. We note, first of all, that SS_x and SS_y can only be positive. Thus the sign of r is determined entirely by the term in the numerator, SS_{xy}. From Eq. 5.15 we see that the same can be said of b_1. In other words, r has the same sign as b_1, and will furthermore be zero if b_1 is zero. It is therefore, like b_1, a measure of the correlation between x and y. It differs, however, in that it is scaled in such a way that it is useful for comparison of various cases.

An analysis of Figure 5.6 will help us to see how r is scaled. If there were no correlation between x and y, we would expect that b_1 would be equal to zero. Then line 2, which is actually $y = \bar{y}$, would be the best representation of the data we could give. The sum of squares of the deviations about this line is

$$SS_y = \Sigma\,(y_i - \bar{y})^2$$

If, on the other hand, there were some correlation between x and y, we would expect that the sum of the squares of the deviations about the line given by Eq. 5.9 (indicated in the figure as line 1) would be smaller. We have seen that this sum of squares may be written (Eq. 5.18)

$$SS = SS_y - \frac{SS_{xy}^2}{SS_x}$$

Since SS_{xy}^2 and SS_x can only be positive, we see that SS is equal to SS_y minus a positive quantity. We conclude that $SS \le SS_y$. By squaring Eq. 5.27, we get

$$r^2 = \frac{SS_{xy}^2}{SS_x SS_y}$$

and insertion of Eq. 5.18 gives

$$r^2 = 1 - \frac{SS}{SS_y} \tag{5.29}$$

With $SS \le SS_y$, we see that $0 \le r^2 \le 1$, or in other words, $-1 \le r \le +1$.

If x and y are perfectly correlated, then $r = 1.00$. By this we mean that for an increase in x there is a corresponding increase in y. Similarly, a value $r = -1.00$ means that there is a perfect negative correlation. A value $r = 0.00$ indicates no correlation whatsoever between x and y.

In most real cases, none of these extremes is observed; one must make a subjective judgment to indicate whether a value of, say 0.5 or 0.6, is significant. There is no concrete rule that can be given which will cover all situations. It should be noted, however, that the actual decrease in the sum of squares deviation corresponding to $r = 0.5$ is only $r^2 = 0.25$ or 25% compared with no correlation. This is not generally considered to be a very impressive improvement.

Finally, a caveat should be given regarding the conclusions that might be drawn from these analyses. In no way have we proven a cause-and-effect relationship between x and y simply through demonstration of a strong correlation between them. We have all heard horror stories about the misuse of statistics, such as the relating of the number of illegitimate children in the Orient to the number of Protestant ministers in New England! The fact that $\beta \neq 0$ does not necessarily mean that the two variables under consideration are related, but only that they follow the same trend for whatever reason. If we have good reason to believe that there should be a relationship between two variables, analysis by the t-test or with the correlation coefficient may be useful. It should be recognized, however, that the results could be quite misleading if applied indiscriminately.

5.2.3 Generalized Linear Least Squares Methods

The least squares procedure outlined in Section 5.2.1 can be generalized in many ways; we will briefly mention some of the more important ones in this section.

First, we considered only random errors in the dependent variable y in the discussion given above. There are many cases where this is a good approximation, but in some instances the variability of x values may exceed that of the y values and one should then minimize the deviations along the abscissa rather than along the ordinate. It may be possible to merely interchange the dependent and independent variables for some applications, in which case the analysis above could still be applied. If random errors of approximately the same magnitude occur in both x and y, the best procedure involves minimization of the perpendicular distance of each point from the line. The calculations are then considerably more involved and we will not go into them here, but the reader should be aware that such considerations may be important in some special cases and that solutions may be found in some of the more extensive treatments.

A second point of considerable importance is the weight that is given to the data points in least squares analysis. In the foregoing treatment all data points were treated equally. In general, however, we may know some points more accurately than others, in which case they should be weighted more heavily. For *weighted least squares analysis,* use is made of the weighted sum of squares

$$SS_w = \Sigma\, w_i(y_i - b_0 - b_1 x_i)^2 \tag{5.30}$$

in place of Eq. 5.9. If a standard deviation or variance for each quantity y_i is determined by the experiment, we may want to use $w_i = 1/s_i$ or $w_i = 1/s_i^2$ in the analysis. When uncertainties are not available for individual points, it is often assumed that the smaller the point, the less accurately it is known, and then $w_i = 1/y_i$ can be used. If we carry through the derivation above with weighting factors included, we have

$$b_0 = \frac{\Sigma w_i x_i^2 \, \Sigma w_i y_i - \Sigma w_i x_i \, \Sigma w_i x_i y_i}{D} \tag{5.31}$$

in place of Eq. 5.16, and

$$b_1 = \frac{\Sigma w_i \, \Sigma w_i x_i y_i - \Sigma w_i x_i \, \Sigma w_i y_i}{D} \tag{5.32}$$

in place of Eq. 5.14. Here the denominator is defined by $D = \Sigma w_i \, \Sigma w_i x_i^2 - (\Sigma w_i x_i)^2$. Similarly, Eq. 5.17 for the standard deviation of the fit must be replaced by

$$s_w = \sqrt{\frac{SS_w}{n - 2}} \tag{5.33}$$

and the standard deviations of the parameters are

$$s_{b_0} = \sqrt{\frac{s^2 \, \Sigma w_i x_i^2}{D}} \qquad s_{b_1} = \sqrt{\frac{s^2 \, \Sigma w_i}{D}} \tag{5.34}$$

Of course, these formulas all reduce to those given in Section 5.2.1 if the weighting factors w_i are all equal to unity.

We are often confronted with a variable that depends on several others, each of which may be regarded as independent. If y is linearly dependent on each of several variables $x_1, x_2, x_3, \ldots$, we may write

$$y = b_0 + b_1 x_1 + b_2 x_2 + b_3 x_3 + \cdots \tag{5.35}$$

Again the analysis is a little more involved, but in fact quite straightforward. One simply defines a sum of squares of the deviations, analogous to Eq. 5.9, and minimizes the result with respect to each of the parameters $b_0, b_1, b_2, b_3, \ldots$ to be determined. The procedure is referred to as *multilinear regression*.

The analysis of Section 5.2.1 is not restricted simply to equations of the form given in Eq. 5.7 or 5.35. For example, thermodynamic studies show that the change in free energy ΔG° is related to the equilibrium constant by

$$\Delta G^\circ = -RT \ln K_{eq}$$

Furthermore,

$$\Delta G^\circ = \Delta H^\circ - T\,\Delta S^\circ$$

so the equilibrium constant for a chemical reaction is related to the temperature by

$$\ln K_{eq} = -\frac{\Delta H^\circ}{RT} + \frac{\Delta S^\circ}{R}$$

Thus we often measure K_{eq} as a function of T. A plot of $\ln K_{eq}$ versus $1/T$ is expected to be linear. Here $\ln K_{eq}$ plays the role of y and $1/T$ plays the role of x, and the analysis given above can be applied directly to determine the best values of $\Delta H°$ and $\Delta S°$ that fit the experimental data.

If more than one independent variable is involved, we might consider the possibility of an equation such as

$$y = ax_1 + bx_1x_2 + cx_2 \tag{5.36}$$

We can still use procedures similar to those given above to find optimum values of the parameters a, b, and c. Equations of this general form may be particularly useful in fitting chemical data where equilibria are concerned, since the equilibrium concentrations generally depend on products of the general form $x_1^n x_2^m$.

Let us now examine in a little more detail the situation where a single independent variable is involved, but let us assume that the data points show a certain curvature. It is apparent that finding a simple line through the points will be inadequate; we should fit the data with an expression of higher order. A generalization of this nature is called *curvilinear least squares analysis* or *curvilinear regression*.

We will illustrate the procedure using a quadratic function

$$y = b_0 + b_1x + b_2x^2 \tag{5.37}$$

to fit the data. The deviations between the y values and those calculated for the corresponding value of x are given by

$$d_i = y_i - b_0 - b_1x_i - b_2x_i^2$$

and the sum of squares of the deviations for weighted least squares analysis is therefore

$$\mathrm{SS}_w = \Sigma\, d_i^2 = \Sigma\, w_i(y_i - b_0 - b_1x_i - b_2x_i^2)^2 \tag{5.38}$$

We must minimize SS_w with respect to each of the three parameters b_0, b_1, and b_2. Differentiation gives

$$\frac{\partial \mathrm{SS}_w}{\partial b_0} = \Sigma\, 2w_i(y_i - b_0 - b_1x_i - b_2x_i^2)(-1)$$

$$\frac{\partial \mathrm{SS}_w}{\partial b_1} = \Sigma\, 2w_i(y_i - b_0 - b_1x_i - b_2x_i^2)(-x_i)$$

$$\frac{\partial \mathrm{SS}_w}{\partial b_2} = \Sigma\, 2w_i(y_i - b_0 - b_1x_i - b_2x_i^2)(-x_i^2)$$

These three derivatives may be set equal to zero to give

$$\begin{aligned} b_0 \Sigma\, w_i + b_1 \Sigma\, w_ix_i + b_2 \Sigma\, w_ix_i^2 &= \Sigma\, w_iy_i \\ b_0 \Sigma\, w_ix_i + b_1 \Sigma\, w_ix_i^2 + b_2 \Sigma\, w_ix_i^3 &= \Sigma\, w_ix_iy_i \\ b_0 \Sigma\, w_ix_i^2 + b_1 \Sigma\, w_ix_i^3 + b_2 \Sigma\, w_ix_i^4 &= \Sigma\, w_ix_i^2y_i \end{aligned} \tag{5.39}$$

These equations are often called the *normal equations*. As we compare them with those derived in Section 5.2.1, we see that a pattern begins to emerge. The first equation is derived from the fitting equation by inserting the various x_i, y_i pairs and summing the result. The second equation is obtained by similarly inserting x_i, y_i pairs, but we then multiply by x_i and then sum the result. The third equation is obtained from the fitting equation by again inserting x_i, y_i pairs, but multiplying by x_i^2 before summing. Equations 5.39 can be conveniently represented in matrix notation by

$$\mathbf{XB} = \mathbf{Y} \tag{5.40}$$

with

$$\mathbf{X} = \begin{pmatrix} \Sigma\, w_i & \Sigma\, w_i x_i & \Sigma\, w_i x_i^2 \\ \Sigma\, w_i x_i & \Sigma\, w_i x_i^2 & \Sigma\, w_i x_i^3 \\ \Sigma\, w_i x_i^2 & \Sigma\, w_i x_i^3 & \Sigma\, w_i x_i^4 \end{pmatrix} \qquad \mathbf{B} = \begin{pmatrix} b_0 \\ b_1 \\ b_2 \end{pmatrix} \qquad \mathbf{Y} = \begin{pmatrix} \Sigma\, w_i y_i \\ \Sigma\, w_i x_i y_i \\ \Sigma\, w_i x_i^2 y_i \end{pmatrix}$$

We can readily include cubic, quartic, and even higher-order terms; the resulting normal equations are simply written by extension of the pattern above. They may be solved for the b's using standard techniques such as those outlined in Chapter 3. Formally, we write

$$\mathbf{B} = \mathbf{X}^{-1}\mathbf{Y} \tag{5.41}$$

For least squares fitting of a polynomial of degree p, the standard deviation of the fit is given by

$$s = \sqrt{\frac{\mathrm{SS}_w}{n - p - 1}} \tag{5.42}$$

where $n - p - 1$ is the number of degrees of freedom, and the standard deviation of the parameters is given by

$$s_{b_0} = s\sqrt{x_{11}^{-1}} \quad s_{b_1} = s\sqrt{x_{22}^{-1}} \quad s_{b_2} = s\sqrt{x_{33}^{-1}} \quad \cdots \tag{5.43}$$

where the x_{ii}^{-1} are the diagonal elements of the inverse matrix $\mathbf{X}^{-1}$.

We would expect that the sum-of-squares deviation SS would get smaller and smaller as we add more and more terms, and supposedly we could always get an exact fit if we were to push this procedure to the limit where we have as many parameters as data points. Our equation probably then has little if any physical meaning. It is likely to give anomalous values between the data points and certainly outside the range where data are available. A question we must raise, then, is the following: Since we can always include additional terms in our fitting equation, at what stage does the improvement cease to be meaningful?

In answering this question one should certainly examine any physical model that may be available to describe the system. If there is some theoretical justification for additional terms, one might wish to include them in the analysis, but if there is no reason to expect them, the question becomes moot. We may in some instances not be concerned with fitting our data to any particular theoretical model, but rather wish to obtain an equation that simply represents the values in the best possible way so that it may be used for predictive purposes.

A simple way to assess the quality of fit by a given equation is through an *analysis of residuals*, $r_i = y_i - y_{ic}$. Here y_i is the observed value and y_{ic} is the corresponding value calculated using the least squares equation. These residuals r_i may be converted into standardized residuals r_i' by dividing them by the standard deviation of the fit: $r_i' = r_i/s$. The standardized residuals r_i' have zero mean and unit standard deviation. If the fit is proper, they will be normally distributed about zero with 95% of the values in the range -1.96 to $+1.96$. A plot of the standardized residuals will quickly reveal any clustering of the points, which is an indication that the model is not appropriate and that more flexibility may be required.

A rigorous approach that may be used to decide if the inclusion of additional parameters in the analysis is justified is based on an F-test. Let us consider the quantity

$$F_{q,n-p} = \frac{(\mathrm{SS}_{p-q} - \mathrm{SS}_p)/q}{\mathrm{SS}_p/(n-p)} \tag{5.44}$$

Here we have assumed that there are n data points. The denominator represents the variance when p free parameters are used, the degrees of freedom then being $n - p$. The term SS_{p-q} represents the sum of the squares of the deviations between the calculated values and the data being fit when q fewer free parameters are used. It will be observed that the calculated F value represents the relative *improvement* in the variance with q additional parameters, and if this is small, we conclude that the two variances represent the same population and the improvement is not significant. Only if the calculated F value exceeds the critical value from the table for q degrees of freedom in the numerator and $n - p$ degrees of freedom in the denominator would we reject the null hypothesis and conclude that the improvement in the fit to the data is significant.

5.2.4 Nonlinear Least Squares Analysis

The least squares methods we have discussed up to this point have an important feature in common: the equations that were fitted to the data were all linear functions of the parameters that were to be determined. As such, they could all be dealt with using *linear* least squares methods. The equation to be fitted to a set of data points is not always linear in the parameters to be determined, in which case the methods described above are no longer applicable. By a very general method known as the *nonlinear least squares fitting procedure,* it is still possible to determine an optimum set of parameters in many cases.

Let us assume, for purposes of illustration, that we have an arbitrary equation

$$y = f(a, b, c, x) \tag{5.45}$$

which we wish to fit to a set of data, where a, b, and c are the constants to be determined. We begin the procedure by making the best estimates we can of a, b, and c. Let us call them a_0, b_0, and c_0. We do not know the optimum values of a, b, and c at this point, but we can assume that they differ from our estimates by certain small corrections which we call α, β, and γ. That is, we have

$$a = a_0 + \alpha \qquad b = b_0 + \beta \qquad c = c_0 + \gamma \tag{5.46}$$

We can expand the function 5.45 in a Taylor's series about the point a_0,b_0,c_0 (because of the way the terms are defined in Eq. 5.46, this is the equivalent of a Maclaurin expansion in α, β, and γ):

$$\begin{aligned} y = y_0 &+ \left(\frac{\partial y}{\partial a}\right)_0 \alpha + \left(\frac{\partial y}{\partial b}\right)_0 \beta + \left(\frac{\partial y}{\partial c}\right)_0 \gamma \\ &+ \left(\frac{\partial^2 y}{\partial a^2}\right)_0 \alpha^2 + \left(\frac{\partial^2 y}{\partial b^2}\right)_0 \beta^2 + \left(\frac{\partial^2 y}{\partial c^2}\right)_0 \gamma^2 \\ &+ \text{other second-order terms} \\ &+ \text{higher-order derivatives} \end{aligned}$$

Here y_0 is the function defined by Eq. 5.45, but evaluated using the estimates a_0, b_0, and c_0, and $(\partial y/\partial a)_0$ is the partial derivative of y with respect to a evaluated using a_0, b_0, and c_0, and so forth.

If we have made good initial estimates of the parameters, the corrections α, β, and γ will be small. In this case the series converges rapidly and we can neglect derivatives of second and higher order. We can therefore write

$$y \approx y_0 + \left(\frac{\partial y}{\partial a}\right)_0 \alpha + \left(\frac{\partial y}{\partial b}\right)_0 \beta + \left(\frac{\partial y}{\partial c}\right)_0 \gamma$$

For each of the data points x_i, y_i this expression becomes

$$y_i \approx y_{0_i} + \left(\frac{\partial y}{\partial a}\right)_{0_i} \alpha + \left(\frac{\partial y}{\partial b}\right)_{0_i} \beta + \left(\frac{\partial y}{\partial c}\right)_{0_i} \gamma$$

We now have a set of approximate equations that are linear in the *corrections* α, β, and γ. We can therefore construct a set of equations defining the deviations as before,

$$d_i = y_i - y_{0_i} - \left(\frac{\partial y}{\partial a}\right)_{0_i} \alpha - \left(\frac{\partial y}{\partial b}\right)_{0_i} \beta - \left(\frac{\partial y}{\partial c}\right)_{0_i} \gamma$$

If we sum the squares of these deviations assuming unit weighting factors, and minimize with respect to α, β, and γ, we have

$$\Sigma\left(\frac{\partial y}{\partial a}\right)_{0_i}^2 \alpha + \Sigma\left(\frac{\partial y}{\partial a}\right)_{0_i}\left(\frac{\partial y}{\partial b}\right)_{0_i} \beta + \Sigma\left(\frac{\partial y}{\partial a}\right)_{0_i}\left(\frac{\partial y}{\partial c}\right)_{0_i} \gamma = \Sigma\left(\frac{\partial y}{\partial a}\right)_{0_i}(y_i - y_{0_i})$$

$$\Sigma\left(\frac{\partial y}{\partial a}\right)_{0_i}\left(\frac{\partial y}{\partial b}\right)_{0_i} \alpha + \Sigma\left(\frac{\partial y}{\partial b}\right)_{0_i}^2 \beta + \Sigma\left(\frac{\partial y}{\partial b}\right)_{0_i}\left(\frac{\partial y}{\partial c}\right)_{0_i} \gamma = \Sigma\left(\frac{\partial y}{\partial b}\right)_{0_i}(y_i - y_{0_i})$$

$$\Sigma\left(\frac{\partial y}{\partial a}\right)_{0_i}\left(\frac{\partial y}{\partial c}\right)_{0_i} \alpha + \Sigma\left(\frac{\partial y}{\partial b}\right)_{0_i}\left(\frac{\partial y}{\partial c}\right)_{0_i} \beta + \Sigma\left(\frac{\partial y}{\partial c}\right)_{0_i}^2 \gamma = \Sigma\left(\frac{\partial y}{\partial c}\right)_{0_i}(y_i - y_{0_i}) \qquad (5.47)$$

These linear equations can be solved for the corrections α, β, and γ, using the techniques of Chapter 3. The corrections are not exact because we discarded higher-order derivatives in the Taylor's series expansion. Therefore, we add these corrections to the initial values a_0, b_0, and c_0 to get better estimates, and the cycle is repeated. In each succeeding cycle the corrections should become smaller until finally they are negligible and the fitting procedure is converged. If the initial estimates were close to the true parameter values, the convergence should be smooth and rapid. Otherwise, it may be slow, and it may even diverge or converge to a false minimum.

The nonlinear least squares procedure allows us to fit an equation of almost any arbitrary form to a set of data, although, as mentioned above, it may not converge smoothly if we are too far off base. In general, we need to be guided by some theoretical considerations which suggest the form of the equation that should be used.

Example 5.1

Use the linear least squares method to find the best line that passes through the following data points, assuming unit weighting factors:

x_i	y_i
2.3	3.7
2.7	4.2
3.9	5.5
3.3	4.4

Solution. The quantities needed for the calculation are

$$\Sigma x_i = 12.2 \qquad \Sigma y_i = 17.8$$

$$\Sigma x_i^2 = 38.68 \qquad \Sigma x_i y_i = 55.82$$

$$n = 4 \qquad \bar{x} = 3.05 \qquad \bar{y} = 4.45$$

Using Eq. 5.14, we find that the slope is

$$b_1 = \frac{\Sigma x_i y_i - \dfrac{\Sigma x_i \Sigma y_i}{n}}{\Sigma x_i^2 - \dfrac{(\Sigma x_i)^2}{n}} = \frac{4(55.82) - (12.2)(17.8)}{4(38.68) - (12.2)^2} = 1.0408$$

The intercept, by Eq. 5.13, is

$$b_0 = \bar{y} - b_1\bar{x} = 4.45 - (1.0408)(3.05) = 1.2755$$

Thus the best line through the data is

$$y = 1.2755 + 1.0408x$$

Example 5.2

For the data of Example 5.1, find the standard deviation of the fit to the data, the standard deviation of the slope, and the standard deviation of the intercept.

Solution. We calculate the sum of squares of the deviations as follows:

$$SS_x = \Sigma (x_i - \bar{x})^2 = 1.4700$$

$$SS_y = \Sigma (y_i - \bar{y})^2 = 1.7300$$

$$SS_{xy} = \Sigma (x_i - \bar{x})(y_i - \bar{y}) = 1.5300$$

From Eq. 5.19 we have

$$SS = SS_y - b_1 SS_{xy} = 1.7300 - (1.0408)(1.5300) = 0.1376$$

so the standard deviation of the fit to the data by Eq. 5.17 is

$$s = \sqrt{\frac{SS}{n-2}} = \sqrt{\frac{0.1376}{4-2}} = 0.2623$$

From this we find the standard deviation s_{b_1} of the slope (Eq. 5.20),

$$s_{b_1} = \sqrt{\frac{s^2}{SS_x}} = \sqrt{\frac{0.0688}{1.4700}} = 0.2163$$

and finally that of the intercept (Eq. 5.21):

$$s_{b_0} = \sqrt{\frac{s^2 \Sigma x_i^2}{n SS_x}} = \sqrt{\frac{(0.0688)(38.68)}{(4)(1.4700)}} = 0.6727$$

Example 5.3

For the analysis of Examples 5.1 and 5.2, calculate the correlation coefficient relating x and y. Does a t-test at the 0.05 level indicate a significant correlation between the variables? What is the 95% confidence interval for b_1?

Solution. By Eq. 5.27 and the sum of squares deviations calculated in Example 5.2, we find

$$r = \frac{SS_{xy}}{\sqrt{SS_x SS_y}} = \frac{1.5300}{\sqrt{(1.4700)(1.7300)}} = 0.9594$$

which indicates a fairly strong correlation. The t-value, from Eq. 5.26, is

$$t = \frac{b_1}{s}\sqrt{SS_x} = \frac{1.0408}{0.2623}\sqrt{1.4700} = 4.8114$$

This may be compared with the critical value $t_{.05/2}$ for a two-sided test for 2 degrees of freedom, 4.303. Therefore, we reject the null hypothesis H_0: $\beta = 0$ and accept the alternative hypothesis H_a: $\beta \neq 0$; that is, the correlation is considered significant at the 0.05 level. The 95% confidence interval, by Eq. 5.25, is

$$b_1 \pm t_{.05/2} \frac{s}{\sqrt{SS_x}} = 1.0408 \pm 4.303 \frac{0.2623}{\sqrt{1.4700}}$$

$$= 1.0408 \pm 0.9308$$

Example 5.4

Spectrophotometric studies of reaction rates are often complicated by the occurrence of two or more reacting materials that absorb in the same wavelength range. The following absorbances have been determined as a function of time in a rate study:

t_i (min)	A_i
0	0.346
4	0.281
11	0.212
22	0.148
40	0.103
55	0.076

Show how the rate constants may be determined, assuming that two concurrent first-order reactions are taking place.

Solution. As a first step we graph the data in the form ln A_i versus t:

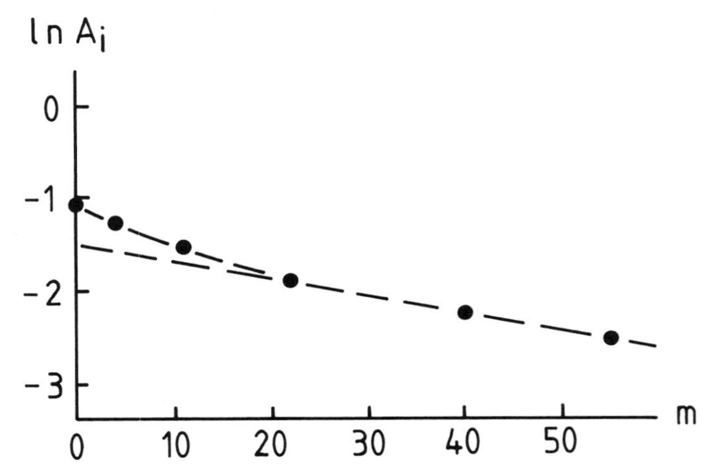

This curve is typical of two parallel reactions with different rate constants. A linear plot is expected for a simple first-order reaction, and we see that this condition is approximated after a long time (i.e., when the more rapid reaction has essentially run its course).

We may analyze the curve by assuming that the observed absorbance is the sum of two component absorbances, each of which decays exponentially with time:

$$A = ae^{-ct} + be^{-dt}$$

The constants a and b may not be particularly meaningful unless we have additional data, since they are functions not only of the initial concentrations, but of the path length and the molar absorptivities as well. The relative values can be determined, however, as well as the two rate constants c and d.

Because the equation we are using is not linear in the parameters we wish to determine, the use of the nonlinear least squares procedure is indicated. The calculations required are tedious and are rarely done by hand. A computer program can be written to carry out the analysis, but we will go through one cycle by hand to clarify the techniques involved.

Our first task is to generate the initial estimates of the parameters; it is important that these be well chosen for uniform convergence. We see that our absorbance at time $t = 0$ is 0.346, so we have the condition

$$a_0 + b_0 = 0.346$$

Extrapolation of the straight-line portion of the curve back to the time $t = 0$ gives

$$b_0 \approx e^{-1.5} = 0.22$$

so

$$a_0 \approx 0.34 - 0.22 = 0.12$$

The slope of the straight-line part of the curve should be the negative of the slower rate constant:

$$-d_0 \approx \frac{-1.5 - (-2.27)}{0 - 40} = -0.019$$

$$d_0 \approx 0.019$$

An estimate of the rate constant of the more rapid reaction can be obtained for any arbitrary point that shows significant deviation from the straight line. Since the other parameters have all been determined, we can solve for c_0. Choosing, for example, the point $t = 4$, we get

$$0.281 = 0.12e^{-4c_0} + 0.22e^{-0.076}$$

$$e^{-4c_0} = 0.64$$

$$-4c_0 = \ln 0.64 = -0.44$$

$$c_0 = 0.11$$

This completes the solution for the initial parameter estimates.

To construct the linear equations analogous to Eq. 5.47, we need not only the fitting equation but also its partial derivatives with respect to each of the parameters to be determined. The partial derivatives are

$$\frac{\partial A}{\partial a} = e^{-ct} \qquad \frac{\partial A}{\partial b} = e^{-dt}$$

$$\frac{\partial A}{\partial c} = -ate^{-ct} \qquad \frac{\partial A}{\partial d} = -bte^{-dt}$$

Using these and the estimates of the parameters given above, we can construct the following table of necessary quantities:

t_i	$\left(\frac{\partial A}{\partial a}\right)_{0_i}$	$\left(\frac{\partial A}{\partial b}\right)_{0_i}$	$\left(\frac{\partial A}{\partial c}\right)_{0_i}$	$\left(\frac{\partial A}{\partial d}\right)_{0_i}$	A_{0_i}	$A_i - A_{0_i}$
0	1.	1.	0.	0.	0.340000	6.00000×10^{-3}
4	0.644036	0.926816	−0.309137	−0.815598	0.281184	-1.83936×10^{-4}
11	0.298197	0.811395	−0.393620	−1.963576	0.214291	-2.29063×10^{-3}
22	0.088922	0.658362	−0.234753	−3.186473	0.155510	-7.51028×10^{-3}
40	0.012277	0.467666	−0.058931	−4.115465	0.104360	-1.35989×10^{-3}
55	0.002358	0.351692	−0.015562	−4.255471	0.077655	-1.65514×10^{-3}

Note that many more digits are retained than could possibly be justified by the data; this is done because the round-off error can be severe in the solution of the linear equations. As a general practice, we carry extra digits until the final result is obtained and then round the answer accordingly.

The linear equations can now be constructed by summing the appropriate quantities from this table. The result is

$$1.51177\alpha + 1.90397\beta - 0.338107\gamma - 1.45471\delta = 4.51005 \times 10^{-3}$$

$$1.90397\alpha + 3.29319\beta - 0.793480\gamma - 7.86828\delta = -2.19164 \times 10^{-3}$$

$$-0.338107\alpha - 0.793480\beta + 0.309326\gamma + 2.08182\delta = 2.82746 \times 10^{-3}$$

$$-1.45471\alpha - 7.86828\beta + 2.08182\gamma + 49.7205\delta = 4.12191 \times 10^{-2}$$

A solution of this set of equations is readily obtained using the techniques of Chapter 3, and the result is

$$\alpha = 0.0291 \qquad \beta = -0.0234$$

$$\gamma = -0.0076 \qquad \delta = -0.0017$$

We recall that these are the corrections to the parameters, not the parameters themselves. The parameters at this stage are given by

$$a = a_0 + \alpha = 0.149$$

$$b = b_0 + \beta = 0.197$$

$$c = c_0 + \gamma = 0.102$$

$$d = d_0 + \delta = 0.0173$$

These parameter values can now be relabeled as estimates, and they can then be used to set up an improved set of linear equations to solve for the next set of corrections. Because the corrections found in the first cycle were fairly small, we have reason to believe that they will become progressively smaller in the subsequent steps, so a uniform, rapid convergence in a few cycles can be anticipated.

5.3 THE PROPAGATION OF ERRORS

The quantities of interest to the chemist are often derived from several factors, each of which is independently measured within certain error limits. The way in which the individual errors are compounded to determine the error of the derived quantity is termed the *propagation of errors*.

Let us assume that we have the quantities $\bar{x} \pm \epsilon_x$, $\bar{y} \pm \epsilon_y$, $\bar{z} \pm \epsilon_z$, ... The uncertainties ϵ_x, ϵ_y, ϵ_z, . . . may be a result of determinate or indeterminate errors, or both, and they may be specified in terms of standard deviations, confidence limits, or other appropriate measures. We assume that some known functional relationship exists by which a derived quantity u is calculated:

$$u = f(x, y, z, \ldots) \tag{5.48}$$

Presumably we get the central value of u by using the central values of the various factors,

$$\bar{u} = f(\bar{x}, \bar{y}, \bar{z}, \ldots)$$

and our task is to find an appropriate measure of the dispersion of u as represented by the quantity ϵ_u:

$$\bar{u} \pm \epsilon_u$$

This problem can be attacked by writing an equation for the total differential of u in terms of the partial derivatives with respect to each of the independent variables:

$$du = \left(\frac{\partial u}{\partial x}\right) dx + \left(\frac{\partial u}{\partial y}\right) dy + \left(\frac{\partial u}{\partial z}\right) dz + \cdots \tag{5.49}$$

If the error limits on the various factors are relatively small, it is possible to replace the differential quantities with the corresponding small finite variations and we have

$$\delta u = \left(\frac{\partial u}{\partial x}\right) \delta x + \left(\frac{\partial u}{\partial y}\right) \delta y + \left(\frac{\partial u}{\partial z}\right) \delta z + \cdots \tag{5.50}$$

This equation states that the total variation in u is equal to the variation in x times the way u varies with respect to x, plus the variation in y times the way u varies with respect to y, and so forth.

If we interpret the variations in the independent variables as the error limits that have been established for them, we may make the replacements

$$\begin{array}{ccccc}
\delta x & \longrightarrow & \bar{x} - x_t & \longrightarrow & \epsilon_x \\
\delta y & \longrightarrow & \bar{y} - y_t & \longrightarrow & \epsilon_y \\
\delta z & \longrightarrow & \bar{z} - z_t & \longrightarrow & \epsilon_z \\
\vdots & & & & \\
\delta u & \longrightarrow & \bar{u} - u_t & \longrightarrow & \epsilon_u
\end{array}$$

where $x_t, y_t, z_t, \ldots$ are the true values of the variables. We now have

$$\epsilon_u = \left(\frac{\partial u}{\partial x}\right)\epsilon_x + \left(\frac{\partial u}{\partial y}\right)\epsilon_y + \left(\frac{\partial u}{\partial z}\right)\epsilon_z + \cdots \tag{5.51}$$

In most cases the true values of the variables are not known, but if we can make estimates of the determinate errors, we can presumably specify $\epsilon_x, \epsilon_y, \epsilon_z, \ldots$ with respect to both magnitude and sign. They can be inserted into Eq. 5.51 to calculate the resulting error in the derived result represented by ϵ_u.

Now let us consider the case of indeterminate errors. We can calculate the maximum possible error in u by assuming that all the component errors are of the same sign, so that they all combine additively to give

$$\epsilon_{u\,\max} = \left|\left(\frac{\partial u}{\partial x}\right)\epsilon_x\right| + \left|\left(\frac{\partial u}{\partial y}\right)\epsilon_y\right| + \left|\left(\frac{\partial u}{\partial z}\right)\epsilon_z\right| + \cdots \tag{5.52}$$

This result is not particularly useful; as pointed out previously, there is little likelihood that the errors will all combine to give the maximum value if they are truly random. A more useful measure of the error in u can be obtained by squaring Eq. 5.51,

$$\epsilon_u^2 = \left(\frac{\partial u}{\partial x}\right)^2\epsilon_x^2 + \left(\frac{\partial u}{\partial y}\right)^2\epsilon_y^2 + \left(\frac{\partial u}{\partial z}\right)^2\epsilon_z^2 + \cdots + 2\left(\frac{\partial u}{\partial x}\right)\left(\frac{\partial u}{\partial y}\right)\epsilon_x\epsilon_y$$
$$+ 2\left(\frac{\partial u}{\partial x}\right)\left(\frac{\partial u}{\partial z}\right)\epsilon_x\epsilon_z + 2\left(\frac{\partial u}{\partial y}\right)\left(\frac{\partial u}{\partial z}\right)\epsilon_y\epsilon_z + \cdots$$

If an average is taken over all the individual observations, the cross terms vanish for variables that are independent and uncorrelated.

As pointed out above, it does not matter how the ϵ's are actually measured as long as they are consistent. If we, for example, use standard deviations to measure the errors, our equation becomes

$$s_u^2 = \left(\frac{\partial u}{\partial x}\right)^2 s_x^2 + \left(\frac{\partial u}{\partial y}\right)^2 s_y^2 + \left(\frac{\partial u}{\partial z}\right)^2 s_z^2 + \cdots \tag{5.53}$$

We will examine the propagation of random errors as determined by Eq. 5.53 for two special cases. First, let us assume that u is a linear combination of x, y, and z:

$$u = ax \pm by \pm cz \tag{5.54}$$

It is immediately apparent that the variance of u is then given by (see Eq. 4.17)

$$s_u^2 = a^2 s_x^2 + b^2 s_y^2 + c^2 s_z^2 \tag{5.55}$$

For the second case, assume that u involves products raised to various powers that we can represent in a general way by

$$u = ax^p y^q z^r \tag{5.56}$$

where p, q, and r may be any integers, positive or negative. The partial derivatives are

$$\left(\frac{\partial u}{\partial x}\right) = apx^{p-1}y^q z^r \qquad \left(\frac{\partial u}{\partial y}\right) = aqx^p y^{q-1} z^r$$

$$\left(\frac{\partial u}{\partial z}\right) = arx^p y^q z^{r-1}$$

and the result can be written in the form

$$\frac{s_u^2}{u^2} = p^2 \frac{s_x^2}{x^2} + q^2 \frac{s_y^2}{y^2} + r^2 \frac{s_z^2}{z^2} \tag{5.57}$$

We see that *absolute errors* should be combined for the addition or subtraction of variables, but *relative errors* are combined for multiplication or division. Complex cases are best handled in steps, broken down in such a way that either Eq. 5.55 or 5.57 applies.

We assumed that the relative errors were small, so that the standard deviations were equivalent to small but finite changes. We can therefore write

$$\left(\frac{s_x}{x}\right)^2 \quad \longrightarrow \quad \left(\frac{\delta x}{x}\right)^2 \quad \longrightarrow \quad (\delta \ln x)^2$$

with similar replacements for each of the other variables. Thus Eq. 5.57 is sometimes written in the alternative form,

$$(\delta \ln u)^2 = p^2(\delta \ln x)^2 + q^2(\delta \ln y)^2 + r^2(\delta \ln z)^2 \tag{5.58}$$

In elementary texts we are generally told to represent the uncertainty in a derived result by retaining the same number of significant digits as contained in the least precisely known factor. This can, of course, lead to a gross underestimation of the error if differences are involved, but even where there is no subtraction the results can be misleading. Suppose that we determine two numbers, 99.9 and 100.1, each to the nearest 0.1. Is it reasonable to say that we know the second number with a precision which is ten times that of the first? Certainly not. The only sure way to avoid problems is by the proper propagation of errors using the formulas given above.

Example 5.5

Assume that you have n values $x_1, x_2, x_3, \ldots, x_n$ which are to be averaged according to the formula

$$y = \frac{x_1 + x_2 + x_3 + \cdots + x_n}{n}$$

What is the standard deviation in y if all the x values have the same standard deviation s?

Solution. For each of the n values of x, we have

$$\left(\frac{\partial y}{\partial x_1}\right) = \frac{1}{n} \qquad \left(\frac{\partial y}{\partial x_2}\right) = \frac{1}{n} \qquad \cdots \qquad \left(\frac{\partial y}{\partial x_n}\right) = \frac{1}{n}$$

Inserting these and $s_{x_1} = s_{x_2} = s_{x_3} = \cdots = s$ in Eq. 5.53 gives

$$s_y^2 = \left(\frac{1}{n}\right)^2 s^2 + \left(\frac{1}{n}\right)^2 s^2 + \cdots + \left(\frac{1}{n}\right)^2 s^2 = \frac{ns^2}{n^2} = \frac{s^2}{n}$$

$$s_y = \frac{s}{\sqrt{n}}$$

This is the same as the standard deviation of means that was derived in Section 3.2.3 (see Eq. 4.20).

5.4 THE SMOOTHING OF EXPERIMENTAL DATA

We frequently obtain experimental data that are subjected to large random errors (i.e., high noise levels). Sometimes it is necessary to smooth the data to a certain extent as a prelude to further analysis, which may include integration, differentiation, or any number of other operations. Some features may be obscured by the noise and become visible only after it is filtered out. Then it may be useful to use one of the smoothing methods described below, but it should be noted that there is an inevitable loss of resolution in this process, and no new information is generated that is not already present in the original signal.

5.4.1 Boxcar and Moving Window Averaging

A very simple and obvious way to smooth data that contains a certain level of random error is to group the data points into subsets, average the points within each subset, and then connect these averages by a smooth curve. One technique for grouping the data consists of assigning the first n points to group 1, the second n points to group 2, the third n points to group 3, and so forth. This process is generally called *boxcar averaging*, and it is indicated schematically in Figure 5.7. The technique is crude, but it can be used to good advantage in those cases where the rate of variation in the desired signal is small compared with the rate at which the data points are collected.

Greater refinement can be achieved by overlap of the groupings of data points by *moving window averaging*, which is also illustrated schematically in Figure 5.7. For example, we may average points 1, 2, and 3 in the first window, points 2, 3, and 4 in the second window, points 3, 4, and 5 in the third window, and so forth. Obviously, the moving window average is preferable to the boxcar average for rapidly varying signals since a higher degree of resolution can be preserved. Better still, however, is an averaging process in which equations are fit to the data points.

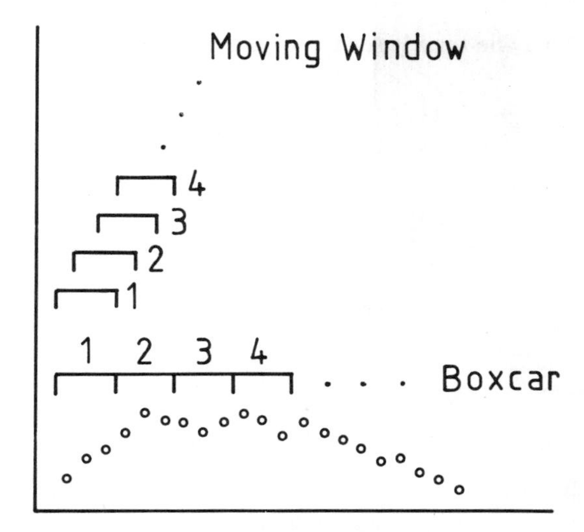

Figure 5.7 Schematic illustration showing the grouping of data for averaging by the boxcar and moving window methods.

5.4.2 The Least Squares Method

In principle we can always smooth experimental data by fitting it with a polynomial of high degree, but this is not practical if a large number of data points is involved, or if the curve is not simply represented. Then a least squares procedure can be used to fit a polynomial *locally* to a limited number of points. This polynomial may be used to compute a "smoothed" point within the local range. The smoothed points may then be joined if desired to obtain a smoothed curve that extends over the entire range of the data.

In a typical application, an odd number of consecutive points is used to generate a localized polynomial of low order. For example, let us assume that our data consist of values of an independent variable y_i corresponding to equally spaced values along the abscissa at $x_i = -2, -1, 0, +1$, and $+2$, and that we desire to fit a polynomial of degree 2,

$$y = b_0 + b_1 x + b_2 x^2 \tag{5.59}$$

to these data. The problem was outlined in Section 5.2.3; the normal equations for unit weighting factors are

$$\begin{aligned} b_0 n + b_1 \Sigma x_i + b_2 \Sigma x_i^2 &= \Sigma y_i \\ b_0 \Sigma x_i + b_1 \Sigma x_i^2 + b_2 \Sigma x_i^3 &= \Sigma x_i y_i \\ b_0 \Sigma x_i^2 + b_1 \Sigma x_i^3 + b_2 \Sigma x_i^4 &= \Sigma x_i^2 y_i \end{aligned}$$

Using the values of x_i above in these equations, we see that

$$\begin{aligned} 5b_0 + 0b_1 + 10b_2 &= \Sigma y_i \\ 0b_0 + 10b_1 + 0b_2 &= \Sigma x_i y_i \\ 10b_0 + 0b_1 + 34b_2 &= \Sigma x_i^2 y_i \end{aligned} \tag{5.60}$$

These three equations may be readily solved for the three unknowns b_0, b_1, and b_2, and the solution is

$$b_0 = \frac{1}{35} \sum_{i=-2}^{+2} (17 - 5x_i^2)y_i$$

$$b_1 = \frac{1}{10} \sum_{i=-2}^{+2} x_i y_i$$

$$b_2 = \frac{1}{14} \sum_{i=-2}^{+2} (x_i^2 - 2)y_i$$

or

$$b_0 = \frac{1}{35} [-3y_{-2} + 12y_{-1} + 17y_0 + 12y_{+1} - 3y_{+2}]$$

$$b_1 = \frac{1}{10} [-2y_{-2} - y_{-1} + y_{+1} + 2y_{+2}]$$

$$b_2 = \frac{1}{14} [2y_{-2} - y_{-1} - 2y_0 - y_{+1} + 2y_{+2}]$$

By insertion of these values into Eq. 5.59, we find the "smoothed" values at each of the points $x_i = -2, -1, 0, +1$, and $+2$ to be

$$y^s_{-2} = \frac{1}{35} [31y_{-2} + 9y_{-1} - 3y_0 - 5y_{+1} + 3y_{+2}]$$

$$y^s_{-1} = \frac{1}{35} [9y_{-2} + 13y_{-1} + 12y_0 + 6y_{+1} - 5y_{+2}]$$

$$y^s_0 = \frac{1}{35} [-3y_{-2} + 12y_{-1} + 17y_0 + 12y_{+1} - 3y_{+2}] \qquad (5.61)$$

$$y^s_{+1} = \frac{1}{35} [-5y_{-2} + 6y_{-1} + 12y_0 + 13y_{-1} + 9y_{+2}]$$

$$y^s_{+2} = \frac{1}{35} [3y_{-2} - 5y_{-1} - 3y_0 + 9y_{+1} + 31y_{+2}]$$

In actual application, we move the interval progressively along through the table of values so that only the equation for y^s_0 is used for the main body of the data; the other four equations are used for the first two and last two end points.

To illustrate the method, we have generated a curve consisting of the superposition of two Gaussian components on an arbitrary background using the program listed below:

```
10 KEY OFF
20 SCREEN 2
30 DIM C(500),S(500)
40 DATA 240,8,15,270,10,80,10,6
50 READ A,B,C,D,E,F,P,N
60 RANDOMIZE TIMER
70 FOR I=1 TO 500
80 T=P+C*EXP(-((I-A)/B)^2/2)+F*EXP(-((I-D)/E)^2/2)
90 FOR K=1 TO N
100 Y=RND-.5
110 IF Y>0 THEN T=T+1 ELSE T=T-1
120 NEXT K
130 C(I)=T
140 NEXT I
150 GOSUB 230
160 FOR J=1 TO 50
170 GOSUB 340
180 FOR I=1 TO 500
190 C(I)=S(I)
200 NEXT I
210 GOSUB 230
220 NEXT J
230 CLS          'PLOT SUBROUTINE
240 FOR I=50 TO 550
250 PSET(I,180)
260 NEXT I
270 FOR I=1 TO 500
280 PSET(I+50,180-C(I))
290 NEXT I
300 LOCATE 8,10
310 PRINT"SMOOTHED ";J;"TIMES";
320 M$=INKEY$ : IF M$="" THEN 320
330 RETURN
340 S(1)=(31*C(1)+9*C(2)-3*C(3)-5*C(4)+3*C(5))/35   'SMOOTH SUBROUTINE
350 S(2)=(9*C(1)+13*C(2)+12*C(3)+6*C(4)-5*C(5))/35
360 FOR I=3 TO 498
370 S(I)=(-3*C(I-2)+12*C(I-1)+17*C(I)+12*C(I+1)-3*C(I+2))/35
380 NEXT I
390 S(499)=(-5*C(496)+6*C(497)+12*C(498)+13*C(499)+9*C(500))/35
400 S(500)=(3*C(496)-5*C(497)-3*C(498)+9*C(499)+31*C(500))/35
410 RETURN
```

The curve is defined at some 500 points along the abscissa and is generated in statement 80 using parameters read in from the DATA statement 40. To impart additional realism, we have then introduced a certain amount of noise by a random walk process in statements 90 to 120. The amount of noise is controlled by the parameter N, which is also read in from the DATA statement 40.

The subroutine starting in statement 230 is used to generate the visual display. Note statement 320 in particular. The function INKEY$ monitors the keyboard for input. The figure can be viewed as long as desired on the video monitor. If no keyboard entry is made, the string variable M$ is a blank that is represented by double quotes "" with no space between. As soon as any key is depressed, however, the variable M$ is no longer blank and control passes to the next statement.

The smoothing of the curve is carried out in the subroutine beginning at statement 340; its operation should be apparent.

Figure 5.8 shows the original curve generated by this program, and after being subjected to varying numbers of smoothing operations. These curves were transferred

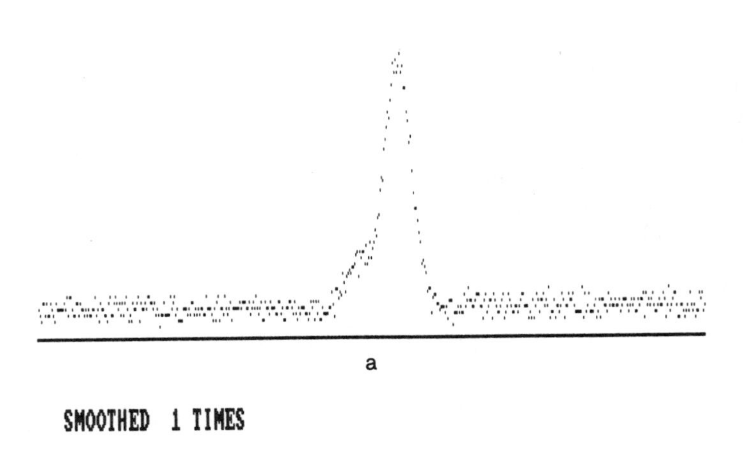

a

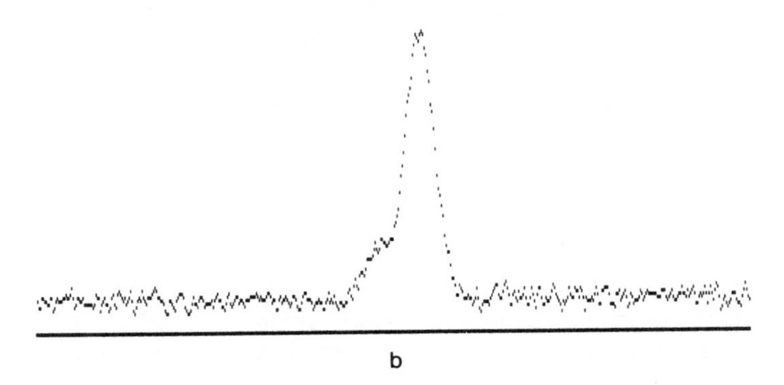

b

Figure 5.8 A "noisy" curve that has been generated and then smoothed using the BASIC program described in the text.

SMOOTHED 5 TIMES

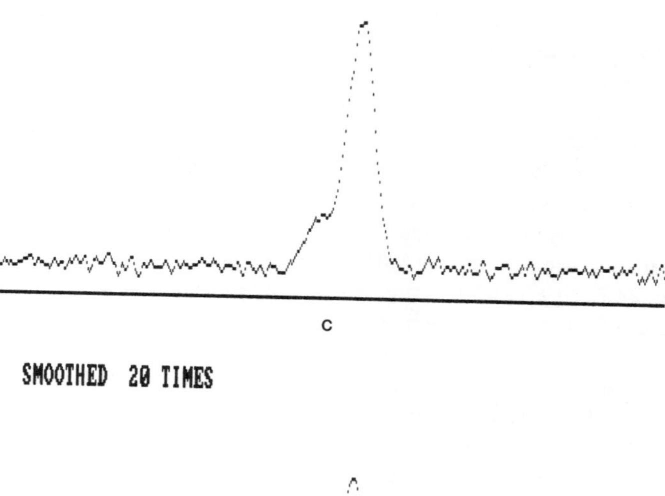

SMOOTHED 20 TIMES

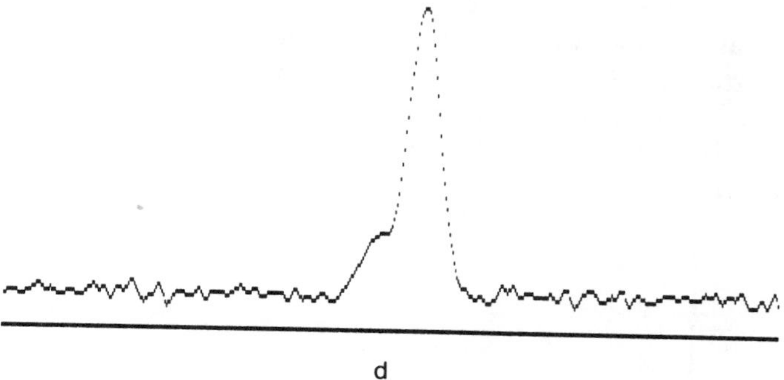

Figure 5.8 (*Continued*)

from the screen to the plotter using PCPLOT® as described in Section 2.4.2. Note that without smoothing the presence of a small second Gaussian component could be easily overlooked unless one were expecting it. It tends to be obscured by the noise. It becomes progressively more distinct as the number of smoothing cycles increases, but there are some obvious practical limitations on how far one might want to push this process.

5.5 LIMITS OF SIGNAL DETECTION

Most modern chemical measurements involve the reading of an instrument of some sort. In accordance with the discussion of Chapter 4, we must recognize that any such reading is subject to a certain level of random error or noise. We will now consider the following question: How can we distinguish between a meaningful response and random fluctuations of background noise? In other words, what is the limit of signal detection?

To keep our discussion within reasonable bounds, we will make some simplifying assumptions. First let us assume that if we make a large number of measurements with our instrument under a given set of circumstances, they will have a normal distribution about the mean. The second assumption we will make is that the standard deviation σ of measurements with a sample in place is the same as that for background measurements which are made in the absence of the sample (or more appropriately in many instances, with a blank).

We designate the mean background response of the instrument by μ_B, and the response of the instrument with sample by μ_S. If we make a large number of measurements of both μ_B and μ_S, the results may be represented by diagrams such as those of Figure 5.9, where the response is plotted along the abscissa and the frequency of a given response is plotted along the ordinate.

Figure 5.9a illustrates the situation where μ_B and μ_S are separated by the distance 6σ. There is only a very small overlapping of the tails of the distribution curves in this particular case (0.13% according to Appendix A.8), which we regard as insignificant. If we make a single measurement of the background and a single measurement of the sample, they will be clearly separated so that the sample response can be stated with certainty. This will be true for any separation of μ_B and μ_S greater than or equal to 6σ. For smaller separations, however, the overlap becomes significant and the distinction is no longer so clear. We take the value 6σ to be the *limit of guaranteed detection*.

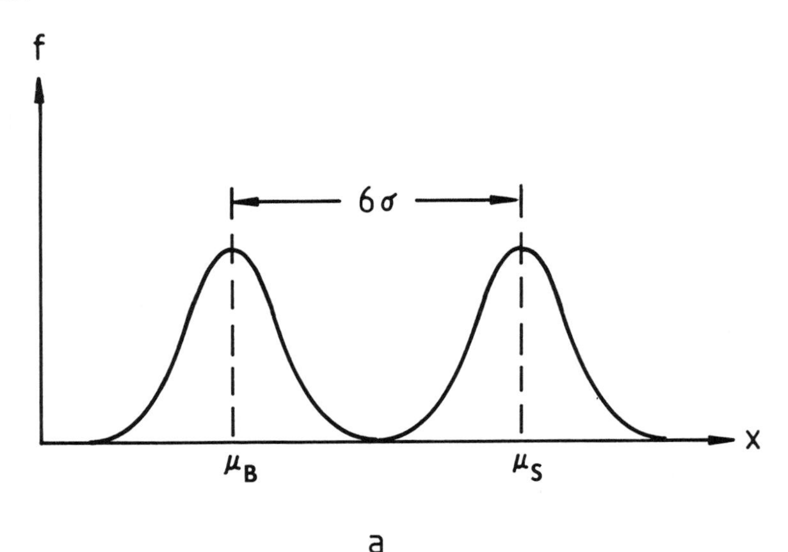

Figure 5.9 Schematic diagram of normally distributed responses for an instrument at (a) the limit of guaranteed detection and (b) the Kaiser detection limit.

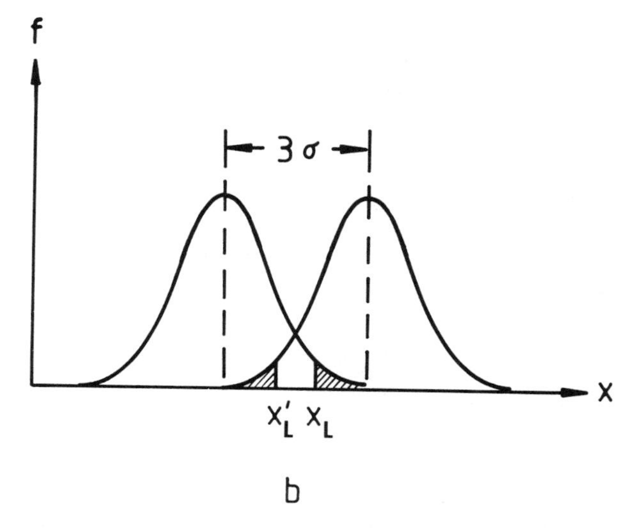

Figure 5.9 (*Continued*)

Figure 5.9b depicts the situation where the two means are separated by the distance 3σ. The overlap is no longer insignificant. We see that 50% of each curve is overlapped by the other. Here a single measurement of μ_B and μ_S is clearly insufficient to distinguish between the response of the sample and the background. Any response that falls in the range μ_S–μ_B could be due to either sample or background, and there is a probability of 0.50 that a single measurement of either will fall within this range. Even so, if a large number of measurements of both are carried out, the means are still distinguishable for this level of separation. This is known as the *Kaiser detection limit*, and it is the criterion recommended by the International Union of Pure and Applied Chemistry.

The above obviously implies that an infinite number of measurements have been made so that the normal distribution is clearly defined. In any real case we must of course rely on finite determinations. The means $\bar{x}_B$ and $\bar{x}_S$ serve as reliable estimates of μ_B and μ_S, respectively. Either s_B or s_S could be used to estimate σ, but as indicated in Section 4.4.2, the pooled standard deviation (Eq. 4.31) is preferable.

If one is willing to accept a certain small probability of error, a lower detection limit can be defined. The range between $\bar{x}_B$ and $\bar{x}_S$ can be subdivided (see Figure 5.9b). The detection limit is determined by a t-test, in which the null hypothesis states that the sample is not present (H_0: $S = 0$), and the alternative hypothesis states that the sample is present (H_a: $S \neq 0$). The small shaded area to the right of x_L represents the probability that H_a is accepted when it is H_0 that is actually correct. This is known as a *Type I error*. Similarly, if we fail to reject the null hypothesis when the sample is present, this is known as a *Type II error*, and its probability is represented by the shaded area to the left of x_L' in Figure 5.9b. If we establish a confidence level for the small area to the right of x_L, the probability that μ_S is significantly larger than μ_B is estimated by (see Eq. 4.32)

$$\frac{\bar{x}_S - \bar{x}_B}{s\sqrt{1/N_S + 1/N_B}} > t_\alpha \tag{5.62}$$

where s is given by Eq. 4.31 and the number of degrees of freedom is $N_B + N_S - 2$ with N_B and N_S being the number of measurements of the background and sample, respectively. From Eq. 5.62, we infer that the smallest difference between μ_S and μ_B which may be considered significant is given by

$$\mu_S - \mu_B = st_\alpha\sqrt{\frac{1}{N_S} + \frac{1}{N_B}} \tag{5.63}$$

It will be observed that the detection limit defined in this way depends on the number of measurements used to determine the response of the instrument with and without the sample.

The result above is often stated in terms of the *signal-to-noise ratio* of the instrument. Obviously, $\bar{x}_S - \bar{x}_B$ is a measure of the size of the response due to the sample (the signal), and

$$s\sqrt{\frac{1}{N_S} + \frac{1}{N_B}}$$

is a measure of the combined effect of random error in x_S and x_B (the noise), so Eq. 5.63 can also be written

$$S/N = t_\alpha \tag{5.64}$$

It should again be carefully noted that there is a relationship between the signal-to-noise ratio and the degrees of freedom $N_B + N_S - 2$. For example, if we wish to establish a detection limit at the 0.05 level with five measurements of the background and five measurements of the sample, we require an instrument with a signal-to-noise ratio of 1.860 (see Appendix A.9 under 8 degrees of freedom).

5.6 THE ANALYSIS OF COMPOSITE SIGNALS

In many cases the data generated by a given instrument represents a composite of several individual signals, and the task of determining just how many may be contained in the composite data set, as well as their size and shape, can be a very complicated task. We will briefly consider two distinct approaches that can be used to shed light on this problem. The first is a straightforward attempt to resolve the composite signal into its component parts through curve fitting, also known as *deconvolution*, and the second makes use of a powerful mathematical technique called *factor analysis*.

5.6.1 Deconvolution

The upper curve of Figure 5.10 is typical of many that are observed in practice. It may be a result of chromatographic separation, x-ray scattering, magnetic resonance or optical spectroscopy, and so forth. It is apparent that there are at least two component signals, since a shoulder is clearly evident. The resolution of the composite signal into its components can be carried out using the least squares methods described in Section 5.2. Each component is described in terms of three parameters, one of which gives its position (x_{0j}), one its width (w_j), and one its height (y_{mj}). In addition, the components are often imposed on a background that is not constant, and additional parameters must be used to define it. We will assume that the background can be described as a sum of constant, linear, and quadratic terms, so the total calculated curve may be written

$$y_{ic} = a + bx_i + cx_i^2 + \sum_j l_j(x_i, x_{0j}, w_j, y_{mj})$$

where $l_j = f(x_i, x_{0j}, w_j, y_{mj})$ is a line shape function for the individual components. The deviations are of the form

$$d_i = y_i - y_{ic} = y_i - a - bx_i - cx_i^2 - \sum_j l_j \tag{5.65}$$

The nonlinear solution technique is required to fit the calculated curve y_{ic} to the observed curve y_i since the individual component line shape functions l_j are not linear in the parameters x_{0j}, w_j, and y_{mj} used to define them. The details of this process will not be reviewed here, since they were outlined in Section 5.2.4.

If reasonable initial guesses can be obtained from the experimental curve, the nonlinear fitting procedure yields optimum values of the various parameters which

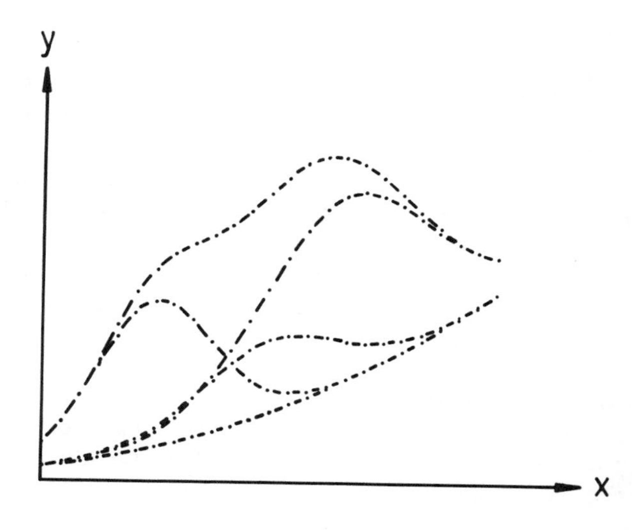

Figure 5.10 A composite curve made up of three Gaussian components superimposed on a curved background having constant, linear, and quadratic terms.

define the background and the individual line shape functions. These can be used to specify the precise location of the components, for the calculation of integrated intensities, and so forth. There is some arbitrariness associated with this process, however. Just how many background terms and how many components should be included in a given case is sometimes questionable. For example, there is no indication whatever of a third component in the curve of Figure 5.10, but the composite was in fact constructed of the three components and the background curve indicated. The general approach that should be used is to assume the smallest number of components clearly indicated by the composite curve and attempt a fit. By examination of the deviations, one can often see that an additional component is required for an acceptable fit, in which case it may be inserted and the fitting process repeated until the desired accuracy is achieved.

Although this sounds simple enough, the actual process is fraught with difficulties. The experimental curve always contains a certain amount of noise, and a plot of the deviations frequently suggests placement of additional components when in fact there is no physical reason to expect them. Care must be exercised to introduce only components for which there is some real justification.

A problem intimately related to the number of components is the question of their individual line shapes. One frequent choice is the *Lorentzian line shape function,*

$$\frac{y_m}{1 + 4(x - x_0)^2/w^2} \tag{5.66}$$

The width w used here is called the *half-width* (or more precisely, the *full width at half maximum*). It is defined as the distance between the two points on either side of the maximum where the intensity is equal to $y_m/2$. Another popular choice is the *Gaussian line shape function,*

$$y_m \exp\left[-2 \ln 4 \frac{(x - x_0)^2}{w^2}\right] \tag{5.67}$$

which is of the same form as the normal distribution function discussed in Chapter 4. The basic difference between the two line shapes is that the Gaussian function is more rounded at the peak but drops rather rapidly in the wings, while the Lorentzian function is more sharply peaked and extends with significant intensity to greater distances from the band center. Based on experience and the theoretical model chosen, either one or the other of these functions may be considered more appropriate for a given case. For example, infrared bands and high-resolution NMR bands generally tend to be more Lorentzian in shape, while chromatographic, broad-line NMR, and many ESR bands tend to be more nearly Gaussian. Ultraviolet and visible spectra may contain either, or a mixture of both.

The proper line shape function to use in a given situation is not always clear, and the results of the analysis can depend on this choice. The Lorentzian and Gaussian functions actually represent two extremes, and real composite curves are often made up of components which are of an intermediate shape. The Lorentzian function is in fact identical with Student's t-distribution for a single degree of freedom, and as we learned in Chapter 4, this distribution tends toward the normal or Gaussian distribution as the number of degrees of freedom increases. Therefore, Student's t-distribution with 2 or 3 degrees of freedom is sometimes selected as the component line shape function.

As indicated above, the number of components that should be used in a given case may be difficult to determine. If the number is not dictated by any theoretical model, we are generally not justified in using more than are evident in the experimental curve. However, the choice can sometimes be aided by the use of derivative spectra. For example, it can be demonstrated that a composite curve of two similar Lorentzian components which are separated by 40% of their half widths gives no indication of the presence of the two components, but a splitting into two peaks is clearly evident in the second derivative. Similar considerations apply to other line shape functions as well.

An additional complication is the fact that the components often tend to be skewed. This can happen in spectroscopy, for example, under rapid scanning conditions where the instrumental response may be too slow. It is also a common problem in chromatography where tailing is frequently observed. In such cases one can use components that have different widths on the two sides of the maximum, but essentially the same result can often be achieved by inclusion of an additional component.

5.6.2 Factor Analysis

Simply stated, factor analysis allows us to determine the number of factors that are contained in any data set which can be expressed in the form

$$D_{ik} = \sum_{j=1}^{n} R_{ij} C_{jk} \tag{5.68}$$

where R_{ij} is known as a *row cofactor* and C_{jk} is known as a *column cofactor*. In matrix form we write

$$\mathbf{D} = \mathbf{RC} \tag{5.69}$$

We will assume that there are as many columns in **R** as rows in **C**; that is, that these matrices are conformable (see Section 1.4). There are numerous situations to which the above applies. As a concrete example of chemical interest, the data matrix **D** may consist of various absorbance readings at wavelengths i for a number of solutions k. At the outset it is possible that neither the number of components contained in the solutions nor the identity of the various possibilities is known. Some cases of special interest include kinetic studies in which intermediates may be changing as a function of time, complex equilibria where the identity and number of species may vary depending on conditions, or the analysis of the effluent from a chromatographic column as the composition of the eluate changes. Factor analysis can be used to determine the number of components involved, as well as their identity. It has also been used to aid the identification of fragmentation patterns in mass spectral analysis, in the characterization of the parameters that occur in various linear free energy relationships, and so forth.

The first step of the analysis is to multiply the data matrix by its transpose

$$\mathbf{r} = \mathbf{D}^{\mathrm{T}}\mathbf{D} \tag{5.70}$$

The matrix $\mathbf{r}$ obtained by this process is known as the *covariance matrix*. It is so named because the off-diagonal elements are a measure of the way in which one column of data varies with another (i.e., the covariance). If there were no correlation between them at all, that particular element of the covariance matrix would be equal to zero. Note that we have indicated a premultiplication by the transpose in Eq. 5.70, which is appropriate if there are more rows than columns in the data set. If not, we postmultiply by the transpose to get the covariance matrix. Stated otherwise, we multiply in such a way as to give a covariance matrix of smaller dimension.

We let λ_1 be the eigenvalue of largest magnitude of the covariance matrix, and represent its associated eigenvector by $\mathbf{C}_1$. They of course satisfy the equation

$$\mathbf{r}\mathbf{C}_1 = \lambda_1\mathbf{C}_1 \tag{5.71}$$

We can arbitrarily choose $\mathbf{C}_1$ and operate on it with $\mathbf{r}$. If $\mathbf{C}_1$ were correctly chosen Eq. 5.71 would be satisfied exactly; in general it is not, but the result can be factored into two terms. The first is a normalization constant from which an approximate value of λ_1 can be calculated, and the second is a new value of $\mathbf{C}_1$, which should be a closer approximation of the real eigenvector. This new value of $\mathbf{C}_1$ can be operated on again by $\mathbf{r}$, and the process continued until λ_1 and $\mathbf{C}_1$ are determined to the desired accuracy (recall the power iteration method of Section 3.6.1).

Once λ_1 and $\mathbf{C}_1$ are known, they are used to construct a *first residual matrix* $\mathbf{r}^1$ defined by

$$\mathbf{r}^1 = \mathbf{r} - \lambda_1\mathbf{C}_1\mathbf{C}_1^{\mathrm{T}} \tag{5.72}$$

Note that postmultiplication of the eigenvector $\mathbf{C}_1$ by its transpose $\mathbf{C}_1^{\mathrm{T}}$ generates a matrix of the same dimensions as the covariance matrix $\mathbf{r}$. When multiplied by λ_1 and subtracted from $\mathbf{r}$, the contribution of the first component is removed. We can now assume an arbitrary eigenvector $\mathbf{C}_2$ and operate on it with $\mathbf{r}^1$:

$$\mathbf{r}^1\mathbf{C}_2 = \lambda_2\mathbf{C}_2 \tag{5.73}$$

Once again Eq. 5.73 will not be satisfied in general by the chosen vector $\mathbf{C}_2$, but the result can be factored to determine an approximate value of λ_2 and an approximate eigenvector $\mathbf{C}_2$ and this process is again continued until λ_2 and $\mathbf{C}_2$ are determined to the desired accuracy. Then the second residual matrix

$$\mathbf{r}^2 = \mathbf{r} - \lambda_1\mathbf{C}_1\mathbf{C}_1^{\mathrm{T}} - \lambda_2\mathbf{C}_2\mathbf{C}_2^{\mathrm{T}} \tag{5.74}$$

is constructed, and from it, values of the third eigenvalue λ_3 and its associated eigenvector $\mathbf{C}_3$ are determined. This process is continued until all the eigenvalues and eigenvectors of $\mathbf{r}$ are determined.

Naturally the eigenvalues and eigenvectors can be equally well determined in a single diagonalization process (for example, by the Jacobi method of Section 3.6.2), but we have purposely chosen to discuss their generation by a sequential process, since they are then ordered naturally in descending order, which is of primary concern for factor analysis. The point is that the *rank* of the matrix **r** is determined by the number of factors involved. As these are removed one by one, we should finally get a residual matrix whose elements are zero. In other words, if there are n factors, the nth residual matrix is the null matrix. Implicit in the above is the obvious assumption that the order of the covariance matrix must be equal to or greater than the number of factors involved.

In any real case the residual matrix is not actually zero but is reduced to a small level. If we carry out the sequential process described above, we will find that the eigenvalues decrease until all the factors have been accounted for; beyond that point they are nearly constant at some small value, and the residual matrix reflects the random error or noise level of the measurement process. The actual criterion for the selection of the number of factors is somewhat arbitrary. Many techniques have been proposed, including the acceptance only of those eigenvalues that are greater than the mean of all eigenvalues; evaluation of the eigenvalues using χ^2 tests can also be used. If a reasonable estimate of the random error of the method can be made, the process can be terminated when the residual matrix elements are decreased to an appropriate level. An interesting point is that once the number of factors has been determined, the residuals can be used to smooth the original data.

The strategy discussed above yields the number of factors, but it does not provide any direct information concerning their identity. The eigenvectors as defined above may be regarded as abstract mathematical entities, but if they are rotated to an appropriate coordinate system, they can reveal the actual nature of the factors involved (e.g., the molar absorptivities ϵ_{ij} and the molar concentrations M_{ij}). A process called *target testing* is used. The process is essentially a trial-and-error test of possible row or column cofactors for internal consistency. For further details, the interested reader is referred to the book by Malinowski and Howery cited at the end of the chapter.

Example 5.6

The absorbances of each of three solutions A, B, and C have been measured at four different wavelengths λ_1, λ_2, λ_3, and λ_4. The results are summarized in the following data table:

	A	B	C
λ_1	0.107	0.017	0.062
λ_2	0.510	0.064	0.260
λ_3	0.135	0.815	0.433
λ_4	0.009	0.029	0.020

Use factor analysis to determine the number of components present in these solutions.

Solution. We first premultiply the data matrix above by its transpose to obtain the following covariance matrix:

$$\mathbf{r} = \begin{pmatrix} 0.289855 & 0.144745 & 0.197869 \\ 0.144745 & 0.669451 & 0.371169 \\ 0.197869 & 0.371169 & 0.259333 \end{pmatrix}$$

The program PIM (power iteration method) of Section 3.6.1 is used to determine the largest eigenvalue and its associated eigenvector; the result is

$$\lambda_1 = 0.9647319 \qquad \mathbf{C}_1 = \begin{pmatrix} 0.3207084 \\ 0.7981849 \\ 0.5099480 \end{pmatrix}$$

The contribution of this factor to the covariance matrix is represented by

$$\lambda_1 \mathbf{C}_1 \mathbf{C}_1^{\mathrm{T}} = \begin{pmatrix} 0.0992264 & 0.2469565 & 0.1577767 \\ 0.2469565 & 0.6146299 & 0.3926775 \\ 0.1577767 & 0.3926775 & 0.2508756 \end{pmatrix}$$

When this is subtracted from the covariance matrix the following first residual matrix is obtained:

$$\mathbf{r}' = \begin{pmatrix} 0.1906286 & -0.1022115 & 0.0400923 \\ -0.1022115 & 0.0548211 & -0.0215085 \\ 0.0400923 & -0.0215085 & 0.0084574 \end{pmatrix}$$

The largest eigenvalue and associated eigenvector of this matrix is found using PIM:

$$\lambda_2 = 0.2538711 \qquad \mathbf{C}_2 = \begin{pmatrix} 0.8665265 \\ -0.4646552 \\ 0.1822843 \end{pmatrix}$$

The contribution of the second factor is

$$\lambda_2 \mathbf{C}_2 \mathbf{C}_2^{\mathrm{T}} = \begin{pmatrix} 0.1906237 & -0.1022177 & 0.4010000 \\ -0.1022177 & 0.0548119 & -0.0215027 \\ 0.4010000 & -0.0215027 & 0.0084355 \end{pmatrix}$$

Subtraction of this from the first residual matrix gives the second residual matrix,

$$\mathbf{r}^2 = \begin{pmatrix} 0.0000049 & 0.0000062 & -0.0000077 \\ 0.0000062 & 0.0000092 & -0.0000058 \\ -0.0000077 & -0.0000058 & 0.0000219 \end{pmatrix}$$

The matrix elements are all very small, which we assume to be indicative of the random error in the raw data. It is concluded that two components are contained in the solutions.

5.7 THE USE OF TRANSFORM TECHNIQUES IN CHEMISTRY

The transformation of functions from one domain to another is an important technique that is used in many different contexts in chemistry, including the solution of differential equations (i.e., by the Laplace transform), x-ray structure determination, spectroscopy, and so forth. An enumeration of all the various usages is far beyond the scope of our present discussion. We will describe but two of the more important transform methods and indicate how they may be applied in spectroscopy to gain certain advantages over conventional techniques.

We have become accustomed to thinking in terms of scanning through a given wavelength range using a single detector in order to record a spectrum. In recent years some multidetector (diode array) spectrometers have been developed so that an optical spectrum can be simultaneously recorded with moderate resolution, but such techniques are not practical except in the visible and near-ultraviolet range. The great advantage of simultaneous recording of a spectrum is that it can be done in much less time. Alternatively, in the same time that a spectrum may be scanned, it can be recorded several times by multidetector methods and averaged to improve the signal-to-noise ratio. There are techniques based on the *multiplex principle* that allow many of the same advantages of multidetection methods to be realized but using only a single broadband detector. The underlying concept can readily be appreciated by analogy with a hypothetical weighing experiment.

Suppose that we have three objects of weights x_1, x_2, and x_3 to be weighed on an analytical balance, and let us further assume that the standard deviation of a given weighing with this balance is σ. The obvious procedure consists of weighing each object in turn, with the results $x_1 \pm \sigma$, $x_2 \pm \sigma$, $x_3 \pm \sigma$. Another technique that may be used is to weigh combinations of weights such as the following:

$$y_1 = x_1 + x_2$$

$$y_2 = x_1 + x_3$$

$$y_3 = x_2 + x_3$$

From the three readings y_1, y_2, and y_3 we can obviously recover the weights x_1, x_2, and x_3 of the three objects from

$$x_1 = \tfrac{1}{2}(y_1 + y_2 - y_3)$$

$$x_2 = \tfrac{1}{2}(y_1 - y_2 + y_3)$$

$$x_3 = \tfrac{1}{2}(-y_1 + y_2 + y_3)$$

Now the standard deviation in each of the weighings y_1, y_2, and y_3 is also σ, and by use of Eq. 5.53 for the propagation of error, the standard deviations in the x_1, x_2, and x_3 values determined from them are

$$\sigma_x = [(\tfrac{1}{2}\sigma)^2 + (\tfrac{1}{2}\sigma)^2 + (\tfrac{1}{2}\sigma)^2]^{1/2} = \frac{\sqrt{3}\sigma}{2}$$

Here we have multiplexed the data; by making the same number of weighings and transforming them, we have recovered the weights of the three individual objects, but the uncertainty in their values is now reduced. This is due to the fact that by multiplexing, each object has been weighed more than once. The same principle applies to the recording of spectra by the Hadamard and Fourier transform techniques, which we will now describe.

5.7.1 The Hadamard Transform

The essential principles of a Hadamard transform spectrometer are illustrated schematically in Figure 5.11. The Hadamard mask is a movable element that can be brought into a number of successive positions where selected portions of the beam are blocked. Ideally, half the positions are blocked in a random fashion, and real spectrometers use a great number of channels. For purposes of illustration we have shown but seven channels, and a suitable mask for this case may be represented by the 13 elements (0101110010111), where 0 indicates that the beam is blocked and 1 indicates that it is allowed to pass. In position 1 the first seven elements are used and the mask consists of 0101110. In position 2 it is shifted such that the pattern is 1011100, and so forth. This constitutes a cyclic permutation of the elements; such masks can be readily designed as long as there are $2N - 1$ elements, where N is any integer. A "mask" for the weighing experiment described above for three elements can be written (11011).

With seven channels, we must make seven photometric readings with the spectrometer above in order to determine the absorbance at the seven different wavelengths. If we use x_i to represent the absorbances at the various wavelengths and y_i to represent the photometric readings, we have for the mask described above,

$$y_1 = \quad x_2 \quad + x_4 + x_5 + x_6$$

$$y_2 = x_1 \quad + x_3 + x_4 + x_5$$

$$y_3 = \quad x_2 + x_3 + x_4 \quad + x_7$$

$$y_4 = x_1 + x_2 + x_3 \quad + x_6$$

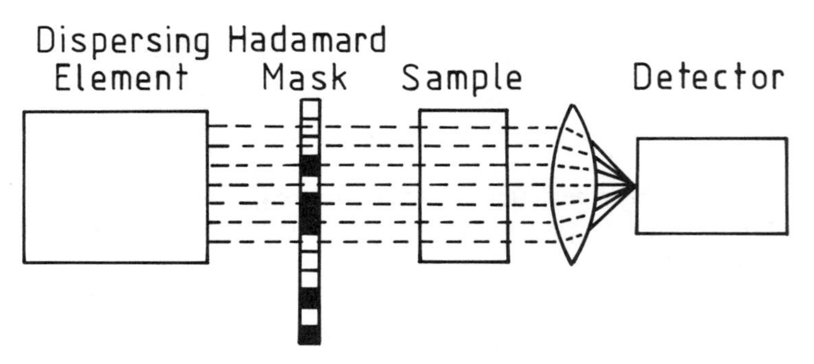

Figure 5.11 Schematic diagram of a Hadamard transform spectrometer.

$$
\begin{aligned}
y_5 &= x_1 + x_2 \qquad\qquad\qquad + x_5 \qquad + x_7 \\
y_6 &= x_1 \qquad\qquad + x_4 \qquad + x_6 + x_7 \\
y_7 &= \qquad\qquad x_3 \qquad + x_5 + x_6 + x_7
\end{aligned}
$$

This may be represented in matrix form by

$$\mathbf{y} = \mathbf{Hx} \tag{5.75}$$

with

$$
\mathbf{H} = \begin{pmatrix}
0 & 1 & 0 & 1 & 1 & 1 & 0 \\
1 & 0 & 1 & 1 & 1 & 0 & 0 \\
0 & 1 & 1 & 1 & 0 & 0 & 1 \\
1 & 1 & 1 & 0 & 0 & 1 & 0 \\
1 & 1 & 0 & 0 & 1 & 0 & 1 \\
1 & 0 & 0 & 1 & 0 & 1 & 1 \\
0 & 0 & 1 & 0 & 1 & 1 & 1
\end{pmatrix}
$$

Formally, we need to invert the matrix **H** to recover the absorbance values **x**:

$$\mathbf{H}^{-1}\mathbf{Hx} = \mathbf{x} = \mathbf{H}^{-1}\mathbf{y} \tag{5.76}$$

The inverse of the matrix **H** may be obtained, except for a constant numerical factor, simply by replacing the 0's by -1's to obtain

$$
\mathbf{H}^{t} = \begin{pmatrix}
-1 & 1 & -1 & 1 & 1 & 1 & -1 \\
1 & -1 & 1 & 1 & 1 & -1 & -1 \\
-1 & 1 & 1 & 1 & -1 & -1 & 1 \\
1 & 1 & 1 & -1 & -1 & 1 & -1 \\
1 & 1 & -1 & -1 & 1 & -1 & 1 \\
1 & -1 & -1 & 1 & -1 & 1 & 1 \\
-1 & -1 & 1 & -1 & 1 & 1 & 1
\end{pmatrix} \tag{5.77}
$$

It is this matrix that is used to give the Hadamard transform, and the absorbance values in this case are determined with an uncertainty which is a factor $\sqrt{7}/4$ times that of an individual photometric reading. The corresponding improvement in the signal-to-noise ratio is therefore $4/\sqrt{7}$, and in the general case it may be represented by the factor $[(N + 1)/2]/N^{1/2}$. This can be substantial for large values of N. Even greater advantages can be realized through Fourier transform methods, which in principle allow the complete multiplexing of signals of all frequencies.

5.7.2 The Fourier Transform

It was shown many years ago (1807–1822) by the French mathematician Fourier that any periodic function may be conveniently represented by a sum of sine and cosine terms. For a function of period $2L$ defined in the fundamental interval $-L$ to $+L$ by $f(x)$, the *Fourier series* expansion may be written

$$f(x) = \frac{a_0}{2} + \sum_{n=1}^{\infty} \left(a_n \cos \frac{n\pi x}{L} + b_n \sin \frac{n\pi x}{L} \right) \tag{5.78}$$

with

$$a_n = \frac{1}{L} \int_{-L}^{+L} f(x) \cos \frac{n\pi x}{L}\, dx \qquad n = 0, 1, 2, 3, \ldots$$

$$b_n = \frac{1}{L} \int_{-L}^{+L} f(x) \sin \frac{n\pi x}{L}\, dx \qquad n = 1, 2, 3, \ldots$$

If the function $f(x)$ is even [i.e., if $f(-x) = f(x)$], then only cosine terms contribute to the sum,

$$f(x) = \frac{a_0}{2} + \sum_{n=1}^{\infty} a_n \cos \frac{n\pi x}{L} \qquad a_n = \frac{2}{L} \int_0^L f(x) \cos \frac{n\pi x}{L}\, dx \tag{5.79}$$

while only sine terms are included if $f(x)$ is an odd function such that $f(-x) = -f(x)$:

$$f(x) = \sum_{n=1}^{\infty} b_n \sin \frac{n\pi x}{L} \qquad b_n = \frac{2}{L} \int_0^L f(x) \sin \frac{n\pi x}{L}\, dx \tag{5.80}$$

Such expansions are possible because the sine and cosine series constitute a *complete orthogonal set* of functions; that is, for an infinite number of terms they completely span configuration space, and when averaged over an entire period,

$$\int \sin mx \sin nx\, dx = \int \cos mx \cos nx\, dx = \int \sin mx \cos nx\, dx = 0 \qquad m \neq n$$

From this property we can readily justify the expansion coefficients given in Eqs. 5.79 and 5.80. For the sine series, for example, we multiply both sides of Eq. 5.80 by $\sin(m\pi x/L)$ and integrate. Because of orthogonality the only term which remains on the right is that for which $m = n$ and we have

$$\int_0^L f(x) \sin \frac{n\pi x}{L}\, dx = b_n \int_0^L \sin^2 \frac{n\pi x}{L}\, dx$$

Since

$$\int_0^L \sin^2 \frac{n\pi x}{L}\, dx = \frac{L}{n\pi} \int_0^{n\pi} \sin^2 y\, dy = \frac{L}{n\pi} \left[\frac{n\pi}{2} \right] = \frac{L}{2}$$

we see that the equation for the b_n coefficient in Eq. 5.80 is justified; a similar derivation can be given for the a_n coefficients of the cosine series.

It should be obvious that the Fourier series can be written in a more compact exponential form through use of the Euler formula (Eq. 1.46):

$$f(x) = \frac{1}{2} \sum_{n=-\infty}^{+\infty} c_n e^{i\omega_n x} \qquad c_n = \frac{1}{L} \int_{-L}^{+L} f(x) e^{-i\omega_n x}\, dx \tag{5.81}$$

where $\omega_n = n\pi/L$ with $n = 0, \pm 1, \pm 2, \ldots$.

As a simple example, let us choose a periodic step function that is equal to $+1$ from $x = 0$ to $x = L$, and equal to -1 from $x = L$ to $x = 2L$. This is shown by the solid heavy line of Figure 5.12. Since this function is odd we represent it by a sine series with

$$b_n = \frac{2}{L} \int_0^L f(x) \sin \frac{n\pi x}{L}\, dx = \frac{2}{L} \left(\frac{L}{n\pi}\right) \int_0^{n\pi} \sin y\, dy$$

$$= \frac{2}{n\pi} [-\cos y]_0^{n\pi} = \frac{4}{n\pi} \qquad \text{for odd values of } n$$

Thus

$$f(x) = \frac{4}{\pi} \sum_{n=1,3,5,\ldots} \frac{1}{n} \sin \frac{n\pi x}{L}$$

In Figure 5.12 the curve with $n = 1$ is shown, together with the sum of the first two terms with $n = 1, 3$ and the sum of the first three terms with $n = 1, 3, 5$. It is apparent that a better approximation of the function $f(x)$ is obtained as additional terms are

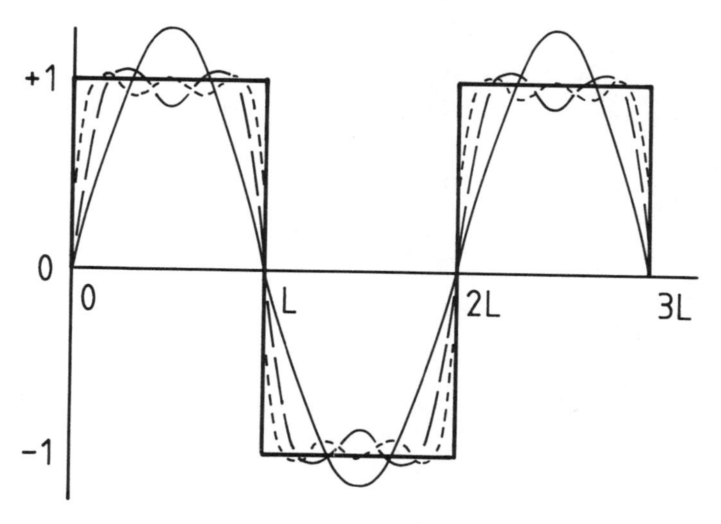

Figure 5.12 Periodic step function and its Fourier series representation with the first (solid curve), first plus second (long dashed curve), and first plus second plus third (short dashed curve) terms.

included in the series. With a sufficient number of terms (infinite in this case), the representation becomes exact.

A Fourier series can be used to synthesize any function as a summation of harmonic waves, and it is a special case of the *Fourier transform*. For spectroscopic applications a signal is generally observed as a function of time. For example, in Fourier transform nuclear magnetic resonance (FT-NMR), a strong radio-frequency (RF) pulse is applied to a sample in a magnetic field, which brings the various magnetic moments into phase. The free induction decay (FID) following this strong RF pulse is a time-dependent signal that can be transformed to determine the natural frequencies and relaxation times of the nuclear moments which contribute to the signal. Similarly, an interferometer can be used to generate a time-dependent signal, and this can be transformed to generate a spectrum of infrared absorption bands or characteristic frequencies that are represented by the interferogram (FT-IR). These and many other signals can be transformed from the *time domain* to the *frequency domain* ($\omega = 2\pi f$) through the Fourier transform

$$H(f) = \int_{-\infty}^{+\infty} h(t) \exp(-i2\pi ft)\, dt \tag{5.82}$$

where $h(t)$ is the time domain signal and $H(f)$ is the corresponding frequency domain signal. The inverse transformation is

$$h(t) = \frac{1}{2\pi} \int_{-\infty}^{+\infty} H(f) \exp(i2\pi ft)\, df \tag{5.83}$$

The Fourier transforms of some simple signals can readily be recognized. For a cosine wave in the time domain, the transform is a single spectral line at the corresponding frequency as shown in Figure 5.13a. If the time domain signal can be recognized as a superposition of two frequencies as in Figure 5.13b, the frequency domain spectrum is again simple. For closely spaced spectral lines in the frequency domain the

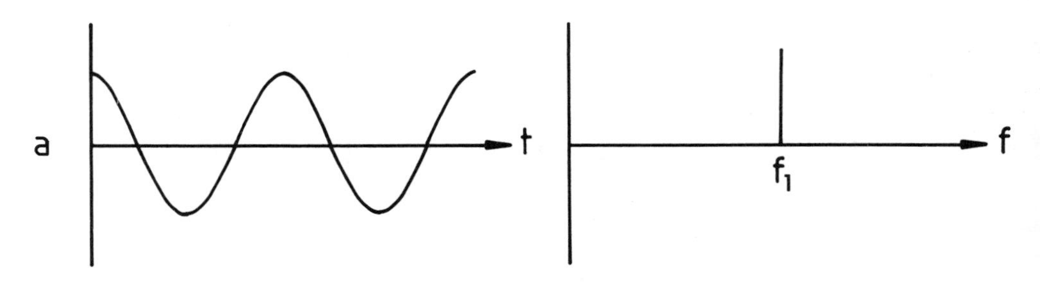

Figure 5.13 Fourier transforms of some simple time domain signals to frequency domain signals.

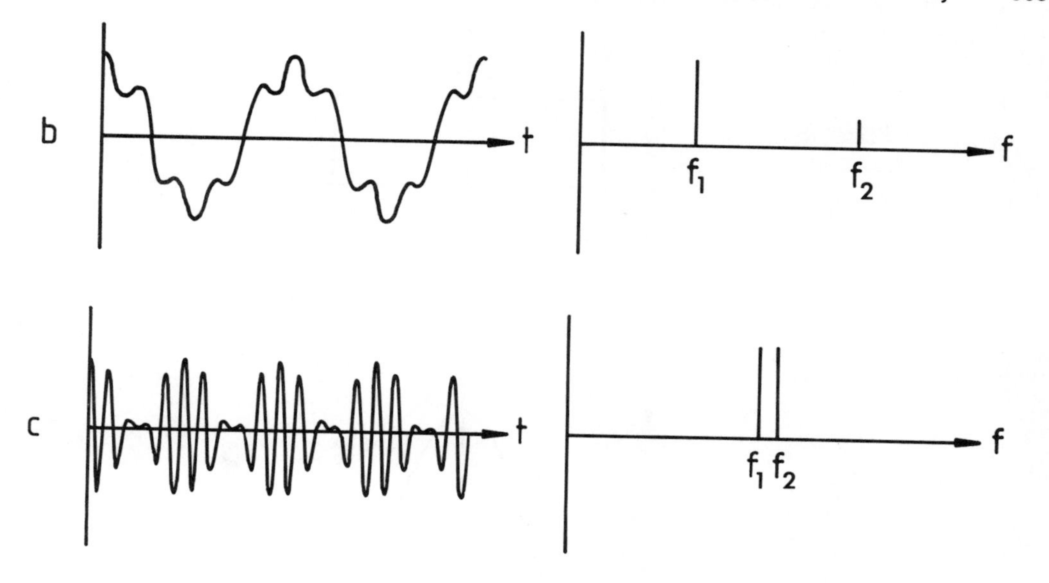

Figure 5.13 (*Continued*)

corresponding time domain signal is composed of beats as illustrated in Figure 5.13c. It was the observation of such beats that led Michelson in 1902 to postulate that the red Balmer line of the hydrogen atom spectrum is actually a doublet (now known to be separated by just 0.014 nm).

Apodization

To obtain a completely accurate frequency domain spectrum, we would need to observe the time domain signal over an infinite period as implied by Eq. 5.82. This is not possible in practice, which has some important consequences. Let us consider the transformation of a simple boxcar function like that illustrated in Figure 5.14a. With $b(t) = 0$ for $|t| > L$ and $b(t) = 1$ for $|t| \leq L$, the transform reduces to

$$
\begin{aligned}
B(f) &= \int_{-L}^{+L} \exp(-i2\pi ft)\, dt \\
&= \frac{-i}{2\pi f} [\exp(-i2\pi ft)]_{-L}^{+L} \\
&= \frac{-i}{2\pi f} [\cos(2\pi ft) - i \sin(2\pi ft)]_{-L}^{+L} \\
&= 2L \frac{\sin(2\pi fL)}{2\pi fL} \equiv 2L \operatorname{sinc}(2\pi fL)
\end{aligned}
\tag{5.84}
$$

This "sinc" function is shown in Figure 5.14b.

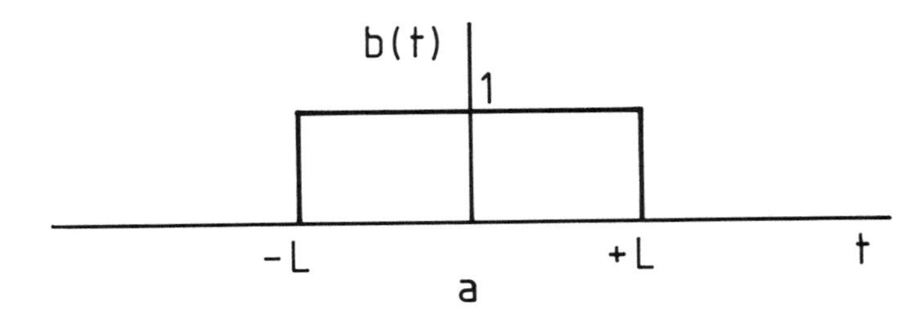

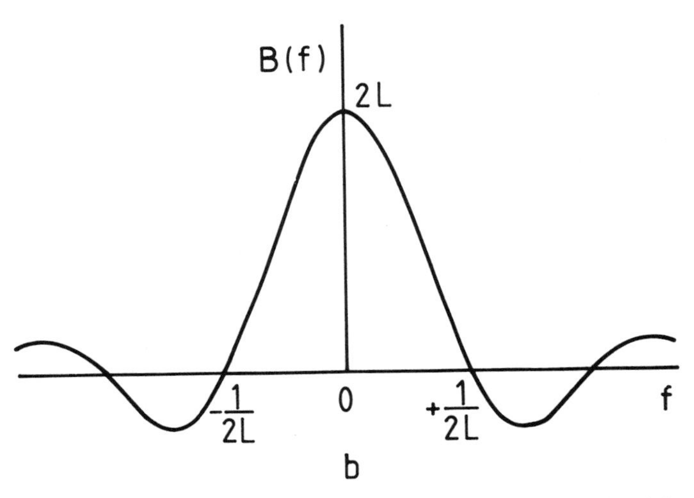

Figure 5.14 (a) The boxcar function $b(t)$ and (b) its Fourier transform $B(f) = 2L \operatorname{sinc}(2\pi f L)$.

The transformation of a product of two functions is called *convolution*. According to the convolution theorem of Fourier transformation,

$$\int_{-\infty}^{+\infty} g(t)h(t) \exp(-i2\pi ft)\, dt = \int_{-\infty}^{+\infty} G(f')H(f - f')\, df' \tag{5.85}$$

In other words, the Fourier transformation is equal to the area under a curve that is a product of the one function times the other function shifted a certain distance from the point of origin. The transformation of the function $h(t)$ weighted by the boxcar function $b(t)$ defined above is tantamount to observation of $h(t)$ for a finite period of time. We have

$$H'(f) = \int_{-\infty}^{+\infty} h(t)b(t) \exp(-i2\pi ft)\, dt = \int_{-\infty}^{+\infty} H(f)B(f - f')\, df'$$

The significance of this result is that transformation of a finite time domain signal does not produce sharp lines in the frequency domain, but gives broadened bands with the sinc line shape. This may not be a serious problem for a simple spectrum with well-separated features, but it can make interpretation difficult in more complex cases. Obviously, the longer the time domain signal, the sharper the frequency domain features.

The weak sidebands of the sinc function can be troublesome, since they can easily be confused with weak bands from other sources which may be close to the center band. The first negative band is in fact some 22% of the intensity of the center band (see Exercise 5.22). Various weighting functions can be used in place of the boxcar function in a convolution process to remove these undesirable sidebands. This process is called *apodization* (possibly derived from a Greek word meaning "without feet"), and one of the more common apodization functions is the triangular function shown in Figure 5.15a. This triangular apodization function may be written $a(t) = 1 - |t|/L$, and its transform is

$$A(f) = L \operatorname{sinc}^2(\pi f L) \tag{5.86}$$

which is shown in Figure 5.15b. Use of this apodization function to weight the spectrum according to the convolution theorem will once again give features in the frequency domain of this same structure. We note that the intensity of the sidebands is

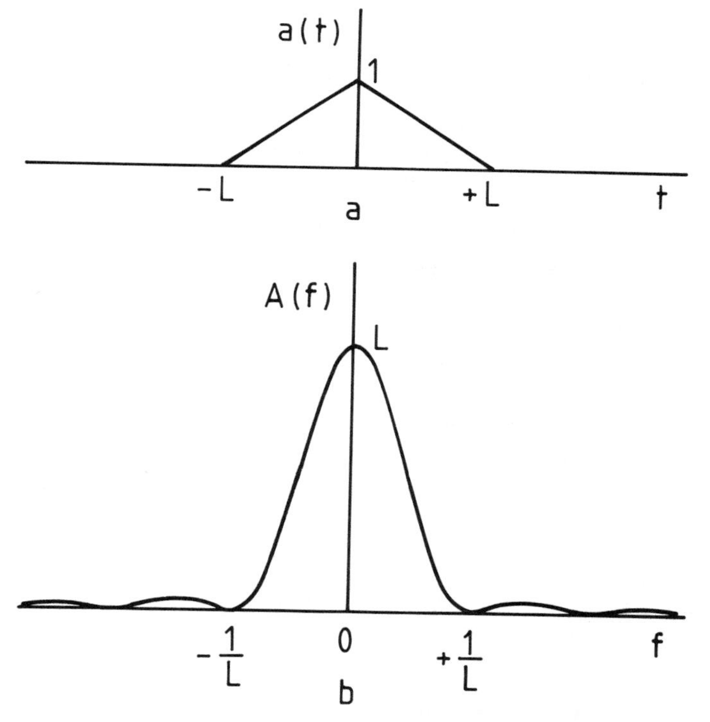

Figure 5.15 (a) The triangular apodization function $a(t)$ and (b) its Fourier transform $A(f) = L \operatorname{sinc}^2(\pi f L)$.

considerably reduced. This has come at the expense of a greater overall broadening, but for many applications this is not too serious.

The Fast Fourier Transform

In actual practice the Fourier transformation is performed digitally from discrete data points, rather than analytically. The *discrete Fourier transform* (DFT) can be represented in a general way for $2N$ data points by

$$H_f = \sum_{t=0}^{2N-1} h_t W^{ft} \qquad f = 0, 1, \ldots, 2N - 1 \tag{5.87}$$

where $W = e^{-i2\pi/2N} = e^{-i\pi/N}$. Here h_t represents the time domain data points, and H_f represents the corresponding frequency domain points. It is apparent that $2N$ (complex) multiplications are required to calculate each H_f point and since there are $2N$ of them, a total of $4N^2$ multiplications are necessary. It may take many thousands of data points to obtain sufficient resolution for certain cases and the computation time is then excessive. In fact, this is such a formidable problem that Fourier transform techniques did not come into immediate prominence with the advent of modern computers, but only after the development of the *Cooley-Tukey algorithm,* also called the *fast Fourier transform* (FFT), which was published in 1965.

To illustrate how the FFT works, we first consider the time domain data h_t to be divided into two subseries f_t and g_t, where f_t is composed of even terms and g_t is composed of odd terms:

$$\begin{aligned} f_t &= h_0, h_2, h_4, \ldots, h_{2t} \\ g_t &= h_1, h_3, h_5, \ldots, h_{2t+1} \end{aligned} \qquad t = 0, 1, 2, \ldots, N - 1$$

The DFT of each of these subseries may be constructed:

$$\begin{aligned} F_f &= \sum_{t=0}^{N-1} f_t \exp(-i2\pi ft/N) = \sum_{t=0}^{N-1} f_t W^{2ft} \\ G_f &= \sum_{t=0}^{N-1} g_t \exp(-i2\pi ft/N) = \sum_{t=0}^{N-1} g_t W^{2ft} \end{aligned} \qquad f = 0, 1, 2, \ldots, N - 1 \tag{5.88}$$

In the sum of Eq. 5.87 we can separate the odd and even terms to get

$$H_f = \sum_{t=0}^{N-1} f_t W^{2ft} + \sum_{t=0}^{N-1} g_t W^{(2t+1)f}$$

or

$$H_f = F_f + W^f G_f \qquad f = 0, 1, 2, \ldots, N - 1 \tag{5.89}$$

F_f and G_f are periodic functions in N such that $F_{f+N} = F_f$ and $G_{f+N} = G_f$. We observe that $W^{f+N} = \exp(-i\pi) W^f = -W^f$, so

$$H_{f+N} = F_f - W^f G_f \qquad f = 0, 1, 2, \ldots, N - 1 \tag{5.90}$$

Thus the entire DFT of $2N$ points can be constructed from the two subsets of N points each using Eqs. 5.89 and 5.90. This has reduced the number of multiplications from $(2N)^2$ to $2(N)^2$, that is, by a factor of 2.

The process above can be repeated. That is, f_t and g_t can each be divided in two, and so forth, any number of times until each array finally contains just a single point. According to Eq. 5.87, the DFT of a single point is just the point itself. From these point transforms the entire frequency spectrum can be constructed by repeated application of equations of the form 5.89 and 5.90. It is found that the number of multiplications required to do this is $2N \log_2 2N$. For example, the reduction in multiplications for a DFT composed of 8192 data points is from $(8192)^2 = 67{,}108{,}864$ to $(8192) \log_2(8192) = 8192 \times 13 = 106{,}496$, which represents a factor of about 630. The actual time saved is not quite the same as this number since the shuffling and recombination of the data points require some time, but the savings is nevertheless quite substantial, and most Fourier transformations are performed using the FFT algorithm, many of which would not be feasible otherwise.

5.8 CHEMICAL GRAPH THEORY

The branch of mathematics that deals with the way in which objects are connected is called *graph theory*. Applications to chemistry follow quite naturally, since the structural formula that we assign to a molecule corresponds closely to the definition of a graph. Many of the characteristic properties of molecules, both physical and chemical, are intimately related to the bonding arrangement or connectivity, which is easily represented. Thus chemical graph theory allows us to make many useful correlations and predictions using only very simple mathematical computations. Our discussion will be confined to *structural graphs,* but it might be noted that various species and the reactions that connect them can be represented by *reaction graphs* and treated in much the same way. The term *graph* as used in this context is quite distinct and should not be confused with common usage, which connotes the plotting of equations or data.

A *graph* consists of a collection of *nodes,* various pairs of which are connected by *edges*. *Adjacent nodes* are connected by a single edge, and a *path* represents a distinct set of edges and nodes that are traversed in moving from any one node to another. The *length* of the path is the number of edges traversed, and the *distance* between two nodes is defined as the shortest path connecting them. If every pair of nodes is joined by one or more paths it is a *connected* (single-component) *graph,* and it may be *cyclic* or *acyclic*. A connected, acyclic graph is sometimes called a *tree*. The *degree* of a given node is the number of edges joining it to other adjacent nodes. If the degree of each node of a component graph is equal to 2, we have a *closed cycle*. Some examples of simple connected graphs follow:

Closed cyclic graph with six nodes.

Acyclic graph (tree) with five nodes, one of which is of degree 4.

We can identify such graphs with molecular structures. If we wish to consider complications such as heteroatoms and multiple bonding, a more elaborate system may need to be devised, but the definitions above are adequate to illustrate some simple principles of chemical graph theory. For example, if we consider all trees consisting of seven nodes and specify that no node can be of higher degree than 4, the possible structures are shown in Figure 5.16. We can consider these graphs to be representations of all the isomeric heptanes of formula C_7H_{16}, where the nodes represent carbon atoms and the hydrogen atoms are suppressed.

Each of the graphs of Figure 5.16 can be represented by an *adjacency matrix* **A**, which has $A_{ij} = 1$ if there is an edge between adjacent nodes i and j, and $A_{ij} = 0$

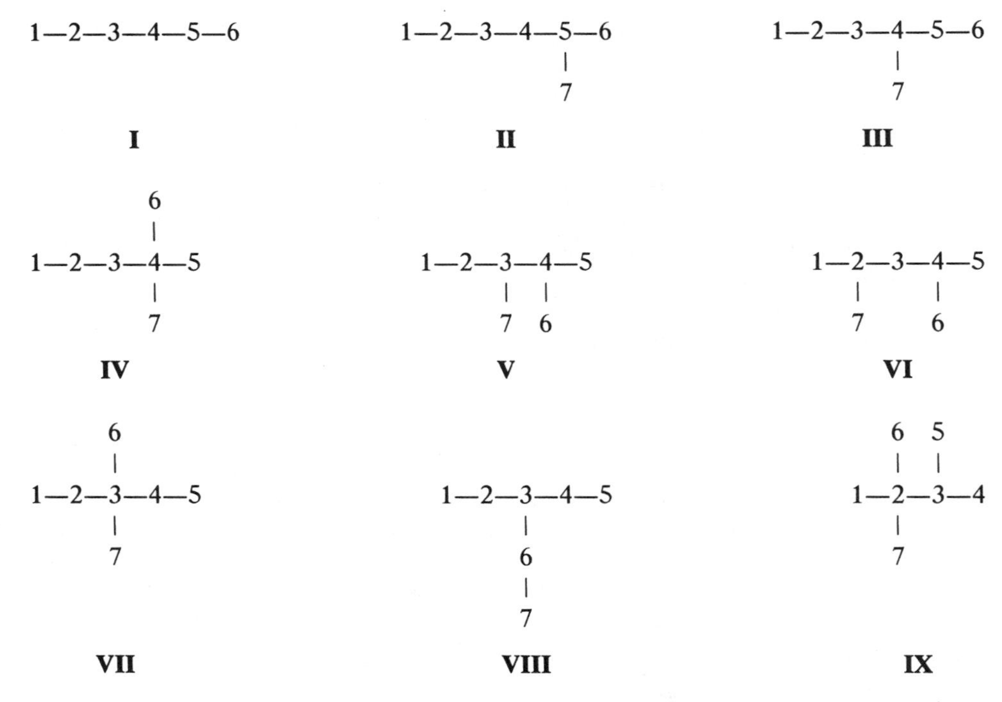

Figure 5.16 The nine acyclic graphs (trees) consisting of seven nodes with maximum degree 4.

otherwise. For example, the adjacency matrix of graph VII is

$$\begin{pmatrix} 0 & 1 & 0 & 0 & 0 & 0 & 0 \\ 1 & 0 & 1 & 0 & 0 & 0 & 0 \\ 0 & 1 & 0 & 1 & 0 & 1 & 1 \\ 0 & 0 & 1 & 0 & 1 & 0 & 0 \\ 0 & 0 & 0 & 1 & 0 & 0 & 0 \\ 0 & 0 & 1 & 0 & 0 & 0 & 0 \\ 0 & 0 & 1 & 0 & 0 & 0 & 0 \end{pmatrix}$$

A particularly interesting and useful result of graph theory is the following: If we calculate $\mathbf{A}^n$, the ij element of the product matrix represents the number of paths of length n between nodes i and j. In our particular case of acyclic graphs, there can be but one path of length n linking any two nodes, and they may be easily located by searching for 1's among the off-diagonal elements of $\mathbf{A}^n$. In the present context, we refer to the number of paths of length n as the *molecular path code*. For the example above, there are $12/2 = 6$ molecular paths of unit length, the division by 2 being necessary to prevent counting a given path twice. We find, either by inspection of structure VII itself or by matrix multiplication using the adjacency matrix given above, that the molecular path code for 3,3-dimethylpentane is

n	1	2	3	4	5	6
M_n	6	8	6	1	0	0

One of the useful facts we learn as we study organic chemistry is that the degree of branching in a molecule has an effect on both the physical and chemical properties. For example, it is known that the more highly branched molecules within a set of isomers tend to be more spherical with correspondingly reduced van der Waals attractions, so that the boiling points are generally lower. Combustion properties are also effected; more highly branched molecules have higher octane numbers.

The molecular path code can be reduced to a single number called the *average molecular path code* ($\bar{A}$),

$$\bar{A} = \frac{\sum_n n \cdot M_n}{\sum_n M_n} \tag{5.91}$$

which is related to the degree of branching. This is but one of many different *topological indices* that have been proposed in attempts to find correlations with physical and chemical properties. The more highly branched the molecule, the smaller its value of $\bar{A}$. The BASIC computer program of Figure 5.17 is designed to calculate the molecular path codes and average molecular path codes for a given set of structures. When

```
10 REM--PROGRAM MPC--CALCULATES MOLECULAR PATH CODES
20 REM--AND DISSIMILARITIES FOR A SET OF STRUCTURES
30 DIM A$(8),A(7,8,8),MPC(18,7),D(18,18)
40 INPUT"INPUT NUMBER OF ATOMS & NUMBERS OF ISOMERS";N,M
50 FOR I=1 TO M : PRINT"ADJACENCY MATRIX NUMBER";I
60 FOR J=1 TO N : PRINT"INPUT ROW NUMBER";J
70 INPUT A$(J) : NEXT J
80 FOR J=1 TO N : A(1,J,1)=VAL(LEFT$(A$(J),1)) 'CONVERT TO NUMERICAL VALUES
90 FOR K=2 TO N : A(1,J,K)=VAL(MID$(A$(J),K,1))
100 NEXT K : NEXT J
110 FOR J=2 TO N-1 : JJ=J-1         'COMPUTE MATRICES A^M WHERE M=2,3,...,N-1
120 FOR K=1 TO N : FOR L=1 TO N
130 A(J,K,L)=0 : FOR KK=1 TO N
140 A(J,K,L)=A(J,K,L)+A(1,K,KK)*A(JJ,KK,L)
150 NEXT KK : NEXT L : NEXT K : NEXT J
160 NMPC=0 : AMPC=0
170 FOR J=1 TO N-1 : MPC(I,J)=0               'LOOP TO COUNT OFF-DIAGONAL 1'S
180 FOR K=1 TO N : FOR L=1 TO N
190 IF A(J,K,L)=1 AND K<>L THEN MPC(I,J)=MPC(I,J)+1
200 NEXT L : NEXT K
210 MPC(I,J)=MPC(I,J)/2 : NMPC=NMPC+MPC(I,J) : AMPC=AMPC+J*MPC(I,J)
220 NEXT J
230 AMPC=AMPC/NMPC : PRINT : PRINT
240 PRINT"FOR ADJACENCY MATRIX" : PRINT
250 FOR J=1 TO N : PRINT A$(J) : NEXT J
260 PRINT : PRINT"THE MOLECULAR PATH CODE IS";
270 FOR J=1 TO N-1 : PRINT MPC(I,J); : NEXT J
280 PRINT : PRINT"THE AVERAGE MOLECULAR PATH CODE IS";AMPC
290 NEXT I
300 FOR I=1 TO M : FOR J=1 TO M : D(I,J)=0 'LOOP TO COMPUTE DISSIMILARITIES
310 FOR K=1 TO N-1
320 D(I,J)=D(I,J)+(MPC(I,K)-MPC(J,K))^2
330 NEXT K
340 D(I,J)=SQR(D(I,J))
350 NEXT J : NEXT I
360 PRINT
370 PRINT"THE DISSIMILARITY MATRIX FOR ABOVE STRUCTURES IS"
380 PRINT : PRINT
390 FOR I=1 TO M : FOR J=1 TO M
400 PRINT USING"###.##";D(I,J);
410 NEXT J : PRINT : NEXT I : PRINT
```

Figure 5.17 BASIC computer program that can be used to calculate the molecular path codes, the average molecular path codes, and the dissimilarity matrix for a set of structures specified by their adjacency matrices.

this program is executed for the set shown in Figure 5.16, the average molecular path codes range from 2.667 for structure I, to 2.000 for structure IX. That of structure VII is 2.095.

Figure 5.18 shows a plot of the boiling points (b.p.) of the isomeric heptanes versus their average molecular path codes. The line shown is a linear least squares fit to the data,

$$\text{b.p.} = 23.892\bar{A} + 33.208$$

and the correlation coefficient of the fit is $r = 0.7298$. Also shown are the research octane numbers (RON), for which

$$\text{RON} = -155.97\bar{A} + 418.23$$

with $r = -0.9897$. Such equations can be useful for qualitative predictions in those cases, like the latter, where experimental data may be lacking.

More sophisticated schemes are used in treating molecules with more complex bonding arrangements, but the basic ideas are similar. Some other applications of chemical graph theory include isomer enumeration, registration of compounds, and substructure searching. The specification of how much similarity there may be between a pair of molecules, based on their molecular path codes and other structural features, is useful when considering problems like drug design, or the prediction of

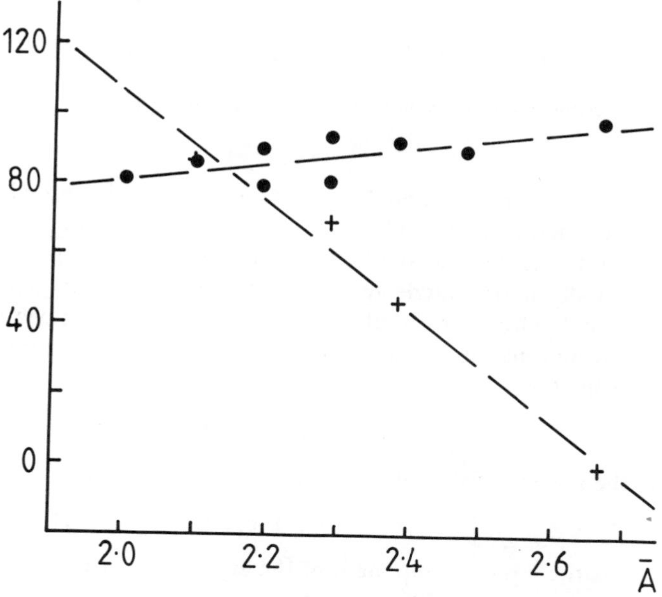

Figure 5.18 Plot of boiling points (circles) and research octane numbers (crosses) for the isomeric heptanes against their average molecular path codes ($\bar{A}$). The lines were determined by the linear least squares method; the equations of them are given in the text.

carcinogenicity. A convenient ranking of molecules can be given in terms of their *dissimilarity,* which is defined for a pair of structures i and j by

$$D_{ij} = \sqrt{\sum_n (M_{in} - M_{jn})^2} \tag{5.92}$$

where the sum is over the corresponding molecular path codes of length n of the two molecules. The program of Figure 5.17 computes a dissimilarity matrix for the set of input structures. As might be anticipated, the greatest dissimilarity occurs in our example set between structures I and IX, for which the value is $D_{\text{I IX}} = 5.83$.

Graph theory can be used to simplify molecular orbital calculations of the type introduced in Section 3.6. As indicated there, one method of solution is by expanding the secular determinant and finding the roots of the resulting characteristic equation of degree n. This is a very tedious procedure for large values of n, but graph theory can provide some shortcuts. For solution of the characteristic equation it is useful to write the secular equations in a somewhat different form. The secular equations for HMO calculations can be simplified by dividing through by $-\beta$ and then by setting $x = -(\alpha - \lambda)/\beta$. In the case of butadiene, which was considered in Example 3.14,

$$\begin{pmatrix} \alpha - \lambda & \beta & 0 & 0 \\ \beta & \alpha - \lambda & \beta & 0 \\ 0 & \beta & \alpha - \lambda & \beta \\ 0 & 0 & \beta & \alpha - \lambda \end{pmatrix} \begin{pmatrix} c_1 \\ c_2 \\ c_3 \\ c_4 \end{pmatrix} = \begin{pmatrix} x & -1 & 0 & 0 \\ -1 & x & -1 & 0 \\ 0 & -1 & x & -1 \\ 0 & 0 & -1 & x \end{pmatrix} \begin{pmatrix} c_1 \\ c_2 \\ c_3 \\ c_4 \end{pmatrix} = 0$$

In a more compact matrix notation this may be written

$$[x\mathbf{I} - \mathbf{A}]\mathbf{c} = 0 \tag{5.93}$$

where $\mathbf{I}$ is the unit matrix and $\mathbf{A}$ is the adjacency matrix. An edge in the current context is associated with π bonding.

We note that for a straight-chain polyene (i.e., for a linear molecule written classically with alternating single and double bonds) the degree of a given atom can be no greater than 2. This means that there can be no more than two off-diagonal elements in the adjacency matrix and so for a given eigenvalue, according to Eq. 5.93, a general term is of the form

$$xc_i = c_j + c_k$$

Taking the absolute value of both sides of this equation gives

$$|xc_i| = |c_j + c_k| \leq |c_j| + |c_k|$$

Now $|c_i|$ is the largest component of the eigenvector, so

$$|x| \leq \left|\frac{c_j}{c_i}\right| + \left|\frac{c_k}{c_i}\right| \leq 2$$

We see that all the roots of the characteristic equation lie in the range $-2 \leq x \leq +2$.

The characteristic equation may be written very simply in terms of molecular graphs using the *Harary–Sachs theorem,* according to which the coefficients of the equation

$$x^n + a_1x^{n-1} + a_2x^{n-2} + \cdots + a_jx^{n-j} + \cdots + a_n = 0$$

are obtained by breaking up the structural graph of the molecule into component graphs S_n of n vertices, where the component graphs contain only nodes of degree 1 or 2 (i.e., a simple linking of nodes by either one edge, or by a cycle). If we let c equal the number of component graphs and r equal the number of closed cycles in S_n, then

$$a_j = \sum_{S_n} (-1)^c(2)^r \tag{5.94}$$

As an example, we will write the characteristic equation of methylene cyclopropene, whose structural graph may be written

$$\begin{matrix} & 1 & \\ & | & \\ & 2 & \\ / & & \backslash \\ 3 & - & 4 \end{matrix}$$

We have

$$S_1 = \{0\} \qquad a_1 = 0$$

$$S_2 = \left\{ \begin{matrix} 1 \\ | \\ 2, \end{matrix} \quad \begin{matrix} & 2 \\ / & \\ 3 & , \end{matrix} \quad \begin{matrix} 2 & \\ & \backslash \\ & 4, \end{matrix} \quad 3 - 4 \right\} \qquad a_2 = (-1)^1 2^0 + (-1)^1 2^0 + (-1)^1 2^0 + (-1)^1 2^0 = -4$$

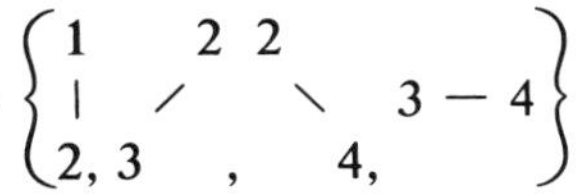

$$S_3 = \left\{ \begin{matrix} & 2 & \\ / & & \backslash \\ 3 & - & 4 \end{matrix} \right\} \qquad a_3 = (-1)^1 2^1 = -2$$

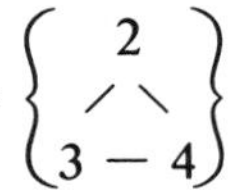

$$S_4 = \left\{ \begin{matrix} 1 \\ | \\ 2 \\ \\ 3 - 4 \end{matrix} \right\} \qquad a_4 = (-1)^2 2^0 = 1$$

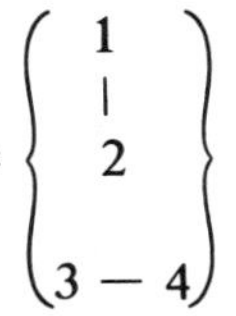

Therefore, the characteristic equation is

$$x^4 - 4x^2 - 2x + 1 = 0$$

Recursion formulas can be developed for the writing of characteristic equations for molecules of various types. In the case of straight-chain polyenes, for example, the secular determinant is of the form

$$\begin{vmatrix} x & -1 & 0 & 0 & \cdots \\ -1 & x & -1 & 0 & \cdots \\ 0 & -1 & x & -1 & \cdots \\ \vdots & & & & \end{vmatrix} = 0$$

It is apparent that if we write $g_n(x)$ for the characteristic polynomial represented by this determinant, the first step in the expansion by minors can be written

$$g_n(x) = xg_{n-1}(x) - g_{n-2}(x)$$

Since

$$g_1(x) = |x| = x \qquad g_2(x) = \begin{vmatrix} x & -1 \\ -1 & x \end{vmatrix} = x^2 - 1$$

we have

$$g_3(x) = x(x^2 - 1) - x = x^3 - 2x$$

$$g_4(x) = x(x^3 - 2x) - (x^2 - 1) = x^4 - 3x^2 + 1$$

etc.

For very large molecules the numerical solution of the characteristic equation can be difficult, but graph theory can be used to help factor the equation and determine the eigenvectors (for further details, see A. C. Tang, Y. S. Kiang, G. S. Yan, and S. S. Tai, *Graph Theoretical Molecular Orbitals,* Science Press, Beijing, China, 1986).

5.9 PATTERN RECOGNITION

The quantities with which the chemist is concerned can often be classified into two or more distinct categories. For example, we may have several paint chips submitted to a forensics laboratory and we wish to categorize them as having originated from plant A or plant B. Other examples include ore samples, which may be distinguished as having come from different mines, or environmental samples, which may need to be classified by their point of origin. We refer to such a category designation for a given member of a data set as its *class*. In general, each member of the data set is characterized by a certain set of variables called *descriptors*. In some cases we may be fortunate enough to find a single descriptor which, when determined, serves to distinguish between the various possible classes. This often proves impossible, however, and then we need to see if there is some combination of the descriptors taken together that will allow us to assign the members of the data set to their respective classes with a certain degree of confidence. The methods by which this may be done are called *pattern recognition,* which is one branch of artificial intelligence that we will describe in a more general way in the following section. There are many techniques used in pattern recognition, which are generally grouped as preprocessing, mapping, classification by discriminants, clustering, and modeling methods. In our brief introduction to the subject, we will limit our discussion to three particularly useful techniques, one for each of the first three categories listed.

We will assume that each member of our data set $\mathbf{D}_j$ may be assigned to a particular class, and that each member is characterized by a set of descriptors d_i:

$$\mathbf{D}_j = (d_1, d_2, \ldots, d_n)$$

For purposes of illustration we will use the data set given in Figure 5.19, for which $n = 3$ and the number of classes is 2. The techniques we will describe can be generalized to include an arbitrary number of descriptors and classes as required.

Pattern recognition methods are of two types. The first employs a data set with known classifications that may be used to develop a discriminant function. We call this a *supervised method,* as opposed to an *unsupervised method*, which makes use of the data itself to develop a classification scheme. The unsupervised methods are generally based on various techniques to detect a clustering of the data points after they have been subjected to certain transformations. Our example data set provides known class designations and we will only be concerned with supervised methods, although they are quite general and can be used for unsupervised analysis as well.

A casual inspection of the three descriptors given in Figure 5.19 reveals no systematic trends. There is very little correlation between them ($r_{12} = 0.1410$, $r_{13} = 0.1431$, $r_{23} = -0.3231$), and no one of them can be easily expressed in terms of another. Furthermore, no one descriptor can be used to classify the data set members very reliably. For descriptor 1 the best classification that can be made assigns data of value greater than 5.7 to class 1 and less than 5.7 to class 2. When this is done only 32 of the 40 points, or 80%, are classified correctly. Similarly, a cutoff for class 1 above 11.5 for descriptor 2 gives 33 correct classifications or 82.5% success, and a value less than 1.92 for class 1 according to descriptor 3 gives 31 correct designations, or 77.5%. Clearly, classification using a single descriptor is of limited utility.

A preprocessing technique in common use is called *autoscaling*. Further analysis of the raw data is often impeded by the fact that the magnitudes and ranges of the various descriptors can be quite different, and this renders pattern recognition difficult. They can be transformed such that the values for each descriptor are centered about zero with unit standard deviation by replacing each descriptor d_{ij} by

$$d'_{ij} = \frac{d_{ij} - \bar{d}_i}{s_i} \tag{5.95}$$

where $\bar{d}_i$ is the mean value for the descriptor i and s_i is its standard deviation. For our example data set of Figure 5.19, we find $\bar{d}_1 = 5.4263$, $s_1 = 1.4844$, $\bar{d}_2 = 10.2426$, $s_2 = 3.0240$, $\bar{d}_3 = 1.9605$, $s_3 = 0.4620$, and the autoscaled descriptors using the transformation indicated by Eq. 5.95 are also shown in Figure 5.19.

	Raw Data			Autoscaled Data		
Class	d_1	d_2	d_3	d_1'	d_2'	d_3'
2	5.1863	5.9455	1.6211	−0.1616	−1.4210	−0.7346
1	6.0223	15.4162	1.0851	0.4015	1.7109	−1.8947
2	3.2775	8.7467	1.9354	−1.4476	−0.4947	−0.0543
1	6.5643	13.6908	1.4739	0.7667	1.1403	−1.0532
1	6.0865	12.2698	1.6026	0.4448	0.6704	−0.7746
1	5.2109	14.7695	1.1944	−0.1451	1.4970	−1.6581
1	6.4188	15.7017	1.6972	0.6687	1.8053	−0.5699
2	3.2637	9.0109	2.0786	−1.4569	−0.4073	0.2556
1	7.7001	6.9308	1.5546	1.5318	−1.0952	−0.8785
2	3.8153	8.4900	2.3042	−1.0853	−0.5796	0.7439
1	8.2741	8.3262	2.6080	1.9185	−0.6337	1.4015
2	3.8590	9.7133	1.9609	−1.0558	−0.1750	0.0009
2	5.6916	4.7188	2.0412	0.1788	−1.8267	0.1747
1	7.0532	16.9060	1.9115	1.0960	2.2035	−0.1060
2	5.4096	11.9686	2.7241	−0.0112	0.5708	1.6527
1	4.8013	12.0930	1.5021	−0.4210	0.6119	−0.9921
1	3.4803	12.2261	1.4722	−1.3109	0.6559	−1.0568
1	6.3709	10.7409	1.8370	0.6364	0.1648	−0.2673
2	6.5997	4.2494	2.4935	0.7905	−1.9819	1.1536
1	5.5432	11.7680	2.3669	0.0788	0.5044	0.8796
1	5.0027	11.1205	1.8663	−0.2853	0.2903	−0.2039
2	3.2032	11.3731	2.1214	−1.4976	0.3738	0.3483
1	7.8629	13.9696	2.7137	1.6415	1.2325	1.6302
2	5.6137	9.8984	2.6740	0.1263	−0.1138	1.5443
2	5.1946	8.5531	1.8148	−0.1561	−0.5587	−0.3153
2	6.8437	6.9790	2.6860	0.9549	−1.0792	1.5703
2	4.0946	11.4486	2.5330	−0.8971	0.3988	1.2391
2	5.2066	8.2822	2.7042	−0.1480	−0.6483	1.6097
1	6.1381	12.0986	1.9133	0.4796	0.6138	−0.1021
1	5.7831	11.7080	1.8742	0.2404	0.4846	−0.1868
2	3.2788	10.1574	1.3094	−1.4467	−0.0282	−1.4092
2	3.8487	6.1398	1.9571	−1.0628	−1.3568	−0.0073
2	3.9281	8.0246	1.5583	−1.0093	−0.7335	−0.8705
1	4.3517	12.1103	1.3969	−0.7239	0.6176	−1.2198
2	5.3709	6.7547	1.7872	−0.0373	−1.1534	−0.3751
1	7.6108	13.1197	1.4102	1.4717	0.9514	−1.1910
1	8.2725	9.6450	2.2652	1.9175	−0.1976	0.6595
1	6.8358	7.7542	1.8347	0.9496	−0.8229	−0.2723
2	4.6448	7.9751	2.5105	−0.5264	−0.7498	1.1904
2	3.3362	8.9097	2.0248	−1.4080	−0.4408	0.1392
$\bar{d}$	5.4263	10.2426	1.9605			
s	1.4844	3.0240	0.4620			

Figure 5.19 Descriptors and class designations for the data set used to illustrate some of the techniques of pattern recognition. The autoscaled descriptors are also given.

Most humans are very adept at spotting patterns and trends in data if they can be reduced to two- or three-dimensional representations, but we have great difficulty with the multidimensional data of the kind under consideration here. A mapping technique that reduces the dimensionality of the representation, while preserving its principal distinguishing characteristics, is the *Karhunen–Loéve transformation,* also known as *principal component analysis*. The first step is the multiplication of the data matrix by its transpose. This can be done with the original data to give the covariance matrix (see Section 5.6.2), but it is now more convenient to use the autoscaled data in which case we obtain the *correlation matrix*. The diagonal elements are all unity and the off-diagonal elements are the correlation coefficients of the data as defined by the various descriptors. For the data of Figure 5.19 we have

$$\begin{pmatrix} 1.0000 & 0.1410 & 0.1431 \\ 0.1410 & 1.0000 & -0.3231 \\ 0.1431 & -0.3231 & 1.0000 \end{pmatrix}$$

We solve for the eigenvalues and their corresponding eigenvectors according to the methods described in Section 3.6 to obtain

$$\lambda_1 = 1.3232 \qquad \lambda_2 = 1.0962 \qquad \lambda_3 = 0.5806$$

$$\mathbf{c}_1 = \begin{pmatrix} 0.0059 \\ -0.7058 \\ 0.7084 \end{pmatrix} \qquad \mathbf{c}_2 = \begin{pmatrix} 0.9018 \\ 0.3098 \\ 0.3012 \end{pmatrix} \qquad \mathbf{c}_3 = \begin{pmatrix} 0.4320 \\ -0.6371 \\ -0.6383 \end{pmatrix}$$

The total variance represented by the autoscaled data set is just the sum of the diagonal elements, or 3.0000. According to the above results, then, the eigenvalue λ_1 represents 44.1% of the total, eigenvalue λ_2 represents 36.5%, and eigenvalue λ_3 represents 19.4%. Thus if we transform the data according to eigenvectors $\mathbf{c}_1$ and $\mathbf{c}_2$, we obtain a set of two-dimensional descriptors that retains 80.6% of the total variance represented by the original data. For example, transformation of autoscaled descriptor 1 by the eigenvector $\mathbf{c}_1$ gives $(0.0059)(-0.1616) + (-0.7058)(-1.4210) + (0.7084)(-0.7346) = 0.4816$, while transformation by the eigenvector $\mathbf{c}_2$ gives $(0.9018)(-0.1616) + (0.3098)(-1.4210) + (0.3012)(-0.7346) = -0.8072$. The point $(0.4816, -0.8072)$ may be mapped in a two-dimensional Cartesian representation, this particular member of the data set belonging to class 2. Continuing in this way we may map all 40 members of the autoscaled data set, and the result is shown in Figure 5.20. This technique has allowed us to reduce the dimensionality of the representation, while retaining most of the variance, such that a map can be made providing an obvious separation of the data into their two distinct classes.

We will now consider a third technique of pattern recognition that uses numerical methods to develop a *discriminant function,* according to which the data may be classified. This process removes the subjectivity that might be required if we were to

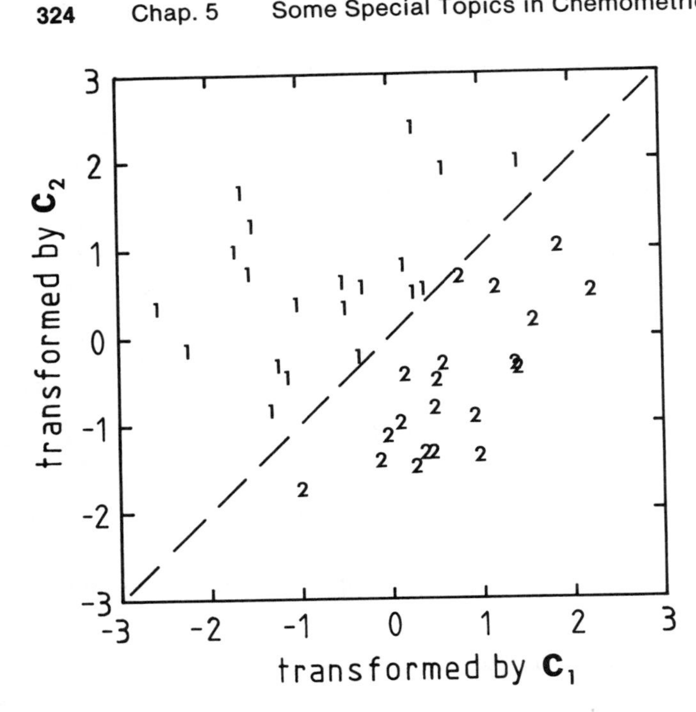

Figure 5.20 Autoscaled descriptors of Figure 5.19 transformed by $\mathbf{c}_1$ and $\mathbf{c}_2$, the eigenvectors corresponding to the two largest eigenvalues of the correlation matrix.

rely completely on our analysis of maps or graphical representations. In certain cases constraints may be imposed and we develop a *parametric* discriminant function. Otherwise, we rely on a *nonparametric* method, such as that described below.

The data may be classified in the n-dimensional vector space it spans by construction of a so-called *decision surface,* such that the points on one side of the surface belong to one class, while those on the other side belong to the other class. The simplest decision surface is a plane in the n-dimensional space, called a *hyperplane*. We define a hyperplanar decision surface by means of a vector that is perpendicular to it. As a matter of convenience we construct the surface such that it passes through the origin. This can be done by addition of an $(n + 1)$st component to the n-dimensional descriptor space. The extra component is usually given the value unity, although it can in fact be assigned any arbitrary value. With this representation, a vector from the origin to any data point defines the perpendicular distance of that point from a surface. Our object is to find a *weight vector* **W** which defines a decision surface such that the dot product with the descriptors of one class are all on one side of the decision surface, while those of the other class are all on the other side.

A popular method of nonparametric discriminant development makes use of the *linear learning machine* to define the decision surface. In this case a certain number of members of the data set may be used as a *training set* to adjust the decision surface so that it properly classifies the members of the set, and the remaining members are then used as a *prediction set* to test the reliability of the decision surface that has been found. As the training set is processed, each member is classified and if it is found to be incorrect, an adjustment of the decision surface is made to correct the classification. This is carried out repetitively in a "learning" process until all of the training set members are classified correctly before it is tested on the prediction set.

The dot product of the weight vector $\mathbf{W}$ with a given descriptor $\mathbf{D}_j$ will be represented by the symbol S:

$$S = \mathbf{W} \cdot \mathbf{D}_j$$

We will arbitrarily assign members of class 1 a negative value of S, and those of class 2 a positive value of S. We proceed through the training process by calculating S for each member of the training set as above. If the sign of S is correct, no adjustment is made, but if the sign of S is wrong, we must modify the weight vector such that the correct sign is obtained. For the modified weight vector $\mathbf{W}'$ we write

$$S' = \mathbf{W}' \cdot \mathbf{D}_j$$

The new weight vector is derived from the original by addition of a certain multiple of the descriptors for that particular member of the training set:

$$\mathbf{W}' = \mathbf{W} + C\mathbf{D}_j$$

Thus

$$S' = (\mathbf{W} + C\mathbf{D}_j) \cdot \mathbf{D}_j = S + C\mathbf{D}_j \cdot \mathbf{D}_j$$

If we use X to represent the square of the magnitude of $\mathbf{D}_j$ (see Section 1.4), we have

$$C = \frac{S' - S}{X}$$

We require a value of C such that S' is opposite in sign to S. It is convenient to set $S' = -S$, so that

$$C = -\frac{2S}{X}$$

For a data point located at a certain distance $|S|$ on the wrong side of the decision surface, this adjustment is tantamount to a reflection that places the point at the corresponding distance $|S|$ on the right side of the surface. By repetitive cycling through the training set, we ultimately define a weight vector such that all members are classified correctly. We then use this weight vector with the prediction set to assign them to their respective classes. The BASIC program LLM of Figure 5.21 is a linear learning machine that is designed to carry out such a pattern recognition calculation on a data set of the format given in Figure 5.19 [the DATA statements containing the classes (CL) and descriptors (D) for the 40 data set members must be included but are not shown]. NTS is the number of members assigned to the training set, and NPS is the number of members assigned to the prediction set. ND is the number of descriptors for each member of the data set (three in the present case). The weight vector W can be given any arbitrary initial value; here it is set equal to zero for all but the $(n + 1)$st component, which is given the value one. TS is an array containing the index for train-

```
10 REM--PROGRAM LLM--LINEAR LEARNING MACHINE
20 DIM CL(40),D(3,40),TS(40)
30 READ NTS,NPS,ND : ND1=ND+1
40 FOR I=1 TO ND1 : READ W(I) : NEXT I                      'READ WEIGHT VECTOR W
50 FOR I=1 TO NTS : READ CL(I)     'READ CLASS & DESCRIPTORS FOR TRAINING SET
60 FOR J=1 TO ND : READ D(J,I) : NEXT J : NEXT I
70 LPRINT "NUMBER OF TRAINING SET MEMBERS CLASSIFIED INCORRECTLY:"
80 LIM=1000 : NT=0 : NM=0                          'INITIALIZE FOR TRAINING LOOP
90 NR=0 : IF NM=0 THEN M=NTS ELSE M=NM
100 IF M<NTS THEN 120
110 FOR I=1 TO NTS : TS(I)=I : NEXT I
120 FOR N=1 TO M : K=TS(N) : S=W(ND1)
130 FOR I=1 TO ND : S=S+W(I)*D(I,K) : NEXT I                      'COMPUTE S=W*D
140 IF CL(K)=1 AND S<0 OR CL(K)=2 AND S>0 THEN 180
150 X=1 : FOR I=1 TO ND : X=X+D(I,K)*D(I,K) : NEXT I 'ADJUST DECISION SURFACE
160 C=-2*S/X : FOR I=1 TO ND : W(I)=W(I)+C*D(I,K) : NEXT I
170 W(ND1)=W(ND1)+C : NR=NR+1 : TS(NR)=K : NT=NT+1
180 NEXT N
190 NM=NR : LPRINT NM;
200 IF NT>LIM OR M=NTS AND NM=0 THEN 220 ELSE 210        'TEST FOR TERMINATION
210 T$=INKEY$ : IF T$="" THEN 90 ELSE LPRINT"ABNORMAL TERMINATION"
220 LPRINT : LPRINT NT;" ADJUSTMENTS TO THE DECISION SURFACE WERE USED"
230 LPRINT"TO CLASSIFY THE";NTS;" MEMBERS OF THE TRAINING SET"
240 LPRINT : LPRINT"THE WEIGHT VECTOR IS:"
250 FOR I=1 TO ND1 : LPRINT W(I), : NEXT I : LPRINT : LPRINT
260 FOR I=1 TO NPS : READ CL(I)  'READ CLASS & DESCRIPTORS FOR PREDICTION SET
270 FOR J=1 TO ND : READ D(J,I) : NEXT J : NEXT I
280 NI1=0 : NI2=0 : NP1=0 : NP2=0                'INITIALIZE FOR PREDICTION LOOP
290 FOR I=1 TO NPS : S=W(ND1)
300 FOR J=1 TO ND : S=S+W(J)*D(J,I) : NEXT J
310 ON CL(I) GOTO 320,330
320 NP1=NP1+1 : IF S<0 THEN 340 ELSE NI1=NI1+1 : GOTO 340
330 NP2=NP2+1 : IF S>0 THEN 340 ELSE NI2=NI2+1
340 NEXT I
350 LPRINT"FOR THE PREDICTION SET OF";NPS;"MEMBERS,";NPS-NI1-NI2;"WERE CORRECT"
360 LPRINT "CLASS 1","CORRECT","CLASS 2","CORRECT"
370 LPRINT NP1,NP1-NI1,NP2,NP2-NI2
380 DATA 25,15,3,0,0,0,1
```

```
NUMBER OF TRAINING SET MEMBERS CLASSIFIED INCORRECTLY:
 8  0  1  0  0
 9  ADJUSTMENTS TO THE DECISION SURFACE WERE USED
TO CLASSIFY THE 25  MEMBERS OF THE TRAINING SET

THE WEIGHT VECTOR IS:
-.59702        -.7187338       .3344278      -.1230725

FOR THE PREDICTION SET OF 15 MEMBERS, 15 WERE CORRECT
CLASS 1          CORRECT          CLASS 2          CORRECT
 6                6                9                9
```

Figure 5.21 BASIC program for development of a discriminant function using the linear learning machine, together with output from a run using the autoscaled data of Figure 5.19.

ing set members which are classified incorrectly in a given cycle. It is initially defined to include all members of the training set, and as the cycle is repeated, only those members classified incorrectly in the previous cycle are tested until all have been classified correctly (NM = 0). Then the entire process is repeated until every member of the training set can be correctly classified with no further adjustment of the decision surface. This condition is signaled by a pair of zeros in the output stream from the training loop (this part of the program logic is patterned after a corresponding FORTRAN program given by P. C. Jurs, *Computer Software Applications in Chemistry*, John Wiley & Sons, Inc., New York, 1986). When adjustment of the decision surface for the training set is complete the weight vector is printed, and then it is used to classify the members of the prediction set. The program was executed with the autoscaled data of Figure 5.19 arbitrarily divided such that the first 25 members were in the training set and the final 15 members were in the prediction set. Relatively few adjustments of the decision surface were needed, and the members of the prediction set were all classified correctly as indicated in the output shown. The program can also be executed using the raw data, in which case many more adjustments of the decision surface are required.

Pattern recognition is finding ever-increasing applications in chemistry, not only in terms of classification of samples, but as an aid to structural analysis through identification of functional group characteristics in NMR, IR, and Raman spectra. Studies with mass spectrometry, Auger spectra, γ-ray spectra, and electrochemical data have also been reported, and we will undoubtedly see many fascinating new uses in the future.

5.10 ARTIFICIAL INTELLIGENCE

A large number of the problems we must solve involve reasoning and subjective judgment rather than simple computation and data processing. The science that deals with these matters is called *artificial intelligence* (AI). The field had its origins in mathematical techniques developed about 50 years ago, but it is only with the advent of modern computers that computational capability became powerful enough for significant advancement. AI research was spurred by the development of new programming languages in the early 1970s, such as LISP, which simplifies the process of representation and manipulation of symbolic knowledge. Further improvements in the interactive environment since that time have led to enhanced capabilities, and today AI is regarded as a rapidly maturing discipline.

An interesting paradox with which AI must deal is the simple fact that computers have no inherent reasoning ability. They are very inflexible and can only do what they are told to do through a set of instructions. Computers can be programmed to carry out some highly sophisticated decision-making processes, but it should be recognized that the knowledge base from which they operate is usually rooted in human experience.

One of the problems that must be faced in the construction of an AI program is to define the scope of the process within reasonable limits where an acceptable solution may be expected. For example, it is estimated that the number of legal moves in a chess game produces approximately 10^{120} possible sequences. It is unrealistic even with modern computers to explore each of these and catalog the "perfect" moves. The scope of the problem, or area of *search,* must be limited in such cases.

The second problem of principal concern is the input, management, and efficient utilization of the *knowledge base*. This can consist of numbers or possible ranges of numbers, but for AI applications the data needed often involve attributes, descriptions of objects, rules, or laws. The format for acquisition and storage of such data is a major problem of AI research. Techniques are being continually sought that will automate the process, and expanding the knowledge base through learning experiences such as those alluded to in our brief discussion of pattern recognition is a topic of considerable interest.

Within the field of AI several subdisciplines may be identified, including robotics, machine vision, understanding of natural languages, and expert systems. Each of these has important applications but we will focus the remainder of our discussion on the latter, which has been shown to have some particularly interesting and valuable contributions to make in chemistry.

An *expert system* makes use of an expert or group of experts in a given area to construct a knowledge base that may be used as an aid in various decision-making processes. A carefully designed expert system may contribute to the solution of problems whose complexity is beyond the experience of a given human, not only because many factors can be dealt with simultaneously, but also due to the fact that the knowledge base may represent the combined total of an entire "community" of experts and this presumably exceeds that of any one individual. This allows for the solution of problems that may require the *fusion* of expertise from a variety of fields. Using the expert systems approach, the knowledge base can be captured, refined, and preserved to assure constancy in the future.

Not all problems are likely candidates for solution by expert systems. Some may be so simple that little expertise is required, and the time and money required to develop the software that mimics the human decision-making process may not be justified. On the other hand, a very complex problem whose methods are vague may be so poorly understood that it may not be possible to construct a useful knowledge base and an expert system will then be of limited value. The best candidates are those that lie between these two extremes, and the best philosophy may be that which is often referred to as an Edisonian approach: Try many things to see how they work. If they show promise, see how they may be improved; otherwise, discard them and find new approaches.

A flexible knowledge base is the key to a good expert system. It is not always possible to deal with concrete rules and laws; in many instances the knowledge base is composed of a set of *heuristics* or judgmental rules, which are usually obeyed. These are used to reason with the facts in much the same way that human experts would, based on their personal experiences.

A second important feature of the expert system deals with the actual processes of problem solving and inference. They may operate in a forward direction, using the data given to solve problems by testing the consequences of various possible conclusions. Alternatively, a backward direction might be operative such that a given goal is stated, and then the conditions that must be met to realize the stated goal may be examined. In either case, the essence of the operation is generally a set of

IF (condition) THEN (action)

rules of varying degrees of complexity.

Finally, the expert system must include an appropriate interface to the user, such that the results are clearly stated and, particularly, so that the sequence of decisions upon which the final conclusions are based is made available. It is only by examination of such results that a meaningful critique can be given, which is essential to the improvement of the system.

Expert systems have been developed for medical diagnosis and treatment, molecular biology and genetic engineering, instrumental configuration, diagnosis and control, mineral exploration, chemical synthesis and analysis, and so forth. For purposes of illustration we will briefly describe two highly successful systems that have been designed to aid the chemist in the solution of structural problems.

The program DENDRAL (for *dendritic algorithm*) is an outgrowth of pioneering work of Lederberg beginning in about 1964. Originally designed to enumerate the various structures that may be constructed from a given set of atoms, it was subsequently expanded to incorporate mass spectral data. Constraints can be placed on possible structures by specification of subgroups or structural units which are known to be present or absent, and then possible solutions are generated and tested using a portion of the program called CONGEN (for *constrained structure generation*). Similar techniques have been used with other spectroscopic data, such as ^{13}C NMR.

Another program making use of expert system principles for the solution of structural problems is called PAIRS (for *program for the analysis of infrared spectra*). It has been developed over a period of years by Woodruff and various coworkers, and different versions of it are available through the Quantum Chemistry Program Exchange. Much of the current analysis of IR spectra is based on comparison of an experimental spectrum against catalogs of known spectra. Such catalogs often contain spectra for tens of thousands of compounds, and they are difficult to maintain and update. The basic philosophy of PAIRS is to solve structural problems not by brute force comparison with an extensive catalog, but rather by using the same approach that an expert would. That is, an analysis is made of the "fingerprint" region of the spectrum for characteristic bands which are indicative of certain structural features. For example, aldehydes have a strong carbonyl peak in the range 1765 to 1660 cm^{-1}, and two peaks of moderate intensity at about 2820 and 2720 cm^{-1}, the latter generally being quite sharp. The experimental spectrum is input and examined for a strong peak in the region 1765 to 1660 cm^{-1}. If one is found, a certain probability that aldehyde is present is assigned and then the other spectral regions are searched for the additional characteristic features. If these also appear, the probability of an aldehyde functional group is correspondingly increased. The program proceeds through numerous "decision trees" of this type for the various possible functional groups before final probabilities are given which can aid the chemist in the identification of the structure.

The PAIRS knowledge base contains refinements based on the nature of the sample (e.g., whether the spectrum was run in the vapor phase, in mineral oil, etc.). It is easily updated by the user, and recent versions of the program also include the possibility of automatic addition to the knowledge base by allowing the program itself to develop a set of rules for analysis from representative spectra of known compounds. Techniques of automated knowledge base generation offer unlimited possibilities for future development of expert systems, and it is likely that we will see many new applications to chemical problems in the next few years.

SUGGESTED ADDITIONAL READING

1. A. T. Balaban (Ed.), *Chemical Applications of Graph Theory,* Academic Press, Inc. (London) Ltd., London, 1976.
2. Yonathan Bard, *Nonlinear Parameter Estimation,* Academic Press, Inc., New York, 1974.
3. R. J. Bell, *Introductory Fourier Transform Spectroscopy,* Academic Press, Inc., New York, 1982.
4. William E. Blass and George W. Halsey, *Deconvolution of Absorption Spectra,* Academic Press, Inc., New York, 1981.
5. Joseph J. Breen and Philip E. Robinson (Eds.), *Environmental Applications of Chemometrics,* ACS Symposium Series 292, American Chemical Society, Washington, D.C., 1985.
6. Peter R. Griffiths (Ed.), *Transform Techniques in Chemistry,* Plenum Press, New York, 1978.
7. Peter C. Jurs, *Computer Software Applications in Chemistry,* Wiley-Interscience, New York, 1986.
8. Peter C. Jurs and Thomas L. Isenhour, *Chemical Applications of Pattern Recognition,* Wiley-Interscience, New York, 1975.
9. Bruce R. Kowalski (Ed.), *Chemometrics: Theory and Applications,* ACS Symposium Series 52, American Chemical Society, Washington, D.C., 1977.
10. Edmund R. Malinowski and Darryl G. Howery, *Factor Analysis in Chemistry,* Wiley-Interscience, New York, 1980.
11. Thomas H. Pierce and Bruce A. Hohne (Eds.), *Artificial Intelligence Applications in Chemistry,* ACS Symposium Series 306, American Chemical Society, Washington, D.C., 1986.
12. Muhammad A. Sharaf, Deborah L. Illman, and Bruce R. Kowalski, *Chemometrics,* Wiley-Interscience, New York, 1986.
13. Nenad Trinajstic, *Chemical Graph Theory,* Volumes I and II, CRC Press, Inc., Boca Raton, Fla., 1983.
14. R. J. Wilson and L. W. Beineke (Eds.), *Applications of Graph Theory,* Academic Press, Inc. (London) Ltd., London, 1979.

EXERCISES

5.1. Verify Eq. 5.4.

5.2. Carry through the optimization of steepest descents begun in Section 5.1.2 for one more cycle, using a step size $s = \frac{1}{4}$. That is, find the minimum gradient at the point (1.0272,2.3286), and then calculate r for lengths 0.25 along the line s until the next minimum is located.

5.3. It is claimed that the *Redlich–Kwong equation* [see O. Redlich and J. N. S. Kwong, *Chem. Rev.*, **44**, 233 (1949)],

$$\left[p + \frac{a}{V_m(V_m + b)T^{1/2}}\right](V_m - b) = RT$$

is one of the most accurate of the two-parameter equations of state for gases. Modify the simplex program given in Figure 5.3 to determine optimum values of a and b by fitting to the pVT data for CO_2 at 330.90 K, and compare the result obtained with those using the van der Waals and Dieterici equations.

5.4. Select data for a given gas at various temperatures from the International Critical Tables. Modify the program of Figure 5.3 to fit an equation of state to the data for which the temperature is also a variable. Compare the results obtained using the van der Waals, Dieterici, and Redlich–Kwong equations. Which appears on the basis of your calculations to be the superior two-parameter equation of state for real gases?

5.5. Modify the simplex program of Figure 5.3 to find optimum values of the parameters A and B that fit the exponential equation $y = Ae^{-Bt}$ to the following data using $AI = 0.5$ and $BI = 0.1$ as the initial estimates:

t_i	y_i
1	0.4441
2	0.3876
4	0.3364
6	0.2454
8	0.1322
10	0.1200
15	0.1176
20	0.0550
30	0.0236
40	0.0109

5.6. A stock solution of NaOH is checked every two days by titration against a potassium hydrogen phthalate standard. In the order taken, the normality of the NaOH was reported as 0.1204, 0.1220, 0.1168, 0.1201, 0.1143, 0.1133, 0.1164, 0.1148. Fit a straight line to these data using the linear least squares method. Calculate the correlation coefficient and the 95% confidence interval of the slope. Should one conclude that there is a significant change in the concentration of the base with time?

5.7. The following data give the yield of a particular reaction as a function of the oxygen pressure under which the reaction was carried out:

P_{O_2} (torr)	Yield (%)
80	21
70	20
50	18
30	19
20	16
10	17
5	15
0	11

Use the linear least squares method to fit a line to these data and compute the correlation coefficient. Would you claim a significant dependence of the reaction yield on oxygen pressure?

5.8. Using the least squares method, find the values of the parameters b_0 and b_1 that give the best fit of the equation $y = b_0 + b_1x$ to the following data, and calculate the standard deviation of each.

x_i	y_i
0.7	7.1
1.1	10.0
2.3	18.4
4.2	33.3
5.5	43.3

5.9. Fit the equation $y = b_0 + b_1x + b_2x^2$ to the data given in Exercise 5.8 using the least squares fitting procedure.

5.10. Calculate the predicted values of y_i according to each of the equations determined in Exercises 5.8 and 5.9, and plot the least squares equations together with the raw data as functions of x.

5.11. Fit the following data with the equation $y = b_0 + b_1x$ and then with the equation $y = b_0 + b_1x + b_2x^2$, and determine whether the fit with the quadratic equation is a significant improvement over the fit with a linear equation at the 0.05 level.

x_i	y_i
0.8	38
1.4	32
2.4	26
3.2	20
4.2	16
5.2	14

5.12. The dependence of the rate r of a chemical reaction on the concentration c of a reactant can be written in general as

$$r = kc^n$$

if the concentrations of all other reactants are made sufficiently large so that they remain effectively constant. Assuming that the following initial values of r_i have been determined as a function of c_i, find the values of k and n that best fit the data.

c_i	r_i
0.2	1.97
0.6	15.85
0.9	34.48
1.1	54.00
1.5	99.61
1.9	157.93

5.13. Fit the equation $y = Ax/(B + x)$ to the following data

x_i	y_i
2	2.1
6	3.5
8	4.2
11	4.3
14	4.6

using the nonlinear least squares method, taking $A_0 = 6$ and $B_0 = 4$ as the initial guesses.

5.14. Repeated determinations of the silver and phosphate ion concentrations of a saturated solution gave $[Ag^+] = (6.85 \pm 0.25) \times 10^{-6}$ and $[PO_4^{3-}] = (3.29 \pm 0.31) \times 10^{-5}$, where the quoted uncertainties are standard deviations. Calculate the value of K_{sp} derived from these measurements, and determine its uncertainty (as a standard deviation).

5.15. Derive Eq. 5.21 from Eq. 5.22.

5.16. Insert the equations for b_0, b_1, and b_2 given in the text in Section 5.4.2 in Eq. 5.59, and verify that Eqs. 5.61 are obtained when the expression is evaluated at $x_i = -2, -1, 0, +1$, and $+2$.

5.17. The Gaussian distribution function is usually written in terms of the standard deviation σ in the form Eq. 4.10. Show that Eq. 5.67 is equivalent with the width w defined as indicated in the text.

5.18. Plot the Lorentzian line shape function (Eq. 5.66), its first derivative, and its second derivative for arbitrarily chosen values of the position, width, and height parameters.

5.19. Plot the Gaussian line shape function (Eq. 5.67), its first derivative, and its second derivative for the same parameters as those chosen in Exercise 5.18.

5.20. Plot the superposition of two Lorentzian functions of identical height and width, with their positions chosen such that the separation is 40% of their width. Also plot the first and second derivatives of the composite curve. For which case(s) is the presence of two components evident?

5.21. The absorbances at seven different wavelengths were determined for each of four different solutions of varying concentrations with the following results:

	A	B	C	D
λ_1	0.504	0.003	0.502	0.001
λ_2	0.203	0.019	0.023	0.180
λ_3	0.504	0.050	0.002	0.504
λ_4	0.022	0.141	0.074	0.087
λ_5	0.003	0.399	0.195	0.200
λ_6	0.005	0.123	0.066	0.066
λ_7	0.079	0.007	0.077	0.002

Perform a factor analysis of this data to determine the number of components present in the solutions.

5.22. Show that the Fourier transform of the boxcar function (Eq. 5.84) has its first zero crossing at $2L$, and that the first negative side lobe is 22% of the center height.

5.23. The following physical and chemical data have been determined for the isomeric acyclic hexanes:

Compound	H_f	ΔH_v	$S°$	C_p^0	T_c	$-\chi_m \times 10^6$	T_b
n-Hexane	36.27	7627.2	92.91	34.08	234.2	74.6	68.95
2-Methylpentane	17.38	7676.6	91.03	33.99	224.3	75.26	60.271
3-Methylpentane	17.41	7743.9	91.52	33.49		75.52	63.282
2,2-Dimethylbutane	1.61	7271.0	85.82	33.81		76.24	49.74
2,3-Dimethylbutane	2.22	7120.0	87.43	33.32	226.8	76.22	58

where H_f is the heat of fusion, ΔH_v is the heat of vaporization, $S°$ is the standard entropy at 298 K, C_p^0 is the standard heat capacity at constant pressure at 298 K, T_c is the critical temperature, χ_m is the molar diamagnetic susceptibility, and T_b is the normal boiling point. Use the program MPC (Figure 5.17) to determine the average molecular path codes for each of the molecules listed, and then use the least squares method to find the best linear equation relating the various properties to the average molecular path codes. For which properties is there a strong correlation?

5.24. Use the Harary–Sachs theorem to write the secular equation for the π-electron energy levels of benzene according to the HMO theory.

5.25. Use the recursion formula developed in Section 5.8 to write the characteristic equation for 1,3,5-hexatriene, and show that it is the same as that obtained by expansion of the secular determinant.

5.26. Solve for the values of x that satisfy the characteristic equation of 1,3,5-hexatriene. Note that for *even alternate hydrocarbons,* of which this is an example, the roots always occur as $\pm$ pairs, so that only the positive (or negative) roots need be found.

5.27. A given data set may be characterized by the following descriptors:

d_1	d_2	d_3	d_4
4.267	5.438	0.521	0.577
8.496	1.784	0.954	3.121
8.381	2.453	0.272	3.449
8.240	1.796	1.913	2.310
0.672	1.120	0.736	1.227
1.243	6.218	1.228	3.053
2.580	1.383	0.291	1.388
2.823	3.668	0.757	1.733
6.902	5.193	0.375	4.322
1.855	5.347	2.350	1.358
7.922	3.853	0.559	1.068
7.686	0.354	2.959	4.123
7.801	4.168	0.945	3.235
6.449	4.279	2.846	0.867
7.511	2.439	0.599	0.827
7.999	6.710	0.445	2.270
5.897	2.400	1.527	0.439
8.379	6.419	2.522	1.118
2.466	4.614	2.319	2.595
1.210	6.619	1.550	3.016

Autoscale the set of descriptors and perform a Karhunen–Loéve transformation. Plot the transformed data, select appropriate groupings, and on this basis assign each of the data points to either class 1 or class 2.

5.28. Using the classification scheme selected in Exercise 5.27, run LLM (Figure 5.21) on the original data set, using the first 15 members as the training set and the last 5 members as the prediction set. (Note that you will need to redimension the array D to contain four descriptors.)

APPENDICES

APPENDIX A.1 DIFFERENTIATION FORMULAS

1. $\dfrac{da}{dx} = 0$

2. $\dfrac{d(ay)}{dx} = a\,\dfrac{dy}{dx}$

3. $\dfrac{d(x^n)}{dx} = nx^{n-1}$

4. $\dfrac{d(f \pm g)}{dx} = \dfrac{df}{dx} \pm \dfrac{dg}{dx}$

5. $\dfrac{dy(u)}{dx} = \dfrac{du}{dx}\,\dfrac{dy}{du}$

6. $\dfrac{dy}{dx} = \dfrac{1}{\dfrac{dx}{dy}}$

7. $\dfrac{d(f \cdot g)}{dx} = f\,\dfrac{dg}{dx} + g\,\dfrac{df}{dx}$

8. $\dfrac{d\left(\dfrac{f}{g}\right)}{dx} = \dfrac{g\,\dfrac{df}{dx} - f\,\dfrac{dg}{dx}}{g^2}$

9. $\dfrac{d(\log_b u)}{dx} = \dfrac{1}{u}\log_b e\,\dfrac{du}{dx}$

10. $\dfrac{d \ln u}{dx} = \dfrac{1}{u} \dfrac{du}{dx}$

11. $\dfrac{d \ln x}{dx} = \dfrac{1}{x}$

12. $\dfrac{d(a^u)}{dx} = a^u \ln a \dfrac{du}{dx}$

13. $\dfrac{d(e^u)}{dx} = e^u \dfrac{du}{dx}$

14. $\dfrac{de^x}{dx} = e^x$

APPENDIX A.2 INTEGRATION FORMULAS

1. $\int af(x)\,dx = a \int f(x)\,dx$

2. $\int f(ax)\,dx = \dfrac{1}{a} \int f(x)\,dx$

3. $\int x^n\,dx = \dfrac{x^{n+1}}{n+1} + C \qquad (n \neq -1)$

4. $\int \dfrac{dx}{x} = \ln x + C$

5. $\int [f(x) \pm g(x)]\,dx = \int f(x)\,dx \pm \int g(x)\,dx$

6. $\int a^u\,du = \dfrac{a^u}{\ln a} + C$

7. $\int e^u\,du = e^u + C$

APPENDIX A.3 COMMONLY USED BASIC COMMAND MODE STATEMENTS

AUTO n,m	Automatically generates line numbers beginning with n and using the increment m (Default: n = 10, m = 10).
CLEAR	Clears all program variables without erasing the program currently in memory.
CONT	Continues program execution after a break.
DELETE n—m	Deletes program lines n through m, inclusive.
EDIT n	Displays line n and enters editing mode.

LIST n—m	Lists program lines n through m, inclusive, on the video monitor (Default: Lists entire program).
LLIST n—m	Lists program lines n through m, inclusive, on the line printer (Default: Lists entire program).
LOAD filename	Loads a program file into memory.
NEW	Clears memory of all program lines and variables.
RENUM i,j,k	Renumbers program lines beginning at j with the new number i using increment k (Default: i = k = 10, j = first program line).
RUN filename	Loads a program file into memory and executes it.
RUN n	Executes the program in memory beginning at line n (Default: n = first program line). All variables are set to an initial value of zero.
SAVE filename	Saves the program in memory under the filename specified.

APPENDIX A.4 TYPICAL RESERVED WORD LIST FOR THE BASIC INTERPRETER OF IBM-PC-COMPATIBLE COMPUTERS

ABS	CONT	EOF	INP
AND	COS	EQV	INPUT
ASC	CSNG	ERASE	INPUT#
ATN	CSRLIN	ERDEV	INPUT$
AUTO	CVD	ERDEV$	INSTR
BEEP	CVI	ERL	INT
BLOAD	CVS	ERR	IOCTL
BSAVE	DATA	ERROR	IOCTL$
CALL	DATE$	EXP	KEY
CALLS	DEF	FIELD	KILL
CDBL	DEFDBL	FILES	LEFT$
CHAIN	DEFINT	FIX	LEN
CHDIR	DEFSNG	FNxxxxxx	LET
CHR$	DEFSTR	FOR	LINE
CINT	DELETE	FRE	LIST
CIRCLE	DIM	GET	LLIST
CLEAR	DRAW	GOSUB	LOAD
CLOSE	EDIT	GOTO	LOC
CLS	ELSE	HEX$	LOCATE
COLOR	END	IF	LOF
COM	ENVIRON	IMP	LOG
COMMON	ENVIRON$	INKEY$	LPOS

LPRINT	PEEK	RND	TAN
LSET	PEN	RSET	THEN
MERGE	PLAY	RUN	TIMER
MID$	PMAT	SAVE	TIME$
MKDIR	POINT	SCREEN	TO
MKD$	POKE	SGN	TROFF
MKI$	POS	SHELL	TRON
MKS$	PRESET	SIN	USING
MOD	PRINT	SOUND	USR
MOTOR	PRINT#	SPACE$	VAL
NAME	PSET	SPC	VARPTR
NEW	PUT	SQR	VARPTR$
NEXT	RANDOMIZE	STEP	VIEW
NOT	READ	STICK	WAIT
OCT$	REM	STOP	WEND
OFF	RENUM	STR$	WHILE
ON	RESET	STRIG	WIDTH
OPEN	RESTORE	STRING	WINDOW
OPTION	RESUME	STRING$	WRITE
OR	RETURN	SWAP	WRITE#
OUT	RIGHT$	SYSTEM	XOR
PAINT	RMDIR	TAB	

APPENDIX A.5 FREQUENTLY USED BASIC LIBRARY FUNCTIONS

In the following, X and Y are constants, variables, or arithmetic expressions, S is a string variable, A is the ASCII code for a character, and L and M are constants, variables, or arithmetic expressions which may be evaluated and rounded to integral values. Some of these functions have additional options which are defined in computer manuals.

ABS(X)	Returns the absolute value of X.
ATN(X)	Returns the arctangent (in radians) of X.
CDBL(X)	Converts X to a double-precision number.
CHR$(A)	Returns the string character defined by A.
CINT(X)	Converts X to an integer by rounding.
CLS	Erases the video monitor screen.
COS(X)	Returns the cosine of X (X is in radians).
CSNG(X)	Converts X to a single-precision number.
EXP(X)	Returns the natural exponential e^X.

FIX(X)	Returns the integer equivalent to truncated X [e.g., FIX(2.5)=2, FIX(−2.5)=−2].
INKEY$	Returns the string character read from the keyboard.
INT(X)	Returns the largest integer not greater than X [e.g., INT(2.5)=2, INT(−2.5)=−3].
LEFT$(S,L)	Returns a character string consisting of the L leftmost characters of S.
LOCATE(X,Y)	Moves the cursor to the position defined by the row X and column Y.
LOG(X)	Returns the natural logarithm (base e) of X.
MID$(S,M,L)	Returns a character string consisting of L characters from S starting at position M.
PSET(X,Y)	Turns on the pixel defined by the row X and column Y.
RIGHT$(S,L)	Returns a character string consisting of the L rightmost characters of S.
RND(X)	Returns a pseudorandom number. If X ≥ 0 or if the argument is omitted, the number returned is in the range 0 to 1. For X ≤ 0, or by use of the statement RANDOMIZE M, the random number generator is reseeded. The formula INT(RND*(N+1)) may be used to generate pseudorandom numbers in the range 0 to N.
SGN(X)	Returns −1, 0, or +1 according to whether X is negative, zero, or positive.
SIN(X)	Returns the sine of X (X is in radians).
SPACE$(L)	Returns a string consisting of L spaces.
SQR(X)	Returns the positive square root of X for X ≥ 0.
STR$(X)	Returns a string representation of the number X.
STRING$(L,A)	Returns a string consisting of L identical characters specified by A.
TAN(X)	Returns the tangent of X (X is in radians).
VAL(S)	Returns the numerical value corresponding to the string representation of a number.

APPENDIX A.6 LAGRANGE FORMULAS*

Lagrange interpolation formulas for equally spaced abscissas

Two point: $f(x_0 + ph) = (1 - p)f(x_0) + pf(x_1)$

Three point: $f(x_0 + ph) = \frac{p(p-1)}{2} f(x_{-1}) + (1 - p^2)f(x_0) + \frac{p(p+1)}{2} f(x_1)$

Four point:
$$f(x_0 + ph) = \frac{-p(p-1)(p-2)}{6} f(x_{-1}) + \frac{(p^2-1)(p-2)}{2} f(x_0) - \frac{p(p+1)(p-2)}{2} f(x_1) + \frac{p(p^2-1)}{6} f(x_2)$$

Five point:
$$f(x_0 + ph) = \frac{(p^2-1)p(p-2)}{24} f(x_{-2}) - \frac{(p-1)p(p^2-4)}{6} f(x_{-1}) + \frac{(p^2-1)(p^2-4)}{4} f(x_0) - \frac{(p+1)p(p^2-4)}{6} f(x_1) + \frac{(p^2-1)p(p+2)}{24} f(x_2)$$

Lagrange differentiation formulas for equally spaced abscissas

Two point: $f'(x_0 + ph) = \frac{1}{h}[f(x_1) - f(x_0)]$

Three point: $f'(x_0 + ph) = \frac{1}{h}\left[\left(p - \frac{1}{2}\right)f(x_{-1}) - 2pf(x_0) + \left(p + \frac{1}{2}\right)f(x_1)\right]$

Four point:
$$f'(x_0 + ph) = \frac{1}{h}\left[-\frac{3p^2 - 6p + 2}{6} f(x_{-1}) + \frac{3p^2 - 4p - 1}{2} f(x_0) - \frac{3p^2 - 2p - 2}{2} f(x_1) + \frac{3p^2 - 1}{6} f(x_2)\right]$$

Five point:
$$f'(x_0 + ph) = \frac{1}{h}\left[\frac{2p^3 - 3p^2 - p + 1}{12} f(x_{-2}) - \frac{4p^3 - 3p^2 - 8p + 4}{6} f(x_{-1}) + \frac{2p^3 - 5p}{2} f(x_0) - \frac{4p^3 + 3p^2 - 8p - 4}{6} f(x_1) + \frac{2p^3 + 3p^2 - p - 1}{12} f(x_2)\right]$$

*From *Handbook of Mathematical Functions,* edited by M. Abramowitz and I. A. Stegun, National Bureau of Standards, Applied Mathematics 55. Used by permission.

APPENDIX A.7 ABSCISSAS AND WEIGHTING FACTORS FOR GAUSSIAN QUADRATURE*

n	Abscissa (z_i)	Weighting Factor (w_i)
2	±0.5773503	1.0000000
3	0.0000000	0.8888889
	±0.7745967	0.5555556
4	±0.3399810	0.6521452
	±0.8611363	0.3478548
5	0.0000000	0.5688889
	±0.5384693	0.4786287
	±0.9061798	0.2369269
6	±0.2386192	0.4679139
	±0.6612094	0.3607616
	±0.9324695	0.1713245
7	0.0000000	0.4179592
	±0.4058452	0.3818301
	±0.7415312	0.2797054
	±0.9491079	0.1294850
8	±0.1834346	0.3626838
	±0.5255324	0.3137066
	±0.7966665	0.2223810
	±0.9602899	0.1012285

*Reprinted from A. N. Lowan, N. Davids, and A. Levinson, "Table of the Zeros of the Legendre Polynomials of Order 1-16 and the Weight Coefficients for Gauss Mechanical Quadrature Formula," *Bulletin of the American Mathematical Society*, Volume 48 (1942), pp. 739-743, by permission of the American Mathematical Society.

APPENDIX A.8 AREAS OF A NORMAL DISTRIBUTION CURVE

x/s	p	x/s	p
0.0	0.0000	1.8	0.4641
0.1	0.0398	1.9	0.4713
0.2	0.0793	2.0	0.4773
0.3	0.1179	2.1	0.4821
0.4	0.1554	2.2	0.4861
0.5	0.1915	2.3	0.4893
0.6	0.2257	2.4	0.4918
0.7	0.2580	2.5	0.4938
0.8	0.2881	2.6	0.4953
0.9	0.3159	2.7	0.4965
1.0	0.3413	2.8	0.4974
1.1	0.3643	2.9	0.4981
1.2	0.3849	3.0	0.4987
1.3	0.4032	3.2	0.4993
1.4	0.4192	3.4	0.4997
1.5	0.4332	3.6	0.4998
1.6	0.4452	3.8	0.4999
1.7	0.4554	4.0	0.49997

APPENDIX A.9 CRITICAL VALUES OF STUDENT'S t FOR $\alpha = 0.05$*

Degrees of Freedom	One-Sided Test	Two-Sided Test
1	6.314	12.706
2	2.920	4.303
3	2.353	3.183
4	2.132	2.776
5	2.015	2.571
6	1.943	2.447
7	1.895	2.365
8	1.860	2.306
9	1.833	2.262
10	1.813	2.228
11	1.796	2.201
12	1.782	2.179
13	1.771	2.160
14	1.761	2.145
15	1.753	2.132
16	1.746	2.120
17	1.740	2.110
18	1.734	2.101
19	1.729	2.093
20	1.725	2.086
21	1.721	2.080
22	1.717	2.074
23	1.714	2.069
24	1.711	2.064
25	1.708	2.060
26	1.706	2.056
27	1.703	2.052
28	1.701	2.048
29	1.699	2.045
30	1.697	2.042
∞	1.645	1.960

*Condensed from M. Merrington, *Biometrika*, **32**, 300 (1941); used by permission of the Biometrika Trustees.

APPENDIX A.10 CRITICAL VALUES OF F FOR $\alpha = 0.05$*

Degrees of Freedom in the Denominator	Degrees of Freedom in the Numerator													
	1	2	3	4	5	6	7	8	10	12	15	20	40	∞
1	161.45	199.50	215.71	224.58	230.16	233.99	236.77	238.88	241.88	243.91	245.95	248.01	251.14	254.32
2	18.51	19.00	19.16	19.25	19.30	19.33	19.35	19.37	19.40	19.41	19.43	19.45	19.47	19.50
3	10.13	9.55	9.28	9.12	9.01	8.94	8.89	8.85	8.79	8.74	8.70	8.66	8.59	8.53
4	7.71	6.94	6.59	6.39	6.26	6.16	6.09	6.04	5.96	5.91	5.86	5.80	5.72	5.63
5	6.61	5.79	5.41	5.19	5.05	4.95	4.88	4.82	4.74	4.68	4.62	4.56	4.46	4.37
6	5.99	5.14	4.76	4.53	4.39	4.28	4.21	4.15	4.06	4.00	3.94	3.87	3.77	3.67
7	5.59	4.74	4.35	4.12	3.97	3.87	3.79	3.73	3.64	3.57	3.51	3.44	3.34	3.23
8	5.32	4.46	4.07	3.84	3.69	3.58	3.50	3.44	3.35	3.28	3.22	3.15	3.04	2.93
10	4.96	4.10	3.71	3.48	3.33	3.22	3.14	3.07	2.98	2.91	2.85	2.77	2.66	2.54
12	4.75	3.89	3.49	3.26	3.11	3.00	2.91	2.85	2.75	2.69	2.62	2.54	2.43	2.30
15	4.54	3.68	3.29	3.06	2.90	2.79	2.71	2.64	2.54	2.48	2.40	2.33	2.20	2.07
20	4.35	3.49	3.10	2.87	2.71	2.60	2.51	2.45	2.35	2.28	2.20	2.12	1.99	1.84
40	4.08	3.23	2.84	2.61	2.45	2.34	2.25	2.18	2.08	2.00	1.92	1.84	1.69	1.51
∞	3.84	3.00	2.60	2.37	2.21	2.10	2.01	1.94	1.83	1.75	1.67	1.57	1.39	1.00

*Condensed from M. Merrington and C. M. Thompson, *Biometrika*, **33**, 73–88 (1943); used by permission of the Biometrika Trustees.

APPENDIX A.11 CRITICAL VALUES OF CHI-SQUARE FOR $\alpha = 0.05$*

Degrees of Freedom	χ^2	Degrees of Freedom	χ^2
1	3.84	16	26.30
2	5.99	17	27.59
3	7.81	18	28.87
4	9.49	19	30.14
5	11.07	20	31.41
6	12.59	21	32.67
7	14.07	22	33.92
8	15.51	23	35.17
9	16.92	24	36.42
10	18.31	25	37.65
11	19.68	26	38.89
12	21.03	27	40.11
13	22.36	28	41.34
14	23.68	29	42.56
15	25.00	30	43.77

*From C. M. Thompson, *Biometrika*, **32**, 188–189 (1941); used by permission of the Biometrika Trustees.

APPENDIX A.12 PARAMETERS FOR STATISTICAL ANALYSIS OF SMALL SAMPLES*

Sample Size n	q ($\alpha = 0.10$)	K_w	t_w ($\alpha = 0.05$)
2	—	0.89	6.4
3	0.94	0.59	1.3
4	0.76	0.49	0.72
5	0.64	0.43	0.51
6	0.56	0.40	0.40
7	0.51	0.37	0.33
8	0.47	0.35	0.29
9	0.44	0.34	0.26
10	0.41	0.33	0.23

*Reprinted with permission from R. B. Dean and W. J. Dixon, *Anal. Chem.*, **23**, 636–638 (1951). Copyright 1951 American Chemical Society.

APPENDIX A.13 TABLE OF Q FACTORS FOR CALCULATING RANGE OF DIFFERENCES*

Degrees of Freedom	k' (Difference in Rank plus One)								
	2	3	4	5	6	7	8	9	10
1	18.0	26.7	32.8	37.2	40.5	43.1	45.4	47.3	49.1
2	6.09	8.28	9.80	10.89	11.73	12.43	13.03	13.54	13.99
3	4.50	5.88	6.83	7.51	8.04	8.47	8.85	9.18	9.46
4	3.93	5.00	5.76	6.31	6.73	7.06	7.35	7.60	7.83
5	3.61	4.54	5.18	5.64	5.99	6.28	6.52	6.74	6.93
6	3.46	4.34	4.90	5.31	5.63	5.89	6.12	6.32	6.49
7	3.34	4.16	4.68	5.06	5.35	5.59	5.80	5.99	6.15
8	3.26	4.04	4.53	4.89	5.17	5.40	5.60	5.77	5.92
9	3.20	3.95	4.42	4.76	5.02	5.24	5.43	5.60	5.74
10	3.15	3.88	4.33	4.66	4.91	5.12	5.30	5.46	5.60
12	3.08	3.77	4.20	4.51	4.75	4.95	5.12	5.27	5.40
15	3.01	3.67	4.08	4.37	4.59	4.78	4.94	5.08	5.20
20	2.95	3.58	3.96	4.24	4.45	4.62	4.77	4.90	5.01
40	2.86	3.44	3.79	4.04	4.23	4.39	4.52	4.63	4.74
∞	2.77	3.32	3.63	3.86	4.03	4.17	4.29	4.39	4.47

*From H. L. Youmans, *Statistics for Chemistry*, Charles E. Merrill Publishing Company, Columbus, Ohio, 1973; used by permission of the author.

ANSWERS TO SELECTED EXERCISES

Chapter 1

1.1 (a) -2 **(d)** 1 **1.2 (b)** -0.5 **(c)** -3.5 **1.3 (a)** Minimum at (0, 0) **(c)** Maximum at (3, 25) **1.4 (b)** $-4x(1 - x^2)$ **1.5 (c)** $(2x - 1)/(x^2 - x)$ **1.6 (d)** 48 **1.9 (c)** $5^x = 1 + 1.6094x/1! + 2.5903x^2/2! + 4.1689x^3/3! + 6.7096x^4/4! + \cdots$ **(d)** Cannot be done since ln(0) is not defined. **1.10 (c)** $14 \ln x + c$ **1.11 (b)** 14 **1.12 (c)** 18,707 cal mol^{-1} **1.13 (c)** 18,741.6 cal mol^{-1} **1.15** 3.11 cal mol^{-1} **1.18** 1.987 cal K^{-1} **1.20 (b)** Substitute $y = \sqrt{x + 1}$, integration gives $-2\sqrt{x + 1} - 2 \ln(1 - \sqrt{x + 1}) + c$ **1.21 (b)** $\ln(x - 2) - 1/(x - 2) + c$ **1.22 (a)** $e^x(x - 1) + C$ **1.23** $W = p_0[V_0 - V(V_0/V)^\gamma]/(\gamma - 1)$ **1.26** 494 atm **1.28 (b)** sinh x **1.29 (a)** $y = x^5 + cx^4$ **1.30 (a)** The auxiliary equation is $m^2 + 4m + 4 = 0$, which factors into $(m + 2)(m + 2) = 0$. This is the special case of equal roots, so the general solution is of the form $y = c_1e^{-2x} + c_2xe^{-2x}$. **(c)** The auxiliary equation has equal roots $m = 2$, so the general solution is $y = 3e^{2x} - 2xe^{2x}$ **1.31** $y = 1 + x^2$ **1.37** $r_c/r_a = 0.2247$

Chapter 2

2.1 (c) Not valid; variable name must begin with an alphabetic character **(e)** Not valid; reserved word **2.2 (d)** Not valid; a constant cannot contain a comma **2.3 (d)** Not valid; only a single variable can occur on the left **(g)** Not valid; two operators cannot be juxtaposed **2.4 (c)** 112.3333 **(f)** 432.1786 **(j)** 6.8 **2.5 (b)** 109 **(d)** 3.2 **2.7 (c)** 0 **(e)** -3 **2.8 (c)** 0 **(e)** -4 **2.11 (b)** 480 **(d)** 140

2.14 **(a)**

$$f(x) = 1 - \frac{1}{3}x - \frac{1 \cdot 2}{3 \cdot 6}x^2 - \frac{1 \cdot 2 \cdot 5}{3 \cdot 6 \cdot 9}x^3 - \frac{1 \cdot 2 \cdot 5 \cdot 8}{3 \cdot 6 \cdot 9 \cdot 12}x^4 - \frac{1 \cdot 2 \cdot 5 \cdot 8 \cdot 11}{3 \cdot 6 \cdot 9 \cdot 12 \cdot 15}x^5 - \cdots$$

(b) Note that the series above starts with the value 1, and each succeeding term is obtained by multiplication of the preceding one by the factor $(3n - 4)x/(3n)$. The following is one way to carry out the calculation:

```
10 REM---PROGR TO TEST EVALN OF CUBE ROOT OF (1-X)
20 REM---BY SERIES EXPANSION OF (1-X)^(1/3)
30 INPUT"INPUT X";X
40 GOSUB 500
50 PRINT"THE CUBE ROOT OF  ";1-X;" IS ";R
60 GOTO 30
500 N=0 : R=1 : T=1       'CUBE ROOT SUBROUTINE
510 WHILE ABS(T/R)>.0001
520 N=N+1
530 T=(3*N-4)*X/3/N*T
540 R=R+T
550 WEND
560 RETURN
```

Chapter 3

3.3 **(b)** ZnS begins to precipitate at $\log[S^{2-}] = -21.8$, and at this point $\log[Pb^{2+}] = -6.1$. Therefore, $10^{-6.1}/10^{-3.0} = 10^{-3.1}$ and quantitative separation is (barely) possible.
3.4 **(b)** $[H^+] + [HCO_3^-] + 2[H_2CO_3] = [NH_3] + [OH^-]$ **3.5** **(c)** 3.1 **(d)** 8.1
3.6 **(c)** 3.7 **3.7** **(b)** $[H^+] = 1.416 \times 10^{-7}$ **(d)** $[H^+] = 2.810 \times 10^{-6}$
3.8 **(c)** $[H^+] = 1.901 \times 10^{-6}$ **3.9** **(a)** $[H^+] = 8.157 \times 10^{-3}$ **(d)** $[H^+] = 5.345 \times 10^{-5}$ **3.11** **(a)** $[H^+] = 2.044 \times 10^{-8}$ **(c)** $[H^+] = 3.264 \times 10^{-5}$
3.13 **(c)** $x_{n+1} = [x_n(7/x_n^3) + 2)]/3,\ x = 1.91293118$ **3.14** **(b)** $[H^+] = 7.129 \times 10^{-5}$
3.16 Note that two estimates of the root are needed to start the process. We take $\sqrt{K_1 C_a}$ to be one estimate and arbitrarily take twice this value to be the other in the following program:

```
10 INPUT"Input Ca, pK1, and pK2";CA,PK1,PK2
20 KW=1E-14 : K1=10^(-PK1) : K2=10^(-PK2)
30 DEF FNHY(Z)=Z*(Z*(Z*(Z+K1)+K1*K2-KW-K1*CA)-K1*KW-2*K1*K2*CA)-K1*K2*KW
40 X1=SQR(CA*K1) : Y1=FNHY(X1)
50 X2=2*X1 : Y2=FNHY(X2)
60 WHILE ABS(X2/X1-1)>.0001
70 Y2=FNHY(X2)
80 X3=X2-Y2*(X2-X1)/(Y2-Y1)
90 X1=X2 : Y1=Y2 : X2=X3
100 WEND
110 PRINT"THE HYDROGEN ION CONCN IS ";X2
```

3.17 (a) 5.56 **3.18 (b)** 8.23 **3.19 (a)** 0.76 **3.21 (c)** 2.407901 **(d)** 2.298283 **3.26 (d)** $z = -1, y = 5, x = -2, w = 4$ **3.27 (a)** The result by exact integration is 6.389056. We generate random x and y values on a grid $0 \leq x \leq 2$ and $0 \leq y \leq 8$, whose total area is $2 \times 8 = 16$. The following program counts the fraction of the points which fall under the curve defined by the integrand, and computes the integral by multiplying that fraction by the total area:

```
10 INPUT"INPUT N";N
20 RANDOMIZE TIMER
30 I=0
40 FOR J=1 TO N
50 X=2*RND
60 Y=8*RND
70 IF Y>EXP(X) THEN 90
80 I=I+1
90 NEXT J
100 PRINT"INTEGRAL = ";I*16/N
110 GOTO 10
```

Chapter 4

4.1 $\bar{x} = 26.98$, $s = 2.81$ **4.4** $\bar{x} = 27.05$, $s = 2.80$ **4.5 (a)** $z = (30 - 26.98)/2.81 = 1.07$, $p = 0.500 - 0.357 = 0.143$ **(d)** 0.602 **4.6** 19 days **4.7 (b)** $\bar{x} = 25.998$, $s_{\bar{x}} = 1.488$ **(d)** for $n = 2$, $s_{\bar{x}} = 2.247/\sqrt{2} = 1.589$ **4.8 (c)** 2.88 **(d)** The critical value $t_{.05/2}$ for 3 degrees of freedom is 3.182, and so we cannot reject the null hypothesis at the 0.05 level. **4.9** The t value is $(25.53 - 25.76)\sqrt{5}/0.13 = -4.1$, which is larger in magnitude (we ignore the sign, since our interest is in the magnitude of the difference) than the critical value $t_{.05/2} = 2.776$ under 4 degrees of freedom. We reject the null hypothesis and conclude that there is a significant difference in the results at the 0.05 level. **4.10** The ratio of variances for the two sets is $F = (0.0840)/(0.0521) = 1.61$, which is less than the critical value of F under 3 degrees of freedom in the numerator and 4 degrees of freedom in the denominator (6.59). Thus we are justified in making the comparison. The pooled standard deviation is $s = 0.256$ and the t value is -2.22. This is less than the critical value $t_{.05/2} = 2.365$ under 7 degrees of freedom and we conclude that the difference is not significant at the 0.05 level. **4.13** The paired differences give $\bar{d} = -5.2$ and $s_d = 9.85$, so the t value is -1.40. It is not entirely clear from the way the question is phrased whether we should apply a one-sided or a two-sided test. Under 6 degrees of freedom we find $t_{.05/2} = 2.447$ and $t_{.05} = 1.943$, and so we should regard the difference as insignificant in either case. **4.16** $\Sigma O_i = 144$, $E_i = 144/10 = 14.4$, $\chi^2 = 16.8$. The critical value of χ^2 under 9 degrees of freedom is 16.91, and so the deviation from the expected results is not significant at the 0.05 level. **4.17** $F = 2.699$, and the differences are not significant at the 0.05 level.

Chapter 5

5.2 $r(1.4768, 2.1097) = 0.7348$ **5.5** $A = 0.505$, $B = 0.126$ **5.6** The least squares fit gives $N = 0.1207 - 4.970 \times 10^{-4}d$, with $r = -0.0760$. The standard deviation of the fit is 2.248×10^{-3} and the 95% confidence interval of the slope is $-(4.97 \pm 4.24) \times 10^{-4}$.

The t value is -2.87, and the critical value of $t_{.05/2}$ under 6 degrees of freedom is 2.447. The correlation is considered significant at the 0.05 level. **5.8** $b_0 = 1.562$, $b_1 = 7.557$, $s_{b_0} = 0.149$, $s_{b_1} = 0.089$ **5.9** $b_0 = 2.209$, $b_1 = 6.889$, $b_2 = 0.1092$ **5.11** The sum of squares deviation with two parameters is SS = 19.76 and with three parameters it is SS = 1.400. This gives (Eq. 5.44)

$$F_{1,3} = \frac{(19.76 - 1.400)/1}{1.400/3} = 39.35$$

The critical F value for 1 degree of freedom in the numerator and 3 degrees of freedom in the denominator is 10.13. We conclude that the difference is significant at the 0.05 level. **5.12** $k = 44.5$, $n = 1.953$ **5.14** $K_{sp} = (1.06 \pm 0.15) \times 10^{-20}$ **5.15** *Hint:* a is the value of y_P corresponding to the particular value of $x_P = 0$. **5.21** Three components **5.24** $x^6 - 6x^4 + 9x^2 - 4 = 0$ **5.25** $x^6 - 5x^4 + 6x^2 - 1 = 0$ **5.26** $x = \pm\, 0.4450$, ± 1.2470, ± 1.8019 **5.27** and **5.28** Several different groupings of the data points are possible, and the outcome with LLM in Exercise 5.28 depends on the selection made in Exercise 5.27.

INDEX

D

E

F

G

H

M

N

O

P

T

U

V

W

Z